Plant Biotechnology:
the genetic manipulation of plants

Second edition

Adrian Slater
Nigel W. Scott
Mark R. Fowler
De Montfort University

OXFORD
UNIVERSITY PRESS

OXFORD
UNIVERSITY PRESS

Great Clarendon Street, Oxford OX2 6DP

Oxford University Press is a department of the University of Oxford.
It furthers the University's objective of excellence in research, scholarship,
and education by publishing worldwide in

Oxford New York

Auckland Cape Town Dar es Salaam Hong Kong Karachi
Kuala Lumpur Madrid Melbourne Mexico City Nairobi
New Delhi Shanghai Taipei Toronto

With offices in

Argentina Austria Brazil Chile Czech Republic France Greece
Guatemala Hungary Italy Japan Poland Portugal Singapore
South Korea Switzerland Thailand Turkey Ukraine Vietnam

Published in the United States
by Oxford University Press Inc., New York

First edition 2003
Second edition 2008

British Library Cataloguing in Publication Data
Data available

Library of Congress Cataloging in Publication Data
Slater, Adrian.
Plant biotechnology: the genetic manipulation of plants / Adrian Slater,
Nigel W. Scott, Mark R. Fowler. — 2nd ed.
p. ; cm.
Includes bibliographical references and index.
ISBN 978-0-19-928261-6 (alk. paper)
1. Plant biotechnology. 2. Plant genetic engineering. I. Scott, Nigel W.
II. Fowler, Mark R. III. Title.
[DNLM: 1. Plants, Genetically Modified—genetics.
2. Crops, Agricultural—genetics.
SB 123.57 S631p 2008]
TP248.27.P55S59 2008
631.5′23—dc22
2008003362

Typeset by Graphicraft Limited, Hong Kong
Printed in Great Britain by Ashford Colour Press Ltd., Gosport, Hants.

ISBN 978-0-19-928261-6

3 5 7 9 10 8 6 4 2

Foreword

The Green Revolution, led by Norman Borlaug, Monkombu Swaminathan, and Gurdev Khush, enabled the world's food supply to be tripled during the last three decades of the 20th century. The extraordinary increase in agricultural productivity was made possible by the adoption of genetically improved varieties coupled with advances in crop management. In many countries food supply increased faster than demand and the technological progress contributed to a decrease in the unit cost of production so that farmers were able to share the benefits of the advances with consumers, by offering food to them at lower prices. Intensive (high-input/high-yield) agriculture has served the populations of the developed countries well but two problems have come to occupy these people. First, the full-scale exploitation of intensive agriculture protocols does deliver high yields of high-quality produce, but the environmental impact of the processes is often high. Second, the farmers have, in fact, the need to dispose of surplus food so that there is a downward pressure on prices in the world market that undermines farmers' incentives.

The position of the people in the low-income countries contrasts starkly with that in the developed countries. The world's population has increased from 2.5 billion to 6.1 billion in the last 50 years and it is unlikely to stabilize before 2100 by which time another 3 billion people will inhabit the planet. Most of the increase will occur in the low-income countries where poverty and hunger are already widespread. Each night 800 million people go to bed hungry and suffer from malnutrition, and one-fifth of humankind (about 1.2 billion people) lives on earnings of less than a dollar per day. We must satisfy the need for more food in an environmentally friendly way but we are confronted by major challenges. Prime agricultural land is being diverted to non-agricultural uses to meet the growing demand from housing, urbanization, and industrialization. Countries inhabited by half the world's population are already experiencing water crises, while the high agrochemical inputs that maximize yields exert a high environmental impact, which is not acceptable. There is a desperate need to produce more food from less land with less water and reduced agrochemical inputs.

The majority of agricultural scientists, led by Borlaug, Swaminathan, and Khush, are convinced that the required crops of high yield, high quality, low cost, and low environmental impact can be delivered by the exploitation of the techniques for plant biotechnology in molecular breeding strategies. The commercial adoption of transgenic crops by farmers has been one of the most rapid cases of technology diffusion in the history of agriculture. Between 1996 and 2002, the area planted commercially with transgenic crops increased from 1.7 million to 58.7 million ha. Some 6 million farmers in 16 countries grow transgenic crops and more than a quarter of such crops are grown in developing countries.

The Norman Borlaug Institute was established to provide an international framework for co-operation in development of molecular breeding strategies. The core research activities provide a perfect environment for training students and research scientists who will respond to the challenges and opportunities outlined above. This book is based upon courses taught by the authors in The Norman Borlaug Institute.

The text defines the concepts and describes the technologies that enable the genetic manipulation of crop plants. It describes in detail the development of the two traits (herbicide and pest resistance) that are most prevalent in commercial genetically modified (GM) crops and examines the reasons for their success. The potential for developments in other crop traits such as disease resistance, abiotic stress tolerance, and improvements of yield and quality are considered and the possibility of using plants as factories for molecular farming is also explored.

The book is strengthened by confronting the wider social aspects of GM crops, and several of the controversies surrounding this new technology are thoroughly aired. The eco-terrorist fringe has constrained the development and exploitation of crop biotechnology but the measured discussion in this book will enable readers to deal with the self-serving campaigners whose actions undermine the undernourished, while they themselves benefit from three full meals each day. Meanwhile someone dies of starvation every 2.1 seconds.

Norman Borlaug himself is convinced that the world has the technology to permit a population of 10 billion people to be sustained but he is concerned that farmers may be prevented from exploiting the new technology by small, vociferous, well-financed groups of anti-science zealots. These affluent campaigners can afford to pay high prices for poorly regulated 'organic' food production. On the other hand the billion chronically poor and hungry people already in the world cannot do so and the crisis seems likely to grow as the population increases. The new technology described in this text will be the salvation of the undernourished, freeing them from obsolete, low-yielding, and more costly production technology.

Professor M.C. Elliott
Director
The Norman Borlaug Institute for Plant Science Research

Preface to the first edition

Plant biotechnology has made tremendous progress in recent years and has enjoyed a previously unknown level of public awareness. Unfortunately, much of this awareness has arisen from the negative publicity that surrounds genetically modified (GM) crops. One only has to think of the media coverage of food safety issues (so-called frankenfoods) or environmental concerns (the Monarch butterfly affair) to gain some appreciation of the public antipathy to this technology. As a result, the GM debate has been fuelled largely by misinformation, and has generated much more heat than light. It is surprising that in the course of this debate, there has not been an accessible textbook available to which serious students of the subject could turn for information and understanding.

The aim of this book is to provide enough information and examples to give the reader a sound knowledge of plant biotechnology in all its various guises, but particularly those related to the genetic manipulation of crop plants. It is not intended to provide an encyclopaedic coverage of the subject; such a task would require a volume far larger than this. As such, this is a textbook, and therefore a learning and teaching aid, rather than an academic treatise. Hopefully, the book also encourages a critical appraisal of plant biotechnology: not just the scientific aspects, but also the economic, social, moral, and ethical issues that surround and, some would say, plague the subject. Although this book is aimed at undergraduate and postgraduate students, we do not assume a huge amount of prior knowledge, and hope that other people will find the book to be accessible, informative, and enjoyable. We hope that this book will make a contribution to the GM crop debate, facilitating a rational exchange of views between informed people.

The first four chapters of the book are designed to provide a more technical introduction to subjects such as gene expression, tissue culture, and plant transformation that enables the remainder of the book to be fully appreciated. They can, of course, be read in their own right and contain information relevant to other areas of biology. The rest of the book looks in detail at various aspects of plant genetic manipulation applied to crop improvement. There are four chapters which deal in depth with the application of plant genetic manipulation to agronomic traits (herbicide, pest, and disease resistance), which can be considered as first-generation plant biotechnology. Three of the remaining chapters deal with more challenging and diverse advances in the areas of stress resistance, crop yield and quality, and molecular farming. It is in these areas that GM crops have the potential to produce real, widespread social and economic benefits, particularly for developing countries. Chapter 12 attempts to give an overview of plant biotechnology, past, present, and future, with reference to the legislative framework and economic, social, moral, and ethical issues.

Throughout the book, case studies are used as extended illustrative examples of particular points that have been made in the main text. These are meant to be read as part of the body of each chapter. However, background information for clarification of advanced concepts, or more advanced information itself, is clearly presented in boxes that can be read as and when required. There is, at the end of each chapter, a Further reading section, which is not intended to be an exhaustive reference section, but gives enough pointers to allow and encourage further investigation. Various web-links to informative websites are also included in the further reading section. The World Wide Web provides an invaluable resource for investigating plant biotechnology, but care should be exercised when accessing information of dubious provenance. We would, however, encourage the use of the excellent online-journal sites to enable, in many cases freely, an interchange of knowledge and ideas, one of the cornerstones of science.

There is a website associated with this book, which will contain hyperlinked chapter web-links, further references, downloadable figures from the book, and update sections. We hope that this website will help to keep the reader of this book fully informed and up to date with developments in plant biotechnology.

We would like to thank all those who have contributed to the making of this book, not least the people whose original investigations are the basis of this book. We apologize unreservedly for any mistakes, all of which are ours, omissions or failure to acknowledge fully. Finally we would like to thank our friends and families, in particular Arlene, Jane, and Hilary, for their continued support and encouragement throughout the writing of this book.

February 2003

Preface to the second edition

When we were invited by Oxford University Press to prepare this second edition, we must admit to approaching the task with mixed feelings.

On the one hand, we have been gratified by the success of the first edition and the complimentary reviews it has received. This confirmed our perception that there had been a conspicuous lack of an accessible textbook on plant biotechnology at a time when the field had reached a level of maturity that demanded one. It has been particularly pleasing to note that the book has been popular not only in the USA (the home of commercial genetically modified (GM) crops) and in Europe (where there is so much antipathy to them), but also in the developing economies where the potential value of this technology is widely appreciated.

On the other hand, the fact that we ourselves have been unable to exploit this book fully in our own teaching has been a source of great regret. Shortly after publication of the first edition, De Montfort University was one of a number of UK universities that was compelled to close its Departments of Biological Sciences and Chemistry, due in part to a lack of demand from UK school leavers to study science. This means that we no longer have the opportunity to fully 'road test' the textbook by teaching an entire module on plant biotechnology, although it is still of value in other areas of our teaching.

De Montfort University also closed the Norman Borlaug Institute for Plant Science Research in which we were based, and which Professor Elliott writes about so passionately in the Foreword. The Institute was an ambitious venture to develop high-yield/high-quality/low-cost/low-environmental-impact crops to address the desperate need to increase world food production. The Institute faced insuperable odds in trying to sustain its funding stream in a post-1992 UK university during the era of public backlash against GM crops. One of the knock-on effects of the rejection of GM food by the UK public described in Chapter 12 was the drying up of public and industrial funding for research directly focused on crop improvement. For good or ill, competitive plant science funding in the UK has been focused on the genomics of model systems such as *Arabidopsis* (see Chapters 1 and 13). At the same time, the distribution of central research funds to UK universities via the Research Assessment Exercise meant that the dice was always loaded against an institute whose primary mission was crop improvement for the benefit of mankind, rather than research publications as ends in themselves. As a result, our teaching and research in plant biotechnology now overlaps with other groups and departments in the university, including forensic science, biomedical science, pharmacy, and textiles, in areas such as food and medicinal plant authentication and the genetic modification of non-food crops for pharmaceutical and fibre production.

One of the unforeseen consequences for us of no longer being directly involved in GM food crop research is that we have been able to cast a more dispassionate eye over progress in the past few years. It is striking, for example, that the rapid rate of adoption of GM crops worldwide (a 60-fold increase in area grown in the first 11 years of commercialization) has not been matched by an increase in the number of GM traits. The 'big two' traits – herbicide tolerance and pest resistance – still predominate, with the most rapid advance being the development of stacked herbicide-/pest-resistant crops. The predominant GM crop species also remain the major US commercial crops (soybean, maize, cotton, and oilseed rape), rather than the most important world food crops (rice and wheat). Of particular note is the fact that several of the important agronomic traits such as disease and stress resistance that we discussed at length in the first edition appear to have foundered at the pre-commercialization stage. Some of the more novel 'molecular farming' traits about which there was considerable excitement a few years ago have also still to make a commercial impact.

So what is new in the field of plant biotechnology that justifies a second edition? There have been some important technological advances in the rationalization of transformation vectors around the Gateway platform, and the clarification of the miRNA mechanisms underpinning sense and antisense transgene silencing. However, the major developments on the horizon are those that encompass 'post-genomics' in their deeper understanding of biological systems and the effect of genetic modification upon them. Indeed, some of the advances such as TILLING could be described as 'post-GMics' in the sense that they point the way forward to selecting mutants in targeted genes without GM technology. For this reason, we have written a new chapter that reviews the current status of plant genomics and its potential impact on crop improvement.

We have also taken this opportunity to improve the format and presentation of the book. The use of colour has emphasized the structural components of chapters such as boxes and case studies. It has also allowed us to show important features in the figures and to highlight keywords in the text. The wider margins also permit key concepts to be drawn out. We trust that these improvements will help to make the text more attractive and accessible.

We would like to dedicate this second edition to all of our colleagues over the years who shared the Norman Borlaug Institute vision of harnessing plant biotechnology for humanitarian goals.

January 2008

Contents

Abbreviations

ABA	abscisic acid
ABRE	abscisic acid-response element
ACC	1-amino-cyclopropane-1-carboxylic acid
ACCase	acetyl-CoA carboxylase
ACMV	African cassava mosaic virus
ACP	acyl carrier protein
AdoMetDC	S-adenosyl-L-methionine decarboxylase
AFLP	amplified-fragment length polymorphism
AGI	*Arabidopsis* Genome Initiative
AlMV	alfalfa mosaic virus
ALS	acetolactate synthase
AMPA	aminomethylphosphonic acid
APHIS	Animal and Plant Health Inspection Service
ArMV	arabis mosaic virus
AuxRE	auxin-response element
BADH	betaine aldehyde dehydrogenase
BAP	6-benzylaminopurine
BNYVV	beet necrotic yellow vein virus
bp	base pair(s)
BSE	bovine spongiform encephalitis
Bt	*Bacillus thuringiensis*
BYDV	barley yellow dwarf virus
bZIP	basic leucine zipper
CaMV	cauliflower mosaic virus
CAT	chloramphenicol acetyltransferase
CDK	cyclin-dependent kinase
cDNA	complementary DNA
CE	coupling element
cM	centiMorgan
CMO	choline monooxygenase
CMS	cytoplasmic male sterility
CMV	cucumber mosaic virus
CoA	coenzyme A
COR	cold-responsive
CP	capsid or coat protein
CPMR	coat protein-mediated resistance

CPMV	cowpea mosaic virus
CpTI	cowpea trypsin inhibitor
CRT	C-repeat
2,4-D	2,4-dichlorophenoxyacetic acid
DAHP	3-deoxy-D-arabino-heptulosonate 7-phosphate
DCL	Dicer-like protein
DEFRA	Department for Environment, Food and Rural Affairs
DFR	dihydroflavonol-4-reductase
DHFR	dihydrofolate reductase
DNA	deoxyribonucleic acid
DP	degree of polymerization
2D-PAGE	two-dimensional polyacrylamide gel electrophoresis
DRE	dehydration-responsive element
ds	double-stranded
DST	downstream element
EDTA	ethylenediaminetetra-acetic acid
EFSA	European Food Safety Authority
ELISA	enzyme-linked immunosorbent assay
EPA	(US) Environmental Protection Agency
EPSPS	5-enolpyruvylshikimate-3-phosphate synthase
ER	endoplasmic reticulum
ERA	environmental risk assessment
EST	expressed-sequence tag
EU	European Union
FDA	(US) Food and Drug Administration
GBSS	granule-bound starch synthase
GFP	green fluorescent protein
GGDP	geranyl geranyl diphosphate
GM	genetically modified
GMO	genetically modified organism
GNA	*Galanthus nivalis* (snowdrop plant) agglutinin
GOX	glyphosate oxidase
GSH	glutathione
GSSG	oxidized glutathione
GST	glutathione S-transferase
GUS	β-glucuronidase
HEAR	high-erucic-acid rape
HIV	human immunodeficiency virus
HPP	hydroxyphenylpyruvate
HPPD	hydroxyphenylpyruvate dioxygenase
hpRNA	hairpin RNA
HR	hypersensitive response
HSA	human serum albumin
HSE	heat-shock element

HSF	heat-shock factor
HSP	heat-shock protein
IAA	indole-3-acetic acid
ICP	insecticidal crystal protein
Ig	immunoglobulin
2iP	N^6-(2-isopentyl)adenine
IPR	intellectual property right
IRGSP	International Rice Genome Sequencing Project
IRRI	International Rice Research Institute
ISAAA	International Service for the Acquisition of Agri-Biotech Applications
JIP	jasmonate-induced protein
kb	kilobase(s)
kDa	kilodalton
K_M	Michaelis–Menten constant
LEAR	low-erucic-acid rape
LRE	light-responsive element
LRR	leucine-rich repeat
LTRE	low-temperature-response element
MAPK	mitogen-activated protein kinase
MAR	matrix-attachment region
MCS	multiple cloning site
miRNA	microRNA
MP	movement protein
mRNA	messenger RNA
MS	Murashige and Skoog (culture medium)
4-MU	methylumbelliferone
4-MUG	4-methylumbelliferryl-β-D-glucuronide
NAA	1-naphthylacetic acid
NADPH	reduced nicotinamide adenine dinuleotide phosphate
NB	nucleotide-binding
NLS	nuclear localization signal
NPC	nuclear-pore complex
NSP	nuclear shuttle protein
NVCP	Norwalk virus capsid protein
ORF	open reading frame
PAMP	pathogen-associated molecular pattern
PCR	polymerase chain reaction
PDR	pathogen-derived resistance (originally known as parasite-derived resistance)
PEG	polyethylene glycol
PEP	phosphoenolpyruvate
PEPC	phosphoenolpyruvate carboxylase
PG	polygalacturonase
PHA	polyhydroxyalkanoate

PHB	polyhydroxybutyrate
Phy	phytochrome
PLRV	potato leafroll virus
PME	pectin methylesterase
Pol	polymerase
PPT	phosphinothricin (also known as glufosinate)
PR	pathogenesis-related
PRSV	papaya ringspot virus
PTGS	post-transcriptional gene silencing
pv.	pathovar
PVX	potato virus X
PVY	potato virus Y
QTL	quantitative trait locus
RAPD	random-amplified polymorphic DNA
rbcL	ribulose-1,5-bisphosphate carboxylase/oxygenase large subunit
rbcS	ribulose-1,5-bisphosphate carboxylase/oxygenase small subunit
RdRp	RNA-dependent RNA polymerase
RFLP	restriction-fragment length polymorphism
RIP	ribosome-inactivating protein
RISC	RNA-induced silencing complex
RNA	ribonucleic acid
RNAi	RNA interference
ROS	reactive oxygen species
rRNA	ribosomal RNA
Rubisco	ribulose-1,5-bisphosphate carboxylase/oxygenase
S	Svedberg unit
SA	streptococcal antigen
SAG	senescence-activated gene
SAGE	serial analysis of gene expression
SAR	systemic acquired resistance
SAUR	small auxin-up RNA
SBE	starch-branching enzyme
scFv	single-chain variable fragment antibody
sg	subgenomic
sIgA	secretory IgA
siRNA	small interfering RNA (or guide RNA)
SNP	single-nucleotide polymorphism
SOD	superoxide dismutase
ss	single-stranded
SSR	simple sequence repeat
TAG	triacylglycerol
TBP	TATA-binding protein
T-DNA	transfer DNA (that part of the Ti plasmid of *Agrobacterium* spp. incorporated into the genome of infected cells; not tDNA)

TetC	tetanus toxin fragment C
TetR	tetracycline repressor
TEV	tobacco etch virus
Ti	tumour-inducing
TILLING	targeted induced local lesions in genomes
TMV	tobacco mosaic virus
TPS	Technology Protection System
tRNA	transfer RNA
TSWV	tomato spotted wilt virus
tTA	tetracycline transactivator
TTSS	type-III protein-secretion system
uORF	upstream ORF
USDA	US Department of Agriculture
UTR	untranslated region
VIP1	VirE2-interacting protein 1
VLP	virus-like particle
VPg	viral genome-linked protein
ZYMV	zucchini yellow mosaic virus

1

Plant genomes: the organization and expression of plant genes

INTRODUCTION

In eukaryotes, genetic information is stored in the form of a polymer called **deoxyribonucleic acid (DNA)**. This DNA stores all the **blueprints**, or instructions, for making and controlling the complex machinery of life. The instructions are stored in the DNA as genes. Genes are the DNA sequences that, in molecular terms, are responsible for making proteins (enzymes and structural proteins) and functional ribonucleic acid (RNA; another class of nucleic acid). In plants (as in most eukaryotes), each gene codes for one protein or functional RNA (although we will see later in the chapter that this is a very simplified view), so not surprisingly for such a complex organism as a plant, there are a very large number of genes. The total amount of DNA in the nucleus of a cell, or in organelles, is called the **genome**. In nuclei, the DNA (the nuclear genome) takes the form of large, linear DNA molecules called **chromosomes**. The size and number of chromosomes varies between different plant species; consequently the size of the genome varies between plant species.

● Plants contain three genomes: the nuclear, mitochondrial, and chloroplast genomes.

Plants, unlike animals, also contain two other genomes (Box 1.1). In common with animals they have a mitochondrial genome, but plants also have a chloroplast genome. The chromosomes in mitochondria and chloroplasts are not linear but circular, and there may be many copies of the genome in each organelle. The nuclear genome contains the majority, but by no means all, of the genetic information. The nuclear genome is also the genome most commonly manipulated in plant biotechnology. This chapter will attempt to give an overview of this genome and explain how the genes in it are arranged and regulated.

DNA, chromatin, and chromosome structure

DNA is the famous double helix, the structure of which was elucidated by Crick and Watson in 1953. It is a linear, double-stranded molecule, the two strands of which run in opposite directions (they are said to be antiparallel), being held together by hydrogen bonding. For this DNA to be packed into a nucleus it must be arranged carefully and higher-order structures imposed on it. The necessity for this packing is not surprising when one considers that the DNA in the nucleus of a typical higher eukaryote might have a total length of several metres.

BOX 1.1 *The chloroplast and mitochondrial genomes of plants*

Plastids and mitochondria are thought to have evolved from prokaryotes that were taken up by eukaryotic cells in symbiotic associations very early in their evolution. Although no longer capable of existing independently of the eukaryotic cell, they retain many features of their prokaryotic ancestors. Both organelles have their own genome, although by comparison with the nuclear genome it is extremely simple. Part of the reason for this simplicity seems to be that during evolution many of the genes from these organelles have been transferred to the nucleus. Thus, many proteins that function in chloroplasts or mitochondria are encoded by nuclear genes, the proteins being transported to the organelle.

The genomes of the organelles have maintained many of the features of those of their prokaryote ancestors. Organelle genomes are circular, like bacterial genomes, and do not form complex structures like the chromatin of the nuclear genome. Some of the genes in chloroplasts and mitochondria are arranged in operons, with a common regulatory element or promoter. The promoters of many chloroplast and mitochondrial genes exhibit features of prokaryotic promoters. Protein synthesis in chloroplasts and mitochondria also resembles that of prokaryotes.

The chloroplast genome

The chloroplast genome is a circular double-stranded DNA molecule.

The chloroplast genome is a circular, double-stranded DNA molecule (or chromosome) located in the stroma of the chloroplast. Chloroplast genomes are highly conserved among plant species. There is more than one copy of the genome in each chloroplast (the exact number varies during development, but mesophyll cells in young leaves contain about 100 copies of the genome).

The details of the chloroplast genome can be looked at with reference to the figure, which shows the chloroplast genome of maize in detail and some comparisons with the chloroplast genomes of other species. Most chloroplast genomes are between 120 and 160 kb in size and contain 120–140 genes. Chloroplast genomes can be categorized into three groups, based on the presence or absence of inverted repeats. The majority of land plants (excluding some legumes and conifers) are characterized by the presence of inverted repeats and have group II genomes.

As can be seen from the table, most of the genes present in the chloroplast genome are involved in either protein synthesis or photosynthesis (although the majority of the genes encoding proteins for photosynthesis are located in the nucleus). Many of these genes are in clusters, allowing expression in the form of large polycistronic primary transcripts, which are processed to oligo- or monocistronic messenger RNAs.

The mitochondrial genome

Mitochondria, which are common to all eukaryotes, also contain their own genome. Plant mitochondrial genomes, which are much larger than those of mammals and yeasts, range in size from about 200 kb in *Brassica* spp. to over 2500 kb in *Cucumis melo* (muskmelon). This variability stems, in part, from the accumulation of large regions of non-coding DNA. Thus, the maize mitochondrial genome is much larger than the chloroplast genome but contains far fewer genes, most of the genes being involved in protein synthesis or respiratory electron transport. Plant mitochondrial DNA also contains chloroplast DNA (termed promiscuous DNA).

BOX 1.1 *Continued*

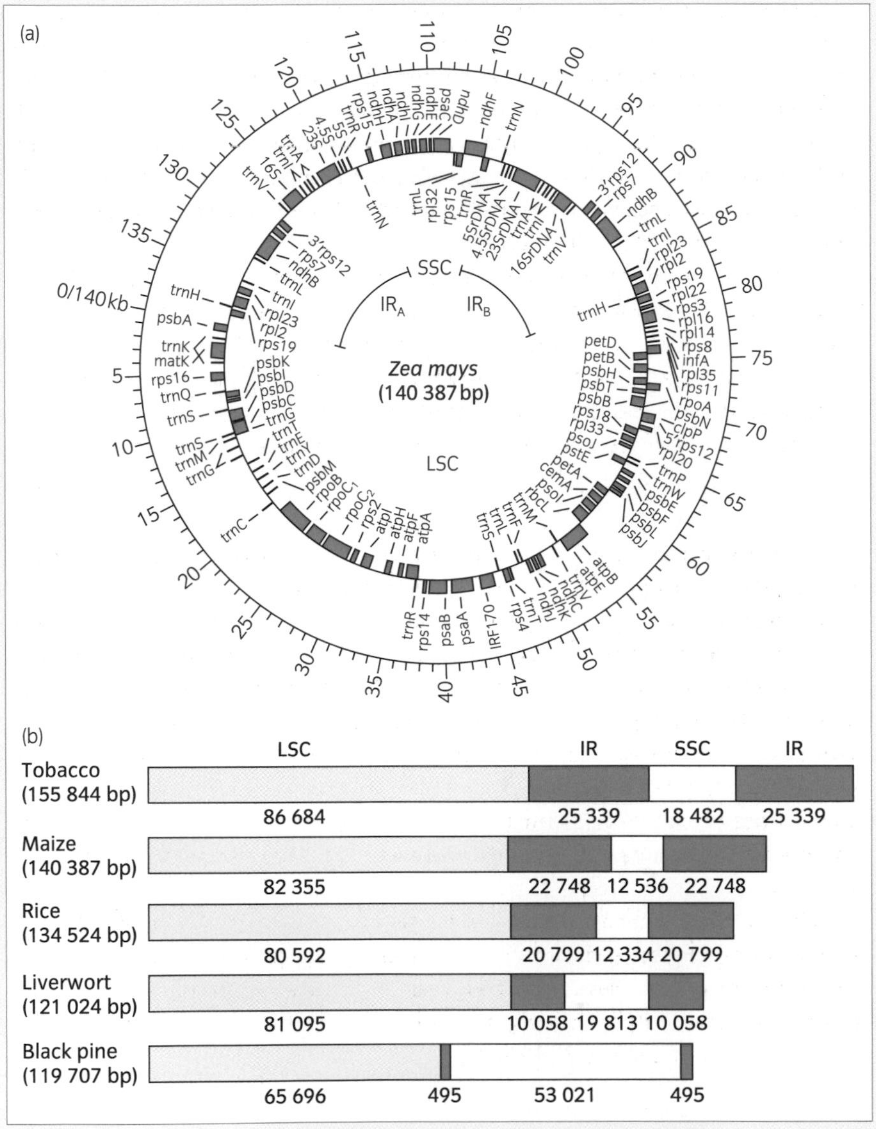

(a) Maize chloroplast genome map. The chloroplast genome of maize is 140 387 base pairs (bp). It contains a pair of inverted repeats (IR_A and IR_B), which are separated by the small (12 536 bp) and large (82 355 bp) single-copy regions (SSC and LSC, respectively). Genes on the outside of the circle are transcribed clockwise, those on the inside anticlockwise. Overall, the A + T content is 61.5%, although this varies from 71.2% in non-coding regions to 45.3% in ribosomal RNA genes. (b) Length and feature analysis of some sequenced chloroplast genomes. (Redrawn with permission from Maier *et al.* (1995).)

(Continued overleaf)

BOX 1.1 *Continued*

The plant mitochondrial genome can, surprisingly, exhibit changes in its structure (although this property is not universal). Complete, circular master chromosomes (or master circles) are postulated to contain all the mitochondrial genes; these, however, have never been isolated or identified. Subgenomic, circular DNA circles lacking a complete set of mitochondrial genes have been identified. These are thought to be formed from the master chromosome by recombination events between repeated DNA sequences in the mitochondrial DNA.

Gene-product function	Acronym[a]	Number of genes in chloroplast genome
Transcription/translation		
rRNA	*rrn*	4
tRNA	*trn*	30
Ribosomal proteins		
— small subunit	*rps*	12
— large subunit	*rpl*	9
RNA polymerase	*rpo*	4
Translation initiation factor	*inf*	1
Photosynthesis		
Photosystem I	*psa*	5
Photosystem II	*psb*	14
Large subunit of Rubisco	*rbcL*	1
Cytochrome	*pet*	4
ATP synthase subunits	*atp*	6
NADH dehydrogenase	*ndh*	11

[a]The acronym can be used to identify the position of the gene in the genome.
Rubisco, ribulose-1,5-bisphosphate carboxylase/oxygenase.

Chromatin

● Nuclear DNA displays extensive secondary structure and is associated with proteins.

The first-order structure of DNA packing consists of DNA wrapped round cores composed of histones, which are basic proteins (Figure 1.1). Each core has two turns of DNA wrapped round it (approximately 150 base pairs (bp)) and is termed a nucleosome. Each nucleosome is separated from the next by spacer DNA, the length of which can vary. This continuous string of nucleosomes is termed the 'beads on a string' structure and is further coiled to produce the 30-nm fibre. The latter fibre has a coiled or solenoid structure, with six nucleosomes to every turn. These 30-nm fibres can be further organized by anchoring the fibre at various points to a protein scaffold and introducing loops. During mitosis, when chromosomes condense and become visible (with the aid of a microscope); these loops can be further coiled.

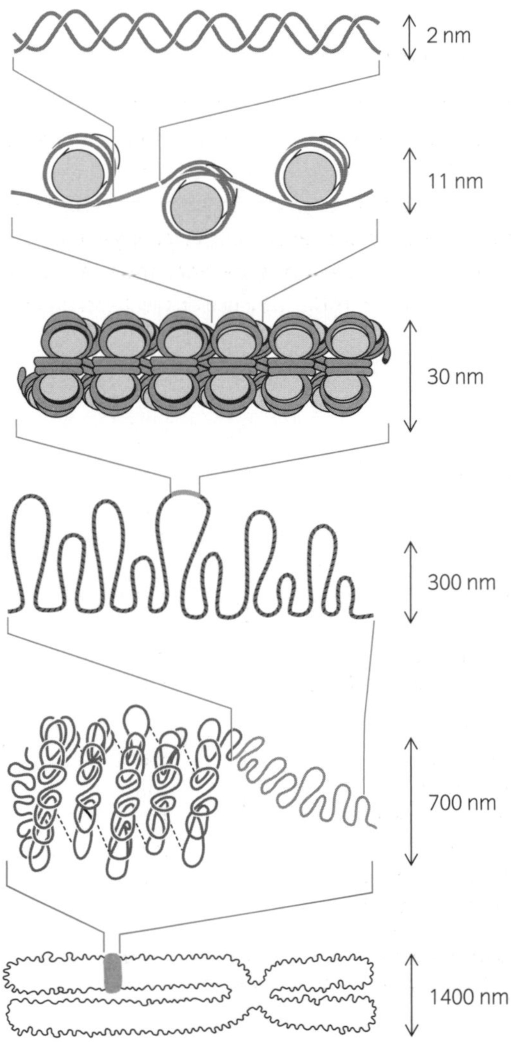

Figure 1.1 The structure of DNA, chromatin, and the higher-order packing of chromatin. DNA is a linear, double-stranded molecule. The nucleosome is the basic unit of chromatin organization and consists of approximately 150 bp of DNA (two turns) wrapped round a core of histone proteins. This core is composed of two molecules each of histones H2A, H2B, H3, and H4. Nucleosomes are separated by spacer DNA to which histone H1 is attached. This structure of a string of nucleosomes considerably reduces the length of the DNA. This structure is further coiled to produce the 30-nm fibre, which has a solenoid structure with six nucleosomes to every turn. This 30-nm fibre can be organized further into loops by anchoring the fibre at AT-rich areas (scaffold-associated regions) to a protein scaffold. In mitosis, when the transcription of the vast majority of genes is halted, the loops can be further coiled and the chromosomes become condensed. (Reproduced with permission from Alberts *et al.* (1989).)

An introduction to gene structure and gene expression

There are three main types of RNA: mRNA, rRNA, and tRNA.

Defined regions of the genome, termed transcription units, serve as templates for the synthesis of another nucleic acid, called ribonucleic acid (RNA). There are three main types of RNA: ribosomal RNA (rRNA), which is a structural component of ribosomes; transfer RNA (tRNA), which is a class of small RNAs involved in protein synthesis (or translation); and messenger RNA (mRNA), which carries the protein-coding information from DNA and is subsequently translated during protein synthesis. Recently other classes of RNA that are extremely important in a diverse range of processes have been discovered and these are described in Box 1.2 and in other chapters (Chapters 4 and 8) where they are relevant to the process being discussed. Although rRNA and tRNA are not translated to give proteins, the DNA regions that code for these RNAs are still termed genes.

Gene structure and expression in a eukaryotic protein-coding gene

The structure of plant protein-coding genes is similar to other eukaryotic protein-coding genes.

As plant biotechnology deals largely with genes that encode proteins, the structure of this type of gene can be looked at in more detail. Figure 1.2 gives a diagrammatic representation of the main features of a eukaryotic protein-coding gene.

BOX 1.2 *Small RNAs*

One of the most important discoveries of recent years is that RNA molecules can regulate gene expression and thereby control or influence a diverse range of processes. Prior to this discovery it was given that RNAs (see An introduction to gene structure and gene expression) could be divided into two general categories: mRNA that carries the protein-coding information from DNA and is translated during protein synthesis, and the structural/decoding rRNA and tRNA. However, the recent elucidation of a variety of gene-silencing phenomena, key among them being RNA interference (RNAi) and post-transcriptional gene silencing (PTGS), has lead to the discovery of a novel class of RNA that is involved in regulating gene expression.

MicroRNA

MicroRNAs (miRNAs) are short (21–24 nucleotides) RNAs encoded in the genome of plants (and most other multicellular organisms) that are involved in the PTGS of endogenous mRNAs, which exhibit a high degree of complementarity to the miRNA. In plants, miRNA genes are generally (although not exclusively) found in regions of the genome not associated with protein-coding genes. Gene families that produce identical or very similar miRNAs can be identified and these gene families may exhibit a degree of redundancy. miRNA genes are generally well conserved across plant species.

Mature miRNAs are produced by a complicated system of transcription and processing of a primary transcript (pri-miRNA) to produce the mature, single-stranded miRNA. This miRNA then interacts with its target mRNA and directs cleavage of the target mRNA. This process is described in more detail below.

BOX 1.2 *Continued*

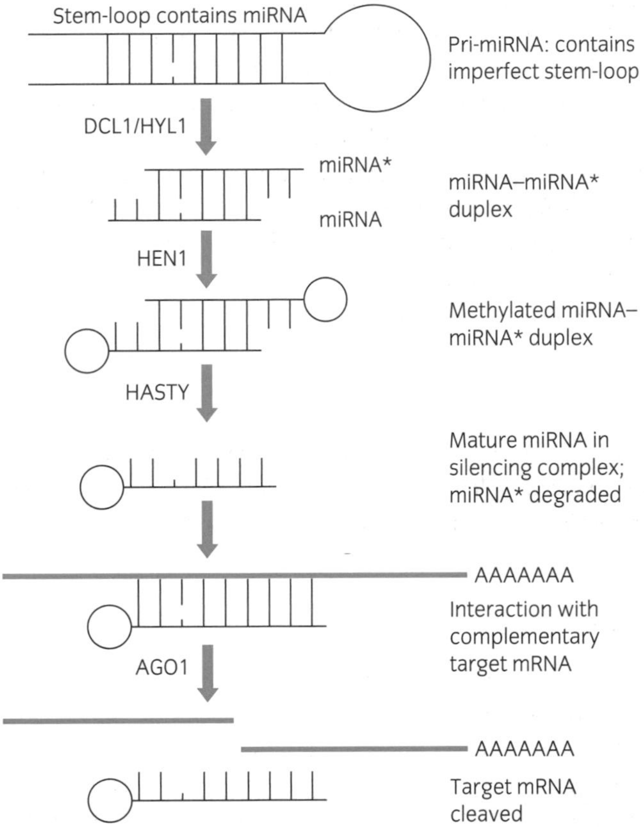

MicroRNA production and action. The primary transcripts from miRNA genes are longer than required to produce the mature miRNA. The transcript must be capable of forming a hairpin, the stem-loop of which contains the miRNA sequence. The miRNA sequence is base-paired with a complementary sequence known as the miRNA*. These pri-miRNAs are produced, like mRNAs, by RNA polymerase II-mediated transcription. The pri-miRNA is cleaved at two positions by a particular class of RNase III-like endonucleases known as Dicer-like (primarily DCL1 in *Arabidopsis*) releasing a miRNA–miRNA* duplex. Exactly how DCL1 recognizes where on the pri-miRNA molecule to cut is not known, but RNA-binding proteins such as HYL1 are thought to facilitate target recognition. The miRNA–miRNA* duplex is then methylated by HEN1, which helps to stabilize the miRNA–miRNA* duplex. To cleave the target mRNA, the miRNA–miRNA* duplex exits the nucleus, this process being facilitated by miRNA exporters such as HASTY (an importin β family member). In the cytoplasm the miRNA–miRNA* duplex is processed, the miRNA strand being incorporated preferentially into what is known as the RNA-induced silencing complex (RISC). The miRNA* strand is degraded. The RISC contains a key component, an Argonaute protein (*Arabidopsis* contains at least 10 Argonaute AGO proteins, but it appears that only AGO1 is required for miRNA function). Argonaute proteins have two domains that are key to their function in a range of gene-silencing phenomena: an RNA-binding domain (the PAZ domain) and the PIWI domain, which is important for mRNA cleavage, being related to RNase H enzymes. It is AGO1 that cleaves the target mRNA resulting in gene silencing, although miRNAs have also been shown to inhibit translation.

(Continued overleaf)

BOX 1.2 *Continued*

miRNAs are involved in regulating many developmental processes in plants, often through targeting transcription factor mRNAs. They are also involved in processes as diverse as responses to stress (both biotic and abiotic), signalling, metabolism, and—interestingly—in regulating the miRNA pathway (both the DCL1 and AGO1 mRNAs are targets for miRNA-directed degradation).

Small interfering RNAs

We have looked at miRNA and seen how it can regulate gene expression by cleaving specific mRNAs, thereby causing gene silencing. However, miRNAs were not the first kind of small RNA to be discovered.

Gene silencing

RNA silencing, or PTGS, was first observed in plants, where it was demonstrated that interactions between multiple foreign transgenes or between transgenes and endogenous genes led to the degradation of specific mRNAs. This degradation was sequence-specific and resulted in the silencing of transgenes and endogenous genes. This had obvious implications for plant biotechnology, some of which are discussed in more detail in Chapter 4. It was also shown subsequently that virus-induced gene silencing (see Chapter 8), whereby viral transcripts (and transcripts of cognate nuclear genes) were degraded, depended on a similar mechanism. With the discovery and elucidation of RNAi it became clear that these phenomena depended on the production of small (21–26-nucleotide), double-stranded RNA molecules as part of the degradation pathway. In many ways these small interfering RNAs (siRNAs, or guide RNAs) are analogous to the miRNAs described above; however, they have some unique aspects.

Unlike miRNAs, which are produced from RNA molecules with an imperfect stem-loop structure, siRNAs are produced from long, double-stranded RNA molecules. This long, double-stranded RNA can arise by several different mechanisms. In transgene-induced gene silencing it may be the product of transcription from antisense transgenes and binding of the transcript to the complementary mRNA. It may also result from transgenes arranged in an inverted repeat arrangement. If the transgene is in the sense orientation, then double-stranded RNA may be produced by the activity of RNA-dependent RNA polymerases copying the transcript (usually it is considered necessary that this is an aberrant transcript; exactly what defines an aberrant transcript is not clear) into double-stranded RNA. In virus-induced gene silencing both viral DNA and RNA can lead to the production of long, double-stranded RNA molecules.

This long, double-stranded RNA is cleaved by the activity of Dicer-like endonucleases to yield multiple siRNA duplexes. In *Arabidopsis* this cleavage usually requires not DCL1 but other members of the Dicer-like family, DCL3 and DCL4. The duplexes produced are of two sizes, small (≈21 nucleotides) and long (24–26 nucleotides), and may reflect the involvement of more than one Dicer-like activity. Like miRNA, one strand of the small class of siRNA is incorporated into the RNA-induced silencing complex (RISC), where it directs the cleavage of target RNAs.

There exists another class of siRNA, so-called *trans*-acting siRNA (ta-siRNA), which is produced by the same basic mechanism as miRNA and siRNA. There are, however, some differences: ta-siRNA results from the DCL4-mediated cleavage of long double-stranded RNAs. These

BOX 1.2 *Continued*

double-stranded RNAs are produced by an RNA-dependent RNA polymerase, which uses part of the ta-siRNA precursor as its template. This template is itself produced by miRNA-mediated cleavage of the ta-siRNA precursor. The DCL4 cleavage products are, like miRNA, incorporated into the RISC and direct cleavage of target RNAs.

RNA-directed methylation

All RNA silencing pathways are capable of directing the methylation of DNA and thereby silencing gene expression at the transcriptional level. The longer (24–26-nucleotide) siRNAs produced by Dicer-like activity on long double-stranded RNAs are thought to be responsible for the methylation of cytosine residues in regions of the target gene. This methylation results in a reduction in transcription from the gene, thus silencing the gene both transcriptionally and post-transcriptionally. A similar effect is proposed to occur with miRNAs.

Interestingly, RNA-directed DNA methylation can occur in the absence of cleavage of a target RNA. In RNA-directed transcriptional gene silencing, transcripts are copied into double-stranded RNA by RNA-dependent RNA polymerase. This double-stranded RNA is cleaved by DCL3 and the resulting siRNA incorporated into an RNA-induced transcriptional silencing complex (RITS complex). The RITS complex, like the RISC, contains a member of the Argonaute family, in this case AGO4. This mechanism is mainly involved in silencing some transposons and sequences with direct repeats.

Each transcription unit is associated with its own regulatory regions, the most important of which is considered to be a flanking region at the 5′ end of a gene; that is, the promoter. One important feature of the promoter is that it contains the binding site for the enzyme RNA polymerase II, which is responsible for making an RNA copy of the DNA in a process known as **transcription**. Transcription is initiated from defined points (transcription start sites) close to the RNA polymerase-binding site (see Figure 1.3).

The transcription unit is the region of the gene that is transcribed by RNA polymerase into pre-mRNA. This pre-mRNA is composed of a 5′ untranslated region (UTR; Figure 1.2, A), the coding region and a 3′ UTR (B). The coding region is composed of exons and introns. Exons are represented in the mature mRNA and carry the information necessary for protein synthesis (they are translated), and are therefore said to be **coding**. Exons are separated by introns (note that not all genes contain introns) that are not present in the mature mRNA and are said to be **non-coding**.

● Protein-coding regions known as exons are interrupted by non-coding introns.

The 3′ end of the gene also contains a polyadenylation, or poly(A), site (AATAAA), which acts as a signal for the adenylation of the RNA. During polyadenylation and processing of the pre-mRNA, the RNA is cleaved, resulting in the pre-mRNA being shorter than the transcription unit. The pre-mRNA is also modified at its 5′ end by the addition of a **cap** of 7-methylguanosine linked to the pre-mRNA by three

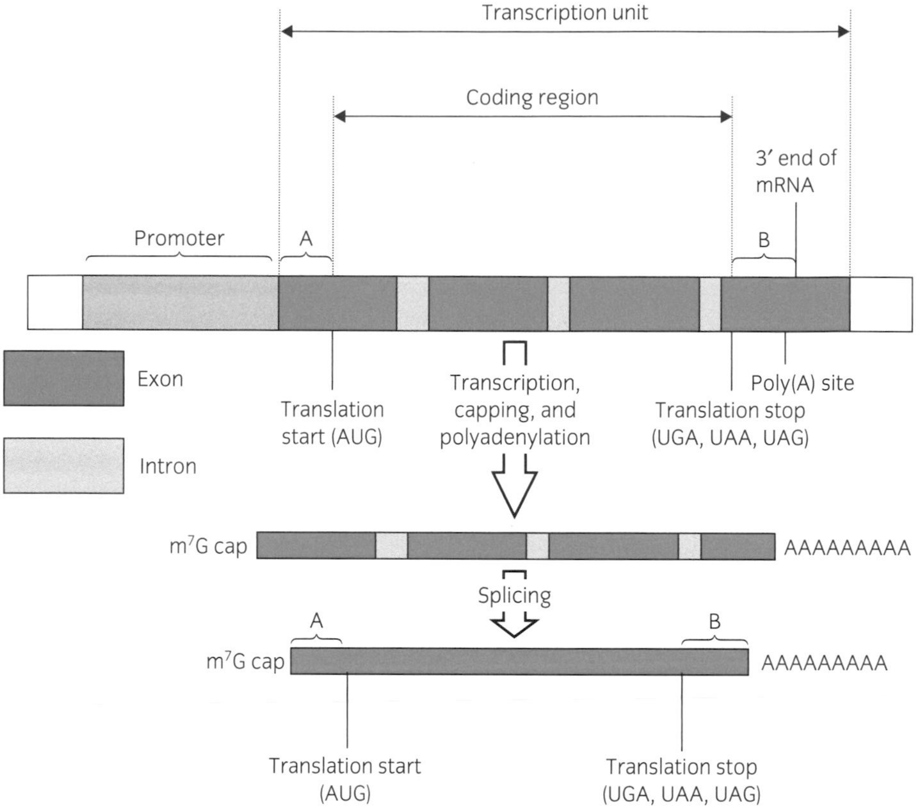

Figure 1.2 The structure and expression of a eukaryotic protein-coding gene. Details are explained in the text.

phosphate groups. The cap and the poly(A) tail tend to make the RNA more stable by preventing nuclease-mediated destruction.

Introns are removed, or spliced out, during mRNA processing. Exon/intron boundaries are delineated by specific base sequences that act as signals for the cleavage and removal of the intron. The left-hand (5′) border (AAG↓GUAAGU, where ↓ indicates the position of cleavage) of the intron is cleaved first, and the free end of the intron loops back and forms a bond with a specific residue in the right-hand border (GCAG↓GU). This structure is called a lariat. This right-hand border is then cleaved and the exons joined together. Plant introns are also characterized by AU-rich regions that may be important for intron splicing in plants. The mRNA is then exported from the nucleus to the cytoplasm where it associates with ribosomes and is translated to produce the protein gene product.

Translation

● mRNA carries the protein-coding information from DNA.

During translation the mRNA serves as a template for ordering the amino acids (Table 1.1) that make up the protein gene product. The process is known as translation because the nucleotide language of DNA and RNA is converted into the amino acid language of proteins (Figure 1.4 gives a simplified overview of translation).

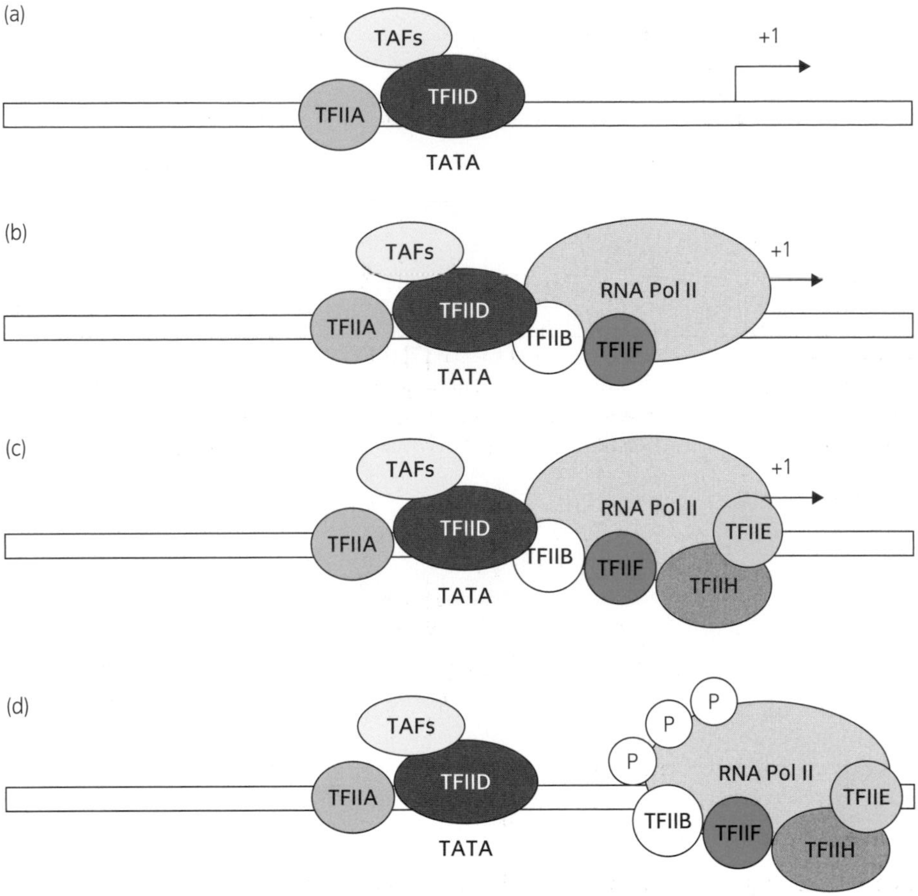

Figure 1.3 A simplified explanation of RNA polymerase II binding to a eukaryotic promoter. (a) General transcription factors bind to the TATA box of the promoter. The first protein to bind, via its TATA-binding protein (TBP) subunit, is TFIID (transcription factor IID), which is itself a multiprotein complex. This interaction requires TBP-associated factors (TAFs) and is promoted by TFIIA. (b) TFIIB binds to the complex and recruits TFIIF, which is bound to RNA polymerase (Pol) II. (c) TFIIE and TFIIH subsequently bind to the complex. An ATP-dependent helicase activity associated with TFIIH unwinds the promoter DNA. In this state the RNA polymerase is said to be in the initiation mode. (d) The C-terminal domain of RNA polymerase is then phosphorylated, causing transition to the elongation mode, and transcription is initiated from the start site (+1). In animals, it is known that several cyclin/cyclin-dependent (protein) kinase (CDK) complexes can phosphorylate the C-terminal domain of RNA polymerase II: CDK7/cyclin H (a component of TFIIH), CDK8/cyclin C, and CDK9/cyclin T. (Data taken with permission from Buratowski (2000).)

The small subunit of a ribosome bound with a tRNA carrying the amino acid methionine associates with the 5′ end of mRNA. The large subunit then binds to the small subunit and translation is initiated from the translation start site (AUG). The ribosome passes along the mRNA in a 5′ to 3′ direction. When a stop codon is reached the ribosome dissociates into its two subunits and the protein is released. Several ribosomes may translate an mRNA at once, forming a polysome.

Table 1.1 Amino acids

Amino acid	Single-letter representation[a]	Three-letter representation[a]	Number of codons[b]
Alanine	A	Ala	4
Arginine	R	Arg	6
Asparagine	N	Asn	2
Aspartic acid	D	Asp	2
Cysteine	C	Cys	2
Glutamic acid	E	Glu	2
Glutamine	Q	Gln	2
Glycine	G	Gly	4
Histidine	H	His	2
Isoleucine	I	Ile	3
Leucine	L	Leu	6
Lysine	K	Lys	2
Methionine	M	Met	1
Phenylalanine	F	Phe	2
Proline	P	Pro	4
Serine	S	Ser	6
Threonine	T	Thr	4
Tryptophan	W	Trp	1
Tyrosine	Y	Tyr	2
Valine	V	Val	4
STOP	—	—	3

[a]Each amino acid can be represented by either a one- or a three-letter code.
[b]Each amino acid can be 'coded' for by up to six codons.

Each amino acid is coded for by a group of three nucleotides called a codon (see Table 1.2). These codons are translated one at a time. Some amino acids are encoded by only one codon, some by multiple codons. Amino acids have to be attached to a tRNA molecule that acts as an adapter to link each mRNA codon with the appropriate amino acid. In order to do this each type of tRNA has to fulfil two tasks. An anticodon (i.e. three complementary nucleotides) must base-pair with the appropriate codon on the mRNA (Figure 1.5). The 3′ end of the tRNA must also bind to a specific amino acid, whence the tRNA is said to be charged.

Protein synthesis actually occurs on specialized multiprotein complexes called ribosomes. Ribosomes are composed of both proteins and rRNA. Ribosomes function to bring the tRNAs carrying amino acids into proximity with the mRNA template and to catalyse the formation of peptide bonds between the amino acids to form the protein.

Initiation of translation

● Translation occurs in the cytoplasm on ribosomes.

The small subunit of a ribosome, which already has a tRNA charged with methionine associated with it, binds to the 5′ cap structure of the mRNA. The small subunit then identifies the start AUG codon on the mRNA, establishing the start position and

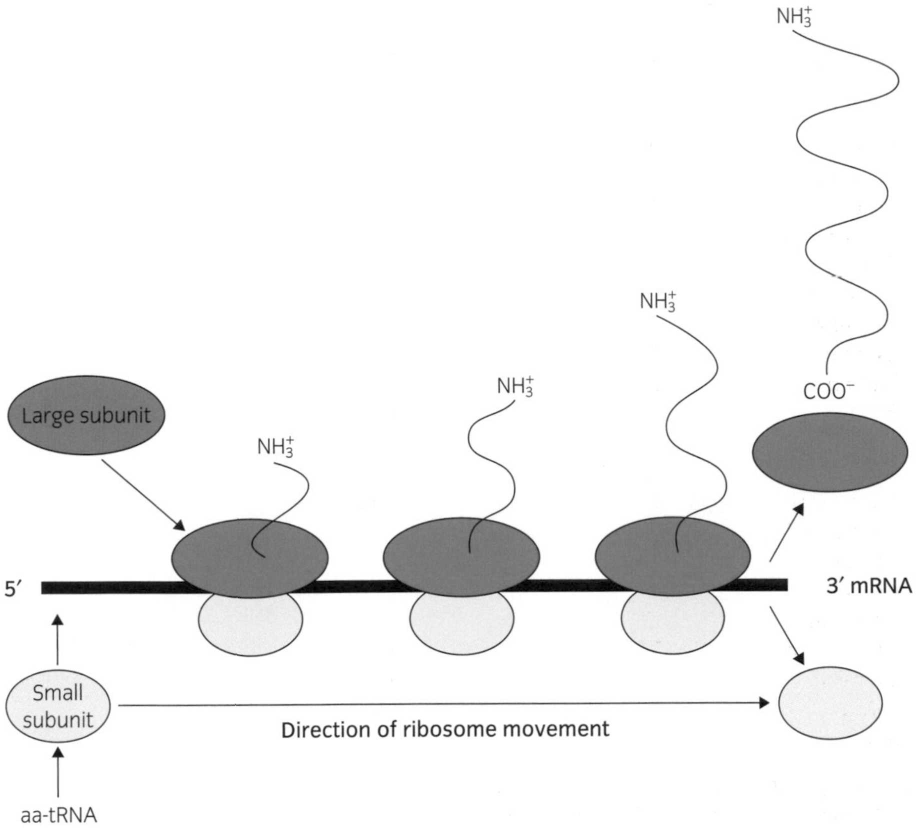

Figure 1.4 A simplified overview of translation. (Adapted with permission from Buchanan *et al.* (2000).) Details are explained in the text.

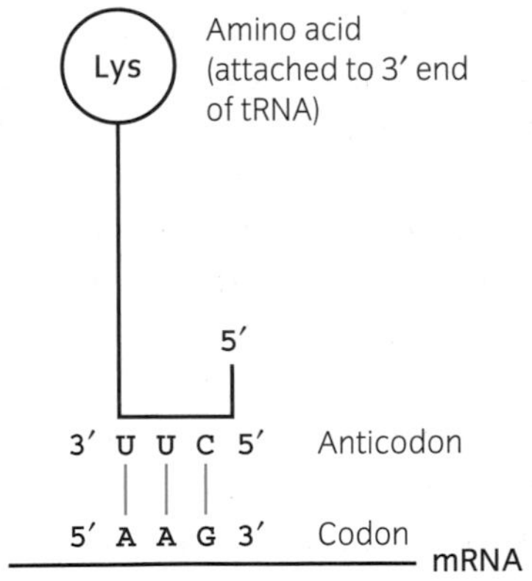

Figure 1.5 Codon/anticodon base-pairing between mRNA and tRNA. The charged (aminoacyl) tRNA carrying an amino acid (in this case lysine) at its 3′ terminus base-pairs with the codon of the mRNA by means of complementary bases, termed the anticodon.

reading frame for translation. This is done in a process known as scanning as the small subunit of the ribosome travels along the mRNA in a 5′ to 3′ direction. Translation is generally initiated from the first AUG codon on the mRNA, but the sequence surrounding this start codon influences the efficiency of initiation of translation. The large subunit of the ribosome then binds to the small subunit.

Table 1.2 The genetic code

		Second base							
		U[a]		C		A		G	
	U[a]	UUU	Phe	UCU	Ser	UAU	Tyr	UGU	Cys
		UUC	Phe	UCC	Ser	UAC	Tyr	UGC	Cys
		UUA	Leu	UCA	Ser	UAA	STOP	UGA	STOP
		UUG	Leu	UCG	Ser	UAG	STOP	UGG	Trp
	C	CUU	Leu	CCU	Pro	CAU	His	CGU	Arg
		CUC	Leu	CCC	Pro	CAC	His	CGC	Arg
		CUA	Leu	CCA	Pro	CAA	Gln	CGA	Arg
First base		CUG	Leu	CCG	Pro	CAG	Gln	CGG	Arg
	A	AUU	Ile	ACU	Thr	AAU	Asn	AGU	Ser
		AUC	Ile	ACC	Thr	AAC	Asn	AGC	Ser
		AUA	Ile	ACA	Thr	AAA	Lys	AGA	Arg
		AUG	Met	ACG	Thr	AAG	Lys	AGG	Arg
	G	GUU	Val	GCU	Ala	GAU	Asp	GGU	Gly
		GUC	Val	GCC	Ala	GAC	Asp	GGC	Gly
		GUA	Val	GCA	Ala	GAA	Glu	GGA	Gly
		GUG	Val	GCG	Ala	GAG	Glu	GGG	Gly

[a]In RNA, T is replaced with U (uracil).

Notes: 61 of the possible 64 codons code for amino acids. Three (UAA, UAG, UGA) cause termination of translation and are termed STOP codons. AUG is the translation initiation codon, but also codes for internal methionines. One amino acid is inserted into the growing protein for every three-nucleotide codon of the mRNA. For amino acids with multiple codons, some species will use particular codons more frequently than others, exhibiting so-called codon preference. This point can be illustrated if we look at the codon usage for proline in tomato and barley. There are four codons for proline (CCU, CCC, CCA, and CCG), which in tomato are used with the following frequencies: CCU 29%, CCC 16%, CCA 50%, and CCG 5%. In barley, the codon usage is CCU 14%, CCC 24%, CCA 31%, and CCG 31%. Thus, in tomato, CCG is the least commonly used codon for proline, whereas in barley it is the most commonly used.

Elongation

The addition of amino acids to the growing protein occurs one amino acid at a time and involves three sites of the ribosome, which are termed P, A, and E (see Figure 1.6 for an overview of translation elongation). The P site is where peptide bonds are formed between the growing protein and the charged tRNA that occupies the P site. The A site, which is empty, binds a charged tRNA if the anticodon of the tRNA matches the codon exposed in the A site. Any tRNA (now uncharged) in the E site is ejected. A peptide bond is formed between the amino acid on the tRNA in the A site and the growing protein, effectively transferring the growing protein to the amino acid attached to the tRNA occupying the A site. This process results in the growing protein being extended by one amino acid.

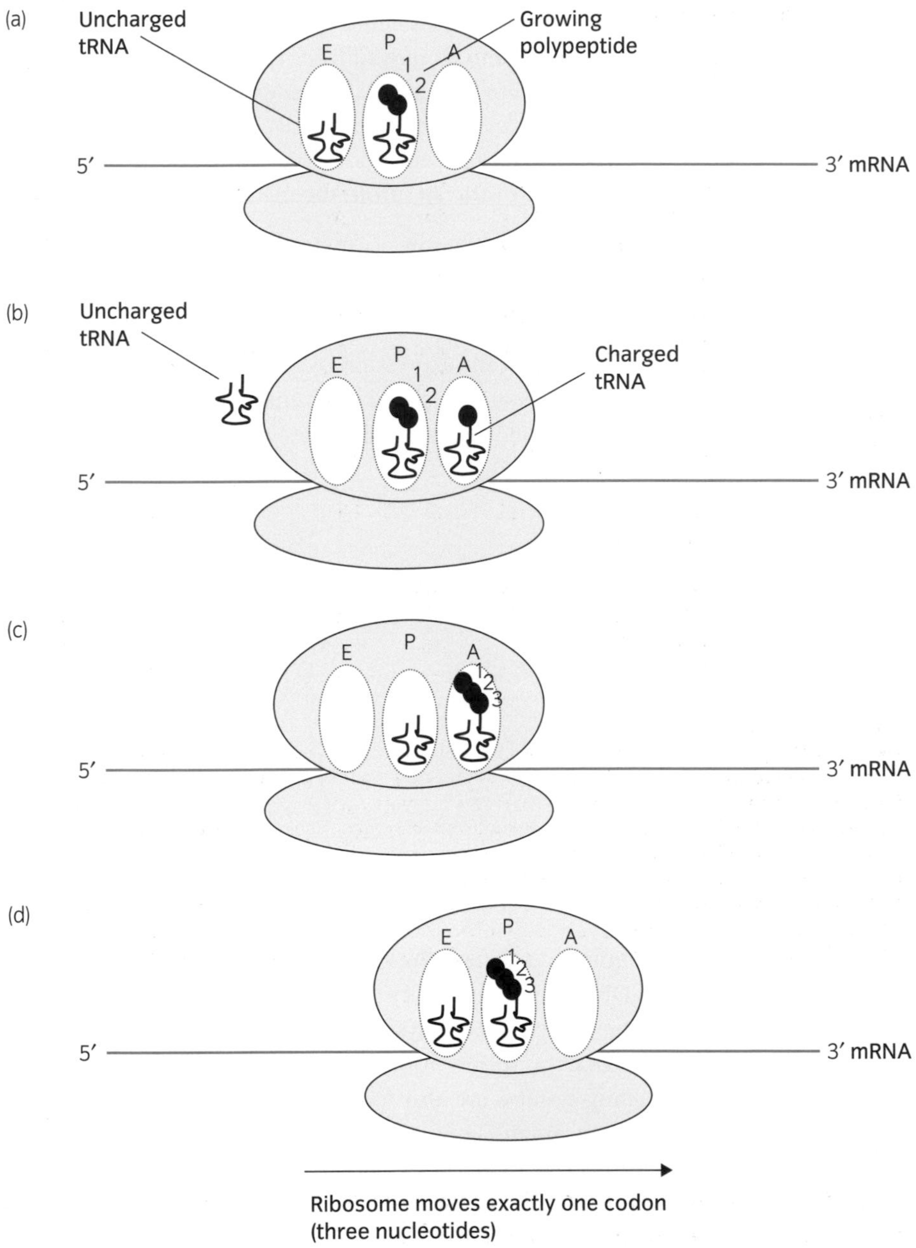

Figure 1.6 Elongation reactions of protein synthesis. (a) A tRNA, to which is attached the growing protein, occupies the P site of the ribosome. The E site is occupied by an uncharged tRNA. The A site is empty, thereby exposing a codon on the mRNA. (b) A charged (aminoacyl) tRNA binds to the A site if its anticodon is complementary to the codon of the mRNA exposed in the A site. The uncharged tRNA that was occupying the E site is ejected. (c) A peptide bond is formed between the growing protein (on a peptidyl-tRNA) and the amino acid on the aminoacyl-tRNA occupying the A site. The growing protein is transferred to the tRNA occupying the A site. The protein is now one amino acid longer. (d) The uncharged tRNA occupying the P site moves to the E site. The peptidyl-tRNA in the A site moves to the P site and the ribosome moves along the mRNA by exactly one codon. This process is called translocation. This translocation exposes the next codon of the mRNA in the now-vacant A site. (Adapted with permission from Buchanan *et al.* (2000).)

The uncharged tRNA now occupying the P site is moved into the now-vacant E site. The tRNA (with the growing protein) is moved from the A site to the P site and the ribosome moves along the mRNA by one codon. This movement exposes another codon in the A site, which then binds another charged tRNA with the appropriate anticodon. This process is repeated until a stop codon is reached, at which point the protein is released and the ribosome dissociates from the mRNA.

Regulation of gene expression

● Gene expression can be regulated at many different points.

It is obviously of vital importance that the expression of genes is carefully controlled to ensure they are expressed in the correct spatial and temporal fashion. In eukaryotes, gene expression can be controlled at many different levels or stages. These can be classified broadly as follows:

1 chromatin conformation;
2 gene transcription;
3 nuclear RNA modification, splicing, turnover, and transport;
4 cytoplasmic RNA turnover;
5 translation;
6 post-translational modification;
7 protein localization;
8 protein turnover.

Chromatin conformation

Chromatin conformation influences gene expression by affecting the ability of RNA polymerase to bind to the DNA. The accessibility of DNA in the nucleosome to RNA polymerase is regulated by acetylation of lysine residues in the core histones. Methylation of DNA can block transcription by altering chromatin conformation. Other chromatin-remodelling systems are also found in plants. These systems can induce reversible changes in chromatin conformation that can control gene expression. Changes in chromatin conformation can, unlike regulation by transcription factors, affect gene expression over large stretches of the chromosome.

Gene transcription

● Control regions in protein-coding genes are located predominantly at the 5′ end of the gene in the promoter.

Most of the controls present in a plant gene are located 5′ of the transcription start site in the region known as the promoter. The promoter contains various sequence elements (short stretches of DNA of a defined sequence) that function to recruit the proteins that regulate transcription of the gene (Figure 1.7). These sequence elements are termed *cis*-elements (because they are on the same DNA strand as the coding region of the gene). The most basic of these sequence elements is the TATA box, which

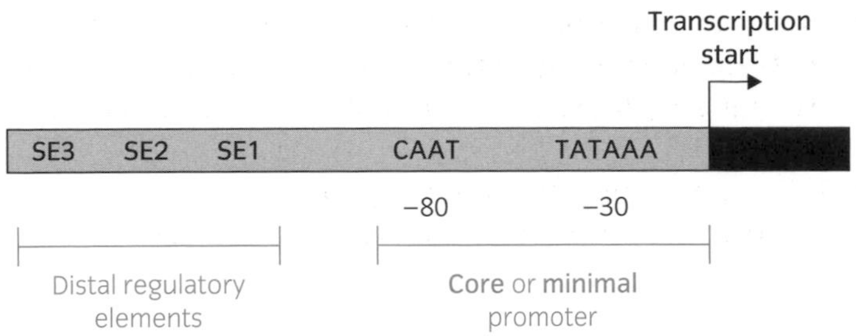

Figure 1.7 A simplified diagram of the promoter from a eukaryotic protein-coding gene. The promoter region contains several sequence elements (*cis*-elements). Some, such as the TATA box [consensus sequence T(C/G)TATA(T/A)A_{1-3}] and the CAAT box , comprise what is known as the minimal or core promoter. Much of the core promoter is involved in binding general transcription factors and RNA polymerase II. Core or minimal promoters are often incapable of initiating transcription by themselves. Promoters also contain distal regulatory elements (sequence elements, or SEs in this figure) that bind transcription factors (*trans*-activating factors), which regulate transcription from the gene. Transcription factors can either promote transcription or inhibit transcription. Several different transcription factors may bind to the promoter of a gene. The regulation of plant gene expression by internal and external stimuli and in particular tissues is mediated by the binding and action of a host of transcription factors. Many genes are capable of responding to more than one stimulus due to the presence of multiple *cis*-elements in their promoter. The presence of multiple *cis*-elements in a promoter and the large number of transcription factors means there is a huge number of potential control 'modules' that can be used to regulate gene expression. Gene expression may also be regulated by elements that are located a great distance from the core promoter. These elements, which may be located either 5′ or 3′ of the gene they regulate, are termed enhancers.

is found about 25–30 bp upstream of the transcription start site. The TATA box functions to direct RNA polymerase II to the correct position on the gene to initiate transcription. Some genes contain multiple TATA boxes, whereas others may contain no recognizable TATA box. The lack of a TATA box was previously thought to be a feature associated with housekeeping genes (i.e. ones that are expressed constitutively) but this has proved not to be the case, with some inducible genes having no TATA box in their promoter. Other motifs can compensate for the lack of a TATA box, and in some classes of genes (such as nuclear genes involved in photosynthesis) a TATA-less promoter may be common. Other sequence elements are often found further upstream from the TATA box. In eukaryotic genes the most common of these are the CAAT and GC boxes, which enhance the activity of RNA polymerase. Sequence elements such as the TATA box are often referred to as core or minimal promoter elements, as they contribute to the binding of RNA polymerase to the promoter and are responsible for the initiation of basal transcription. The core promoter, via interactions with other promoter elements, plays an important role in the regulation of plant gene expression.

Other sequence elements are also found in the promoters of genes. These are involved in binding specific proteins, termed transcription factors (Box 1.3), which are required for transcription by RNA polymerase. These transcription factors (or

BOX 1.3 Plant transcription factors

Plant transcription factors contain several domains: a DNA-binding region, an oligomerization region, and a transcription regulation domain. Most transcription factors have only one type of DNA-binding region and one type of oligomerization region, although some may contain two distinct types. The DNA-binding region contains amino acids (which are often basic) that interact with DNA at the *cis*-sequence elements. These determine the specificity of the transcription factor. The oligomerization region is important in determining protein–protein interactions, as most transcription factors can form hetero- and/or homodimers. The formation of these dimers may affect sequence-element specificity, transactivation efficiency, or DNA-binding affinity. Transcription regulation domains either repress or activate transcription.

Transcription factors are classified, or categorized, on the basis of their structural features. On the whole, plants appear to have evolved independently when it comes to transcriptional control. Some classes of eukaryotic transcription factor appear to be absent from the genome of *Arabidopsis*, whereas many classes are apparently unique to plants. Some conserved classes of transcription factor fulfil different functions in plants as opposed to animals.

The main classes of transcription factor are listed below.

1 Helix–turn–helix (HTH): this type of transcription factor binds DNA as a dimer.

2 Leucine zipper: this type also binds DNA as a dimer. A variation is the basic zipper (bZIP).

3 Zinc finger: binds to DNA as either a dimer or a monomer.

4 Helix–loop–helix (HLH): binds to DNA as a dimer.

5 High-mobility group (HMG)-box motif.

trans-acting factors) are involved in linking various signals (both external and internal) to gene expression and are responsible for determining the level, place, and timing of expression. Similar sequence elements are often found in genes with similar functions or that respond to similar signals.

RNA modification, splicing, turnover, and transport

Many processes must be completed so that functional mRNA can enter the cytoplasm for subsequent translation. All these processes may be used to exert some control over gene expression. Different splice sites may be recognized during splicing of the pre-mRNA, thus generating different transcripts from the same gene. This alternative splicing may be used to generate different transcripts in different tissues, or in response to different stimuli, or to generate multiple transcripts from a single gene (Box 1.4 gives some examples of alternative splicing in plants). Alternative splicing is a common control mechanism in some organisms, and some well-documented examples of biological significance exist in plants.

BOX 1.4 *Alternative splicing*

Alternative splicing is an extremely important control mechanism in humans, where over one-third of the genes may be alternatively spliced. Although predicted to be not as widespread in plants, it is becoming apparent that alternative splicing is used in plants as a control mechanism. In this box we will look at just two examples in a little detail to illustrate the complexity that may be introduced to gene expression by alternative splicing (and processing).

Rubisco activase

The alternative splicing of ribulose-1,5-bisphosphate carboxylase/oxygenase activase, better known as Rubisco activase, is one of the best-studied examples of alternative splicing in plants. Rubisco activase is a nuclear-encoded gene for a chloroplast protein that regulates Rubisco activity in response to light-induced changes in the redox potential and changes in the ADP/ATP ratio.

Alternative splicing results in the production of two polypeptides of different size, the larger isoform having additional amino acids at the C terminus. Both isoforms can regulate Rubisco activity, but the larger isoform is the most sensitive to ADP inhibition and is the only isoform that is redox-regulated. This sensitivity to ADP depends on the presence of two cysteine residues in the additional C-terminal amino acids, which, via a disulphide bond, are involved in the redox regulation of Rubisco activase. The larger isoform can also alter the activity of the smaller isoform, which is proposed to create a 'fine-tuning' mechanism for the regulation of Rubisco activase.

FCA

The *FCA* gene, which encodes an RNA-binding protein, promotes photoperiod-independent flowering in *Arabidopsis*. The *FCA* gene is alternatively processed at two positions to yield four different transcripts: α, β, γ, and δ. Transcript α constitutes less than 1% of the *FCA* mRNA, β about 55%, γ about 35% and δ about 10%. Transcript γ is thought to be the only transcript that encodes the full-length protein. Transcript α results from the inclusion of intron 3 in the transcript and results in the premature termination of translation. Transcript β results from cleavage and polyadenylation within intron 3, generating a protein identical to that from transcript α. Transcript γ results from the correct splicing of intron 3. Transcript δ results from alternative splicing in intron 13. The purpose of the alternative splicing/processing appears to be to regulate the spatial and temporal accumulation of the functional FCA protein.

The addition of a 5′ cap and a 3′ poly(A) tail is known to stabilize RNA (although the most profound effect of these modifications is to increase the efficiency of translation). The half-lives of individual mRNA species may still vary considerably (from minutes to days), with specific sequences or motifs thought to be responsible for rapid mRNA destabilization. In plants, an element termed DST (for downstream element) has been identified in the 3′ UTR of short-lived mRNAs that are induced by the plant hormone auxin (small auxin-up RNA or *SAUR* transcripts). Two conserved subdomains (ATAGAT and GTA) appear to be critical for DST function. In common with mammals, AU-rich areas (which usually contain multiple copies of the sequence motif AUUUA) destabilize plant mRNAs if they are present in the 3′ UTR (these sequence motifs are mentioned in Chapter 4). A variety of environmental and hormonal signals can also influence the half-life of specific mRNAs. For example, the half-life of the *Fed1* gene (encoding the photosynthetic electron carrier ferrodoxin) transcript is much increased in plants in the light compared with those in the dark and α-amylase transcripts are destabilized by the presence of sucrose.

These post-transcriptional processes can have a profound effect on the amount of mRNA present in the cytoplasm of a cell. A long-studied example of this is in the accumulation of legumin mRNA (legumin is a storage protein in seeds) following recovery from sulphur starvation—here the levels of legumin mRNA were increased by more than 20-fold, but transcription was only increased by about twofold.

Introns from plant genes are known to be capable of increasing gene expression from both homologous and heterologous promoters. First intron sequences are often used in monocotyledon transformation experiments to increase transgene expression. The precise mechanism by which introns increase gene expression is unclear, but it is thought to involve an effect on RNA processing/splicing, as mutation of splice sites results in lower expression levels.

Translation

It is becoming increasingly apparent that translational regulation in response to various signals is an extremely important step in regulating gene expression (Box 1.5 gives some more specific examples of translational regulation of gene expression). The efficiency of translation initiation and elongation can be influenced by a variety of factors.

BOX 1.5 *Repression of translation by upstream open reading frames*

Short open reading frames (ORFs) are present in the 5′ UTR of several plant transcription factor gene families. The translation initiation codon of these short ORFs does not facilitate efficient translation initiation. These upstream short ORFs repress translation of downstream ORFs, with the effect that the transcription factor itself (the downstream ORF) is only translated when the competition for ribosomes is not intense. Upstream ORFs (uORFs) only occur in some members of each family and thus provide a mechanism for the differential regulation of transcription factor activity. The plant S-adenosyl-L-methionine decarboxylase (AdoMetDC) transcript also contains uORFs, both a tiny uORF (2–3 codons) and a small, highly conserved uORF (50–54 codons). The product of the small uORF, possibly an RNA-binding protein, represses translation of the AdoMetDC ORF.

Upstream ORFs may also be involved in regulating gene expression in response to specific signals. For example, the 5′ UTR of an *Arabidopsis* bZIP protein mRNA (*ATB2*), which contains four overlapping uORFs, represses translation in response to sucrose, but not other sugars.

The presence of the 5′ cap and 3′ poly(A) tail in mRNA act synergistically to enhance translation. The efficiency of translation initiation depends on several factors, the most important ones being the distance of the AUG translation initiation codon from the 5′ end of the mRNA and the secondary structure around the initiation codon. Computer analysis has allowed consensus sequences for translation initiation codons to be identified in both monocotyledon and dicotyledon plants:

Monocotyledons	**(A/C)**	**(A/G)**	**(A/C)**	C **AUG G** C
Dicotyledons	**A**	**A**	**(A/C)**	A **AUG G** C

Here the translation initiation codon, AUG, is underlined, and bold indicates conserved residues. Further analysis has demonstrated that the nucleotides in positions +4 to +11 are also involved in the regulation of gene expression. Translation may also be promoted by the binding of specific proteins to the mRNA.

Post-translational modification

Many proteins have to be modified after translation in order to become functional. Proteins can undergo a great variety of possible post-translation modifications. Signal peptides (which direct proteins to certain subcellular structures or organelles) may be proteolytically cleaved from the precursor protein. Specific residues may be modified by phosphorylation, glycosylation, or acetylation. Reversible changes in these modifications (particularly phosphorylation and acetylation) are used to regulate the activity and location of a large number of proteins.

Localization

In order to function properly, proteins have to be localized in the correct subcellular compartment. Specific signals (Table 1.3) have been identified that are responsible for targeting proteins to the nucleus, chloroplasts, mitochondria, vacuoles, and peroxisomes, and for retention in the endoplasmic reticulum.

Protein turnover

Most proteins present in the cytosol of cells are stable and have relatively long half-lives (days). However, some proteins (which are usually coded for by short-lived mRNAs) are turned over rapidly due to the presence of particular amino acid residues or motifs, which target these proteins for rapid proteolysis.

The N-terminal amino acid is known to influence protein stability. Although all proteins are synthesized, initially, with a methionine as the N terminal amino acid, this is often subsequently removed. A class of enzymes called aminoacyl-tRNA-protein transferases can also add other amino acids to the N terminus of certain proteins. Some amino acids (such as Met, Gly, Ser, Thr, and Val) are found to stabilize proteins, whereas others (such as Arg, Asn, Asp, Gln, Glu, His, Ile, Leu, Lys, Phe, Trp, and Tyr) result in rapid degradation. The N-terminal amino acid may also be modified by acetylation, which increases the stability of the protein.

Table 1.3 Protein-localization signals

Organelle/subcellular compartment	Signal
Endoplasmic reticulum	HDEL/KDEL/RDEL motif[a]
Vacuole	LQRD (vacuolar targeting sequence)[b]
Nucleus	Nuclear localization signal (NLS)[c]
Peroxisomes/glyoxisomes	SKL motif (S/A/C)(K/R/H)L[d]
Chloroplasts	N-terminal transit peptide[e]
Mitochondria	N-terminal transit peptide[f]

[a]Usually present in the C terminus. Addition of this motif to other proteins may not be sufficient for retention in the endoplasmic reticulum but will slow transport.
[b]May not be sufficient for targeting. Proper targeting may require protein folding to bring 'signal patches' to the surface of the protein.
[c]The NLS usually consists of a small region (located anywhere in the protein) rich in basic amino acids (arginine and lysine). Other types of NLS are also found. NLSs are not absolutely conserved in either size or composition.
[d]The addition of a C-terminal SKLL motif to reporter genes targets the protein to the peroxisome in rice.
[e]The composition of the transit peptide varies, although they usually have no (or few) acidic residues but are rich in serine and threonine. Various computer programs are available to predict chloroplast transit peptides in a protein sequence. The targeting peptide is cleaved from the mature protein. Although the chloroplast has a genome, the majority of chloroplastic proteins are encoded in the nucleus, necessitating transport into the chloroplast.
[f]The composition of transit peptides directing proteins to the mitochondrion is variable, although there may be some regions of homology. They generally lack acidic residues (as is the case with chloroplast transit peptides) and are enriched in alanine, leucine, arginine, and serine. As with chloroplast transit peptides, computer programs are available to predict mitochondrial transit peptides. The transit peptide is cleaved from the mature protein.
(Data taken with permission from Wallace, T.P. and Howe, C.J. (1993) Plant organellar targeting sequences. In *Plant Molecular Biology Labfax* (ed. R.R.D. Croy), pp. 287–292. BIOS Scientific, Oxford.)

Conclusions

In conclusion, gene expression can be controlled at many points. The expression of particular genes may be controlled at multiple steps to ensure correct regulation. Specific stimuli that influence gene expression, such as environmental signals and plant hormones, may also exert their effects at more than one stage in the process of gene expression. This multiplicity of controls allows gene expression to be finely regulated and to respond in a flexible way to the needs of the plant cell. Box 1.6 uses the example of methyl jasmonate to illustrate how one compound can control gene expression in different ways.

Implications for plant transformation

- Transgene expression may need to be controlled carefully.

The introduction and stable integration of transgenes into the genome of the host plant is only the first step in manipulating a plant successfully. The transgenes have to be expressed in the appropriate fashion, both spatially and temporally. The transcript has to be processed properly and the protein product has to be modified appropriately and targeted to the correct cellular compartment. For these prerequisites to be achieved, considerable effort may need to be put into the design of a transgene before it is introduced into the plant.

BOX 1.6 *The control of gene expression by methyl jasmonate*

Methyl jasmonate is a cyclopentanone compound synthesized from linolenic acid. It is one of the main fragrances of jasmine oil, and a flavour component of black tea. However, methyl jasmonate is also involved in a variety of physiological processes, such as embryogenesis and defence against wounding and pathogen attack, as well as osmotic stress. It induces the expression of a large number of sequences, termed jasmonate-induced proteins (JIPs). As well as inducing the expression of some sequences, methyl jasmonate has a profound negative effect on the expression of other genes, particularly those involved in photosynthesis. This may be part of the plant's defence mechanism to wounding, as photosynthesis imposes a heavy energetic burden on the wounded plant.

Transcriptional control

Methyl jasmonate negatively regulates the transcription of a range of nuclear-encoded genes for photosynthetic proteins, such as the ribulose-1,5-bisphosphate carboxylase/oxygenase (Rubisco) small subunit (encoded by the *rbcS* gene) and several light-harvesting chlorophyll–protein complex apoproteins. This negative effect on transcription is not universal, as *JIP* genes continue to be transcribed.

Post-transcriptional control

Methyl jasmonate can, by alternative processing, cause modification of the 5′ end of the transcript of the plastid-encoded Rubisco large-subunit gene. Methyl jasmonate-induced alternative processing of the primary transcript results in a transcript with a 94-base UTR as opposed to a 'normal' 59-base UTR. This extension of the UTR results in an inhibition of translation initiation, due to the presence of a region with homology to the 3′ terminus of both 16S and 18S rRNA. Methyl jasmonate also seems to selectively destabilize some transcripts, such as those for *rbcS*.

Translational control

Methyl jasmonate represses the synthesis of most pre-existing proteins. In the case of proteins involved in photosynthesis, this is because their transcripts are associated with small polysomes (or the non-polysomal fraction) and is indicative of an inhibition of translation initiation. This displacement from polysomes appears to be a specific rather than a general effect, as *JIP* transcripts remain associated with all polysome fractions and are translated normally.

In some cases the decisions are fairly simple. If a high level of expression of a cytoplasmic protein is required, and the timing and location of expression are not important, well-characterized promoters such as the 35S promoter (where S means Svedberg units) from cauliflower mosaic virus could be used (see also Chapter 4). However, in many cases such a simple constitutive pattern of expression is not what is required. Alternative control sequences, either in the form of promoter fragments from other genes or as specific sequence elements, can be used to drive transgene expression in a characteristic fashion (Box 1.7 gives a simple introduction to how suitable promoter fragments or sequence elements can be identified).

BOX 1.7 *Promoter analysis*

Much of plant science depends on understanding promoter function. This is particularly important in plant biotechnology when appropriate promoters have to be added to transgenes to drive expression in the correct temporal and spatial fashion.

Promoters are often studied using a combination of molecular biology and plant transformation methods. A cloned DNA sequence is analysed by computer and the various features (promoter, coding region, poly(A) site, etc.) provisionally identified. The promoter is then fused to a reporter gene (reporter genes are discussed in more detail in Chapter 4), which can be used to monitor the pattern of expression in transgenic plants. In subsequent analyses, the size of the promoter fragment used can be reduced and/or various areas can be mutated and the effect of these changes on the pattern of gene expression monitored. Various molecular techniques, such as DNA footprinting, can be used to determine which areas of the promoter binds proteins (possible transcription factors) and these can be compared with the results from the reporter gene experiments. Depending on the precise experiments used, sequence elements can be mapped on to the promoter fragments down to a resolution of 1 bp.

Increasingly, potential *cis*-elements in cloned, sequenced DNA are identified by computer, using programs that screen libraries of known *cis*-element sequences. Once specific sequence elements have been identified they can be fused to a minimal promoter, and assays performed to investigate whether they confer the expected pattern of expression on the minimal promoter.

The spinach nitrite reductase gene

As an example of promoter analysis we can look at a study of the promoter of the spinach nitrite reductase gene. Nitrite reductase catalyses the reduction of nitrite to ammonia in the chloroplast. The transcript is 2.3 kb long and transcription is induced by nitrate. An analysis of the

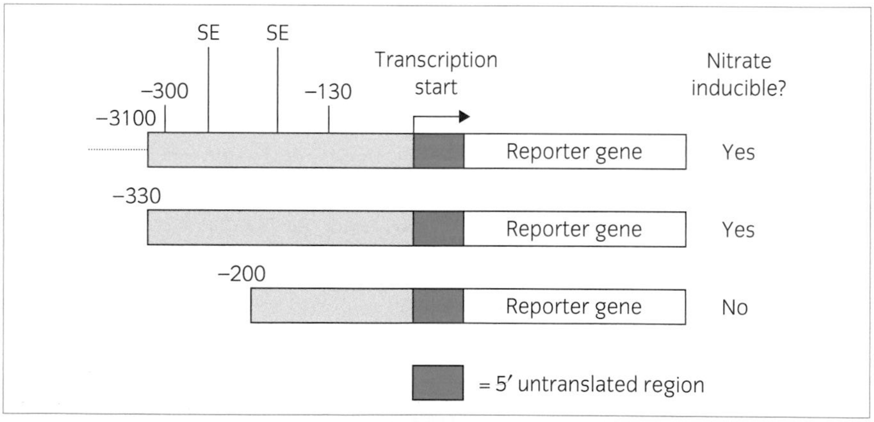

Spinach nitrite reductase gene-promoter analysis. Various promoter::reporter gene fusions are shown. The promoter region used is represented by the lighter-green area, and the figures represent the lengths of promoter fragments used in each construct. The constructs used also contained some of the 5′ UTR. The positions of the two sequence elements (SEs) identified by DNA footprinting are indicated.

BOX 1.7 *Continued*

promoter region has identified several sequence motifs that are responsible for this nitrate inducibility. Various promoter::reporter gene fusions were used to identify which regions of the promoter were responsible for nitrate induction of transcription (see figure). A region between 330 and 200 bp upstream of the beginning of the gene was identified by comparing the nitrate inducibility of reporter gene expression of various constructs. DNA foot-printing assays of the −300 to −130 region identified two GATA sequence elements (SEs) in this region with the sequence 5′-TAGATA-3′ and 5′-TATCTA-3′, which bound proteins in response to nitrate treatment. Additional sequences involved in nitrate inducibility were also found to be located in the 5′ UTR of the gene.

For example, the transgene may be required to be expressed in response to some stimulus, either external or internal, or may need to be expressed in a particular tissue. A few examples of the findings from promoter analyses are given below to illustrate how different promoter elements can be used to drive transgene expression, and Table 1.4 lists some *cis*-elements involved in the control of gene expression in response to various environmental and hormonal signals.

Table 1.4 Examples of plant *cis*-regulatory elements

Stimulus and name[a]	Sequence[b]	Source[c]
Light-response elements (LREs)		
Box I	ACCTAACT	Parsley *chalcone synthase*
Box II	TGTGTGGTTAATATG	Pea *rbcS*-3A
Box III	ACTATTTTCACTATC	Pea *rbcS*-3A
Auxin (AuxREs)		
AuxRE	TGTCTC	Soybean *SAUR*
AuxRE	TGACCTAA	Soybean *GH3*
AuxRE	GTCCCAT	Pea *IAA4/5*
Abscisic acid (ABREs)		
ABRE A	GCC<u>ACGT</u>GGG[d]	Maize *rab28*
ABRE B	TCC<u>ACGT</u>CTC	Maize *rab28*
ABRE	GT<u>ACGT</u>GGCGC	Rice *α-amylase*
CE	ACGCGCCTCCTC	Maize *rab28*
Gibberellic acid	GGCCGATAACAAACTCCGGCC	Barley *α-amylase*
Jasmonic acid		
JASE 1	CGTCAATGAA	*Arabidopsis* spp. *OPR1*
JASE 2	CATACGTCGTCGTCAA	*Arabidopsis* spp. *OPR1*

[a]The name of the *cis*-element is given and/or the stimulus, the transduction of which involves the *cis*-element.
[b]The DNA sequence of the element is written 5′ to 3′.
[c]The plant and gene names are given. Gene names are in italics.
[d]The conserved ACGT core is underlined.
Abbreviations: ABRE, abscisic acid (ABA)-response element; AuxRE, auxin-response element; CE, coupling element; *rbcS*, ribulose-1,5-bisphosphate carboxylase/oxygenase (Rubisco) small subunit; SAUR, small auxin-up RNA.

Examples of promoter elements used to drive transgene expression

Light-regulated gene expression

Analysis of the promoters from light-responsive genes has identified the regions required for light-induced expression. Several consensus sequence elements have been identified in those regions shown to be necessary for promoter activity. These sequence elements are termed light-responsive elements (LREs). The action of LREs is complex, as no one LRE is found in all light-regulated genes and two or more LREs must be present for light-induced expression. Some LREs may also be involved in repression of expression as opposed to induction. It has been demonstrated that pairwise combinations of any LREs are sufficient to confer light-induced expression to minimal promoters.

Abscisic acid-induced gene expression

Abscisic acid (ABA) is a plant hormone that controls the expression of a large number of genes. Promoter analysis has demonstrated that a small fragment from the promoter of some ABA-induced genes is sufficient to confer ABA-inducibility to minimal promoters. ABA induction mediated by the fragment was found to depend on the presence of two sequence elements: one termed the ABA-response element (ABRE), which in common with some other *cis*-elements has an ACGT core; and another termed the coupling element (CE).

Tissue-specific expression

Analysis of the promoters of various genes expressed in specific tissues has allowed the identification of those sequence elements responsible for the pattern of expression. Seed storage proteins have been studied extensively in this respect, and sequence elements (such as the legumin box, 5′-TCCATAGCCATGCAAGCTGCA-3′) responsible for localizing expression have been identified.

Protein targeting

Proteins may also have to be targeted to a particular subcellular location in order to have the desired effect. As will be seen in Chapter 5, the production of herbicide-resistant plants often requires the transgene product to be targeted to the chloroplast, which can be done by adding the transit peptide from a chloroplast-targeted protein. Deliberate mis-targeting of proteins to alternative subcellular locations may be of some benefit in some areas of plant biotechnology, as normal controls that limit accumulation may be circumvented.

Heterologous promoters

Several assumptions are made when considering the use of heterologous promoters to drive transgene expression. The first is that the *trans*-activating factors from one plant will recognize the *cis*-sequence elements of the heterologous promoter. In many cases, this appears to happen, but in some cases (particularly if monocotyledon promoter elements are used to drive transgene expression in a dicotyledon or vice versa) it does not. It is also usually assumed, sometimes incorrectly, that the transgene will be properly spliced and processed and the protein correctly folded and modified. As the controls of gene expression are better understood it is becoming apparent that what the plant biotechnologist can achieve with the use of heterologous promoters and other tools is an approximation of the 'natural' expression profile. Fortunately, in many cases this approximation is close enough for its use in a biotechnology programme.

Genome size and organization

● Genome size does not necessarily reflect organismal complexity.

It is perhaps fitting, especially in the light of recent progress in genome sequencing, to ask what we know about plant genomes at the molecular level and to consider what uses this information can be put to.

The size of the nuclear genome varies among organisms, with eukaryotic cells having an unreplicated DNA content (the *C* value) varying from approximately 10^7 to 10^{11} bp. In the broadest terms, the size of the nuclear genome reflects the complexity of the organism, so that the genome of bacteria is smaller than that of fungi, which in turn is smaller than that of animals and plants. However, this simple relationship does not always hold true, a situation known as the C-value paradox. Higher plants for example, which can be assumed to have a similar degree of complexity and a similar number of genes, exhibit genome sizes that vary by several orders of magnitude (Figure 1.8), and many amphibia have *C* values much larger than that of humans. It is also known that in many organisms only a small percentage of the genome actually encodes proteins. Even accounting for other regions, such as introns and the 5′ and 3′

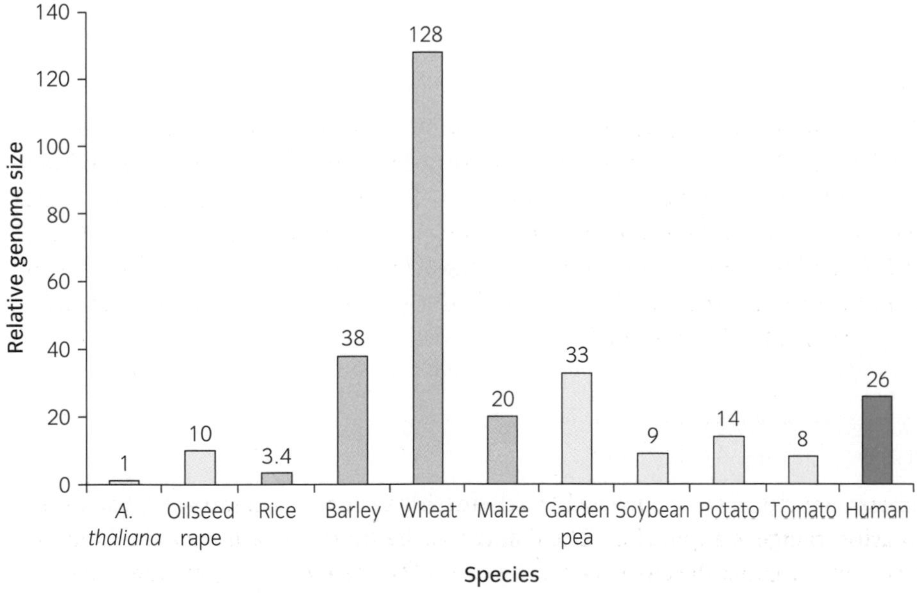

Figure 1.8 Approximate relative genome sizes in a range of crop plants, both monocotyledon (mid-green) and dicotyledon (light green) compared with the human genome (dark green).

regions of genes, this means that the vast majority of the DNA in some organisms is non-coding and apparently functionless.

Much of this non-coding DNA in the nuclear genome comprises repetitive DNA. Repetitive DNA consists of groups or families of similar repeated sequences and can be divided into two types: tandem repeats (or satellite DNA) and dispersed repeats. Tandem repeats are associated with particular locations, and primarily compose the centromeric (where the fibres of the mitotic spindle attach themselves) and telomeric (the ends of chromosomes) regions of chromosomes. Dispersed repeats, which can in some plant species compose a significant proportion of the genome, tend to be found throughout the genome and are often related to transposable elements.

Arabidopsis and the new technologies

Recent years have seen a revolution take place in biology. Perhaps the most publicized example of this revolution is the completion of the sequencing of the human genome, a marvellous technical achievement that should prove invaluable in improving the quality of life. Plant biology has also seen its own revolutions, some of which have admittedly followed from advances in other areas, but some of which are unique to plants (see Box 1.8). These advances in plant science may also yield tangible benefits to humanity very rapidly, given the applied focus of much plant research and the smaller number of ethical/moral issues associated with plant science.

In the light of these advances, it seems appropriate to look at the findings of these new technologies and the uses to which they can be put.

Genome-sequencing projects—technology, findings, and applications

Arabidopsis*: a model plant*

Arabidopsis thaliana is a small, dicotyledonous cruciferous plant that has become *the* model species for most aspects of plant biology. It has achieved this status despite being a weed with no commercial value, although it is a member of the mustard family and is related to crops such as cabbage, cauliflower, sprouts, and oilseed rape (canola).

A. thaliana does, though, have a range of attributes that make it the ideal experimental plant. It has a short life cycle and small stature, and produces large numbers of offspring, making it ideally suited for genetic and mutational analysis. It is also easily transformed and, most importantly for genome-sequencing initiatives, it has one of the smallest genomes in higher plants (125 Mb). *A. thaliana* was therefore chosen to be the first plant to have its complete genome sequenced.

The Arabidopsis *Genome Initiative—successful, international, publicly funded, freely available*

● *Arabidopsis* was the first plant to have its genome sequenced.

December 2000 saw the publication of the complete genome sequence of *A. thaliana*, the culmination of years of effort by a multinational team. The international collaborative

BOX 1.8 *The new technologies*

Genomics is only one of a range of recently developed technologies that can be of use to the plant biotechnologist. Indeed, the true potential of the various genome-sequencing efforts will only be realized when they are seen as part of a co-ordinated effort involving all these new technologies.

Functional genomics

One of the most direct ways of investigating gene function is to mutate the gene so that it no longer produces a functional protein product. In plants, this can be done in a variety of ways, usually involving the introduction of a piece of foreign DNA into the gene. The advantage of this type of insertional mutagenesis is that it is relatively straightforward to locate the point of insertion. As this process is more or less random, large libraries of mutated plants, each of which has a mutation in a particular gene, have been established. These mutants can be used to investigate the function of particular genes in a process known as reverse genetics, where the genotype is known and the resultant phenotype identified.

Transcriptomics

Transcriptomics is a technology developed to enable the genome-wide analysis of gene-expression patterns. This is a highly automated process and allows experimenters to identify all the changes in gene expression that occur, for example, in response to a particular treatment. Identifying genes that show altered expression patterns due to a particular treatment, or at a particular developmental stage, is one way of identifying genes that are potentially useful to the plant biotechnologist. For instance, genes that are found to be expressed in high amounts following an attack by a pest may well be involved in the plant's defence against that pest.

Proteomics

Attempts to study the proteome (the total protein complement encoded by the genome) are in their infancy, and, at present, analysis is restricted to the relatively abundant proteins. The advantage of this technique is that it actually looks at the functional outcome of gene expression (i.e. proteins) and can be used to study post-translational modifications. This technique attempts to analyse the next level of gene expression after transcriptomics.

Metabolomics

This allows the study of the chemical constituents of the cell using analytical chemistry. Different classes of metabolites (such as plant hormones) can be studied, and differences and similarities correlated with desirable traits. One interesting outcome of metabolomics is the demonstration that genetic engineering really is a quite precise science, with far fewer differences between a transformed plant and a control plant than between different varieties of the same plant.

Arabidopsis Genome Initiative (AGI) began sequencing the genome in 1996. A total of 115.4 million bases were sequenced with a previously unmatched accuracy of between 99.99 and 99.999%, thanks to advances in technology. The remaining 10 Mb represents repeats and/or areas that are difficult to sequence. It is surely one of the

Table 1.5 A synopsis of the findings from the *Arabidopsis* genome-sequencing initiative

	Chr1	Chr 2	Chr 3	Chr 4	Chr 5	Total
Length (kb)	29 105	19 647	23 172	17 550	25 935	115 409
Top arm (kb)	14 449	3 607	13 590	3 052	11 132	
Bottom arm (kb)	14 656	16 040	9 582	14 498	14 803	
Number of genes	6 543	4 036	5 220	3 825	5 874	25 498
Exons						
Number	35 482	19 631	26 570	20 073	31 226	132 982
Total length (kb)	8 773	5 100	6 655	5 151	7 571	33 250
Percentage of total length	30	26	29	29	29	29
Average per gene	5.4	4.9	5.1	5.2	5.3	
Average size (bp)	247	259	250	256	242	
Introns						
Number	28 939	15 595	21 350	16 248	25 352	107 484
Total length (kb)	4 829	2 768	3 398	3 031	4 030	18 056
Percentage of total length	17	14	15	17	16	17
Average size (bp)	167	177	159	186	159	
Gene density (kb per gene)	4.0	4.9	4.5	4.6	4.4	

Chr, chromosome.
(Data from The Arabidopsis Genome Initiative (2000).)

best examples of what can be achieved by publicly funded science given the focus and resources.

The statistics of the *A. thaliana* genome are summarized in Table 1.5. There are predicted to be about 25 500 genes in the *Arabidopsis* genome. This is more than for some multicellular eukaryotes (*Caenorhabditis elegans*, a nematode, has about 19 000 genes and the fruit fly *Drosophila melanogaster* has about 13 600 genes) but less than the 30 000–40 000 predicted for humans.

It is important to remember that this analysis is based on computer predictions. The figures in Table 1.5 and comparison with other eukaryotes perhaps give a misleading impression about the relative complexities of these organisms. There is extensive gene duplication in *Arabidopsis*, so that the total number of distinct protein types is only about 11 600, a similar figure to that estimated for both *C. elegans* and *D. melanogaster*.

Functional analysis

About 70% of the predicted genes in *Arabidopsis* can be assigned a function based on their sequence similarity to proteins of known function in all organisms (Figure 1.9). However, it is important to note that only about 9% of the predicted genes have been characterized experimentally. The 30% of genes that cannot be assigned a function comprise both plant-specific proteins and proteins with a high degree of similarity to genes of unknown function from other organisms.

Arabidopsis has a relatively high proportion of genes involved in metabolism and defence. This is likely to reflect the fact that *Arabidopsis* is a sessile autotroph. Within

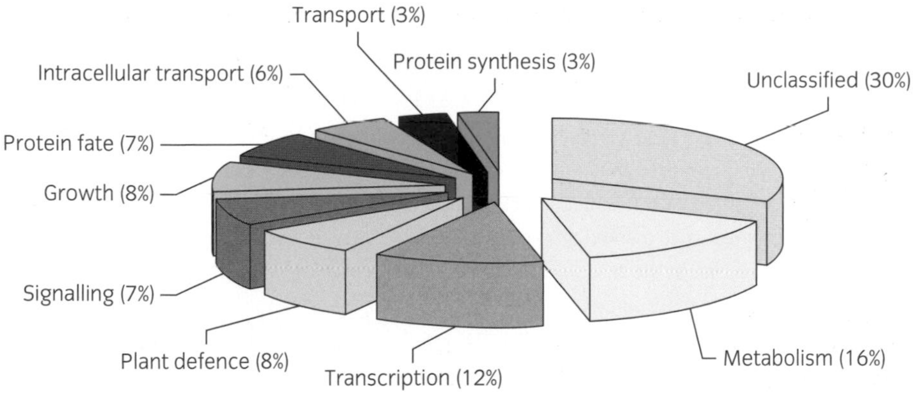

Figure 1.9 Functional analysis of the *Arabidopsis* genome. The proportion of genes assigned to different functional classes is shown. (Data from The *Arabidopsis* Genome Initiative (2000).)

each class of protein, the proportion of *Arabidopsis* proteins having identified equivalents in other eukaryotes varies. Only 8–23% of the proteins involved in transcriptional control have counterparts in other eukaryotes, whereas the figure is 48–60% for proteins involved in protein synthesis.

Biotechnological implications of the AGI

The data that has emerged from the AGI has proved to be extremely useful in many areas of plant biology. Many biotechnology companies are already taking advantage of the results of the *Arabidopsis* genome-sequencing effort. Although *Arabidopsis* is not a crop plant, many of the processes that are of interest to the plant biotechnologist (e.g. disease and pest resistance, and stress tolerance) share common features in both *Arabidopsis* and crop plants. *Arabidopsis* provides the ideal tool for plant scientists to understand these processes, an essential step before they can be successfully manipulated using biotechnology. This aspect is becoming increasingly important as public awareness and concern about plant biotechnology increases and efforts are made to reduce the number of proteins from other, completely different, organisms used in plant biotechnology.

- Plant genome-sequencing programmes will have profound implications for plant biotechnology.

Crop plant genome sequencing

The completion of the AGI was without doubt a major milestone in plant biology, and has already demonstrated the usefulness of *Arabidopsis* as a model system. However, it is also beyond doubt that *Arabidopsis* will not feed the world. Most of the world's population depend on cereals, and rice in particular, to supply their calorific and nutritional needs. Cereals are monocotyledonous plants and therefore exhibit important differences from *Arabidopsis* (a dicotyledon) in some aspects of their development. Particularly important in this respect are the differences in seed/endosperm development between monocotyledons and dicotyledons. Therefore, although *Arabidopsis* is, in many respects, the ideal model system, the elucidation of its genome may not provide clues to some specific problems. Rice also has some

BOX 1.9 *Synteny and colinearity*

Plant genomes show a huge variation in chromosome size and number, which is reflected by a size range from approximately 100 Mb found in some Cruciferae (such as *Arabidopsis*) to approximately 100 000 Mb in some members of the lily family. However, the gene number (and type) does not show such marked differences. Much of the difference in genome size between different plants is due to the amount of repetitive DNA present in the genome (although polyploidy may also account for some of the difference). Changes in genome size can be a (relatively) frequent event during evolution, but these changes do not appear to be associated with large changes in gene content and gene order. This conservation of gene content and order is termed colinearity, although recently the use of the term synteny (derived from the Greek *syn*, meaning together and *taenia*, meaning ribbon) has increased to cover both colinearity and homeology (the residual homology of previously completely homologous chromosomes). Colinearity or synteny is a startling phenomenon. *Arabidopsis* (a dicotyledon) and rice (a monocotyledon) diverged from a common ancestor about 200 million years ago, and yet there is low, but detectable, synteny between the two species. Where synteny becomes much more obvious, and useful, is when the genomes of various cereals are aligned and compared. Agronomically important traits (grain yield, pest resistance, etc.) identified in other cereals, such as maize, can be associated with the rice genome sequence, allowing identification of the genes responsible for the trait. Many cereal quantitative trait loci (QTLs) have already been placed on the rice genome map.

major advantages that made it an ideal choice for the next plant to have its genome sequenced. Perhaps most importantly, the rice genome is only about four times larger than that of *Arabidopsis*, making it one of the smallest cereal crop genomes. Good genetic maps also exist for rice, and genetic mapping experiments had already shown that the gene content and order were conserved in cereals (gene order is explained in Box 1.9).

The rice genome

● Rice was the first crop plant to have its genome sequenced.

A major multinational effort (the IRGSP, or the International Rice Genome Sequencing Project) was established to sequence the genome of rice. In April 2002 draft sequences of the two major subspecies of rice (*indica* and *japonica*) were published and in 2005 the IRGSP published the finished, high-quality sequence of *Oryza sativa* L. ssp. *japonica* cv. Nipponbare in *Nature*.

The usefulness of the rice genome sequence can be illustrated simply by looking at some statistics. Rice is predicted, according to IRGSP estimates, to have 37 544 protein-coding genes; considerably more than estimates of gene number in *Arabidopsis*. In a comparison with the genome of *Arabidopsis*, it was found that, whereas rice had putative homologues to 90% of *Arabidopsis* proteins, only 71% of the predicted rice proteins had homologues in *Arabidopsis*. Therefore, there are significant differences in not only gene number but also gene type between rice and *Arabidopsis* that may reflect the different development and physiologies of monocotyledons and dicotyledons. However, as was expected, there has proved to be considerable conservation of gene type and order in the cereals, with the majority of the known maize, wheat, and barley proteins being present in rice.

Other crops

● The genomes of many crop species are being sequenced.

There was considerable debate as to which other, if any, crop plants should have their genome sequenced. Many considered that the data from *Arabidopsis* and rice genome-sequencing initiatives should be sufficient, especially given the considerable synteny, or colinearity, that exists within monocotyledons and dicotyledons. Others disagreed, and argued that the genomes of other important crops (such as wheat and maize, and also, for example, trees) should also be sequenced. One argument put forward was that technological advances were making genome-sequencing projects cheaper and quicker, but the big disadvantage with these crops is that their genomes are extremely large and complicated (see Figure 1.8), which makes sequencing both time-consuming and difficult as many now-standard sequencing methodologies cannot be applied easily to these genomes.

However, it was perhaps inevitable that the number of crop-plant sequencing efforts would, like sequencing projects in general, mushroom. A brief look at the ERGO™ website (web-link 1.3) reveals a growing list of crop-plant genome-sequencing projects (most of which have yet to be completed), including maize, barley, and wheat.

Findings from genome-sequencing initiatives are not only applicable to projects involving genetic manipulation. The findings should prove invaluable in conventional breeding programmes allowing, for example, specific traits to be followed easily.

SUMMARY

This chapter has looked briefly at various aspects of what might be called the genetic systems of plants. It is not intended to give an all-encompassing description of plant genomes and gene regulation, whole volumes being required to cover such complex subjects. It should, however, provide a framework sufficient to enable the more molecular aspects of the remainder of the book to be understood without reference to other books.

Understanding how gene expression is controlled is key in any strategy to genetically manipulate plants. Transgenes (those foreign genes introduced into the plant genome by plant transformation) are subject to control in the same way as the endogenous genes of the plant. It is interesting to observe the synergy that has developed between the need to understand how transgenes are regulated and the use of plant-transformation strategies as a tool to elucidate the controls of gene expression. As the regulation of gene expression becomes better understood, this information is fed through into improvements to plant-transformation strategies. Some of the most obvious developments are being made in the design of plant-transformation vectors and the modification of the transgene itself so as to ensure optimal expression. Some of these improvements are discussed in subsequent chapters, particularly Chapter 4.

The rice and *Arabidopsis* genome-sequencing projects discussed in this chapter have generated a huge amount of data. Much of these data are of interest because of the light shed on fundamental processes, such as evolution. The data, however, also contribute hugely to our understanding of how plant gene expression is controlled and integrated. Coupled with information from fundamental studies on gene function, these advances are making possible the development of the next generation of genetically modified crops. These new-generation crops, which to a large extent will be based on modifying the regulation or complement of endogenous genes, bring about the possibility of manipulating complex agronomic traits like yield. In Chapter 9 we will see how the manipulation of transcription factor expression, in order to influence the expression of a number of genes, can be used in plant biotechnology. This type of approach will become more common in plant biotechnology as gene-expression controls, genome structures, and gene function become better understood.

FURTHER READING

Visit www.oxfordtextbooks.co.uk/orc/slaterplants2e/ for hyperlinked references that take you directly to online abstracts for the journal articles listed below:

Gene structure and expression

Alberts, B., Bray, D., Lewis, J., Raff, M., Roberts, K., and Watson, J. D. (1989) *Molecular Biology of the Cell*. Garland, New York.

Bailey-Serres, J. (1999) Selective translation of cytoplasmic mRNAs in plants. *Trends in Plant Science* **4**, 142–148.

Bailey-Serres, J., Rochaix, J.-D., Wassenegger, M., and Filipowicz, W. (1999) Plants, their organelles, viruses and transgenes reveal the mechanisms and relevance of post-transcriptional processes. *EMBO Journal* **18**, 5153–5158.

Berk, A. J. (1999) Activation of RNA polymerase II transcription. *Current Opinion in Cell Biology* **11**, 330–335.

Buchanan, B. B., Gruissem, W., and Jones, R. L. (2000) *Biochemistry and Molecular Biology of Plants*. American Society of Plant Physiologists, Rockville, MD.

Buratowski, S. (2000) Snapshots of RNA polymerase II transcription initiation. *Current Opinion in Cell Biology* **12**, 320–325.

Eckardt, N. A. (2002) Alternative splicing and the control of flowering time. *Plant Cell* **14**, 743–747.

Gutierrez, R. A., MacIntosh, G. C., and Green, P. J. (1999) Current perspectives on mRNA stability in plants: multiple levels and mechanisms of control. *Trends in Plant Science* **4**, 429–438.

Kornberg, R. D. (1999) Eukaryotic transcriptional control. *Trends in Cell Biology* **9**, M46–M49.

Li, G., Hall, T. C., and Holme-Davis, R. (2002) Plant chromatin: development and gene control. *BioEssays* **24**, 234–243.

Lorkovic, Z. J., Wieczorek Kirk, D. A. W., Lambermon, M. H. L., and Filipowicz, W. (2000) Pre-mRNA splicing in higher plants. *Trends in Plant Science* **5**, 160–167.

Lukaszewicz, M., Feuermann, M., Jerouville, B., Stas, A., and Boutry, M. (2000) *In vivo* evaluation of the context sequence of the translation initiation codon in plants. *Plant Science* **154**, 89–98.

Maier, R. M., Neckermann, K., Igloi, G. L., and Kossel, H. (1995) Complete sequence of the maize chloroplast genome: gene content, hotspots of divergence and fine tuning of genetic information by transcript editing. *Journal of Molecular Biology* **251**, 614–628.

Miller, W. A., Waterhouse, P. M., Brown, J. W. S., and Browning, K. S. (2001) The RNA world in plants: post-transcriptional control III. *Plant Cell* **13**, 1710–1717.

Reinbothe, S., Mollenhauer, B., and Reinbothe, C. (1994) JIPS and RIPS: the regulation of gene expression by jasmonates in response to environmental cues and pathogens. *Plant Cell* **6**, 1197–1209.

Sawant, S. V., Kiran, K., Singh, P. K., and Tuli, R. (2001) Sequence architecture downstream of the initiator codon enhances gene expression and protein stability in plants. *Plant Physiology* **126**, 1630–1636.

miRNA/siRNA/gene silencing

Jones-Rhoades, M. W., Bartel, D. P., and Bartel, B. (2006) MicroRNAs and their regulatory roles in plants. *Annual Reviews in Plant Biology* **57**, 19–53.

Meins, F., Si-Ammour, A., and Blevins, T. (2005) RNA silencing systems and their relevance to plant development. *Annual Reviews in Cell and Developmental Biology* **21**, 297–318.

Novina C. D. and Sharp, P. A. (2004) The RNAi revolution. *Nature* **430**, 161–164.

Willmann, M. R. and Poethig, R. S. (2007) Conservation and evolution of miRNA regulatory programmes in plant development. *Current Opinion in Plant Biology* **10**, 1–8.

Genome-sequencing projects

Bevan, M., Mayer, K., White, O., Eisen, J. A., Preuss, D., Bureau, T., Salzberg, S. L., and Mewes, H.-W. (2001) Sequence and analysis of the *Arabidopsis* genome. *Current Opinion in Plant Biology* **4**, 105–110.

Goff, S. A., Ricke, D., Lan, T.-H., Presting, G., Wang, R., Dunn, M., Glazebrook, J., Sessions, A., Oeller, P., Varma, H., Hadley, D., Hutchison, D., Martin, C., Katagiri, F., Lange, B. M., Moughamer, T., Xia, Y., Budworth, P., Zhong, J., Miguel, T., Paszkowski, U., Zhang, S., Colbert, M., Sun, W.-L., Chen, L., Cooper, B., Park, S., Wood, T. C., Mao, L., Quail, P., Wing, R., Dean, P., Yu, Y., Zharkikh, A., Shen, R., Sahasrabudhe, S., Thomas, A., Cannings, R., Gutin, A., Pruss, D., Reid, J., Tavtigian, S., Mitchell, J., Eldredge, G., Scholl, T., Miller, R. M., Bhatnagar, S., Adey, N., Rubano, T., Tusneem, N., Robinson, R., Feldhaus, J., Macalma, T., Oliphant, A., and Briggs, S. (2002) A draft sequence of the rice genome (*Oryza sativa* L. ssp *japonica*). *Science* **296**, 92–100.

International Rice Genome Sequencing Project (2004) The map-based sequence of the rice genome. *Nature* **436**, 793–800.

The *Arabidopsis* Genome Initiative (2000) Sequence and analysis of the flowering plant *Arabidopsis thaliana*. *Nature* **408**, 796–815.

Yu, J., Hu, S., Wang, J., Wong, G. K.-S., Li, S., Liu, B., Deng, Y., Dai, L., Zhou, Y., Zhang, X., Cao, M., Liu, J., Sun, J., Tang, J., Chen, Y., Huang, X., Lin, W., Ye, C., Tong, W., Cong, L., Geng, J., Han, Y., Li, L., Li, W., Hu, G., Huang, X., Li, W., Li, J., Liu, Z., Li, L., Liu, J., Qi, Q., Liu, J., Li, L., Li, T., Wang, X., Lu, H., Wu, T., Zhu, M., Ni, P., Han, H., Dong, W., Ren, X., Feng, X., Cui, P., Li, X., Wang, H., Xu, X., Zhai, W., Xu, Z., Zhang, J., He, S., Zhang, J., Xu, J., Zhang, K., Zheng, X.,

Dong, J., Zeng, W., Tao, L., Ye, J., Tan, J., Ren, X., Chen, X., He, J., Liu, D., Tan, W., Tioan, C., Xia, H., Bao, Q., Li, G., Gao, H., Cao, T., Wang, J., Zhao, W., Li, P., Chen, W., Wang, X., Zhang, Y., Hu, J., Wang, J., Liu, S., Yang, J., Zhang, G., Xiong, Y., Li, Z., Mao, L., Zhou, C., Zhu, Z., Chen, R., Hao, B., Zheng, W., Chen, S., Guo, W., Li, G., Liu, S., Tao, M., Wang, J., Zhu, L., Yuan, L., and Yang, H. (2002) A draft sequence of the rice genome (*Oryza sativa* L. ssp *indica*). *Science* **296**, 79–92.

Web-links

Arabidopsis Information Resource, web-link 1.1: www.arabidopsis.org

Biome (searchable links), web-link 1.2: www.intute.ac.uk/healthandlifesciences/

ERGO™ at Integrated genomics, web-link 1.3: http://wit.integratedgenomics.com/ERGO_supplement/index.html

ExPASy, web-link 1.4: www.expasy.ch

Institute for Genomic Research, web-link 1.5: www.tigr.org

Nature Genome Gateway, web-link 1.6: www.nature.com/genomics

NCBI online books, web-link 1.7: www.ncbi.nlm.nih.gov/entrez/query.fcgi?db=Books

2

Plant tissue culture

INTRODUCTION

Most methods of plant transformation applied to genetically modified crops require that a whole plant is regenerated from isolated plant cells or tissues that have been genetically transformed. This regeneration is conducted *in vitro* so that the environment and growth medium can be manipulated to ensure a high frequency of regeneration. In addition to this, the regenerable cells must be accessible to gene transfer by whatever technique is chosen (gene-transfer methods are described in Chapter 3). The primary aim is therefore to produce, as easily and as quickly as possible, a large number of regenerable cells that are accessible to gene transfer. The subsequent regeneration step is often the most difficult step in plant transformation studies. However, it is important to remember that a high frequency of regeneration does not necessarily correlate with high transformation efficiency.

This chapter will consider some basic issues concerned with plant tissue culture *in vitro*, particularly as applied to plant transformation. It will also look at the basic culture types used for plant transformation and cover some of the techniques that can be used to regenerate whole transformed plants from transformed cells or tissue.

Plant tissue culture

Practically any plant transformation experiment relies at some point on tissue culture. There are some exceptions to this generalization (some of which will be looked at in Chapter 3), but the ability to regenerate plants from isolated cells or tissues *in vitro* underpins most plant transformation systems.

● There are important differences between plant and animal cells that allow almost any plant tissue to be regenerated. The key is using the correct culture conditions.

Plasticity and totipotency

Two concepts, plasticity and totipotency, are central to understanding plant cell culture and regeneration. Plants, due to their sessile nature and long life span, have developed a greater ability to endure extreme conditions and predation than have animals. Many of the processes involved in plant growth and development adapt to environmental conditions. This plasticity allows plants to alter their metabolism, growth, and development to best suit their environment. Particularly important aspects of this adaptation, as far as plant tissue culture and regeneration are concerned, are the abilities to initiate cell division from almost any tissue of the plant and to regenerate lost organs or undergo different developmental pathways in response to particular stimuli. When plant cells and tissues are cultured *in vitro* they generally exhibit a very high degree of plasticity, which allows one type of tissue or organ to be initiated from another type. In this way, whole plants can be subsequently regenerated.

This regeneration of whole organisms depends upon the concept that all plant cells can, given the correct stimuli, express the total genetic potential of the parent plant. This maintenance of genetic potential is called totipotency. Plant cell culture and regeneration do, in fact, provide the most compelling evidence for totipotency. In practical terms though, identifying the culture conditions and stimuli required to manifest this totipotency can be extremely difficult and it is still a largely empirical process.

The culture environment

When cultured *in vitro*, all the needs of the plant cells, both chemical (see Table 2.1) and physical, have to met by the culture vessel, the growth medium, and the external environment (light, temperature, etc.). The growth medium has to supply all the essential mineral ions required for growth and development. In many cases (as the biosynthetic capability of cells cultured *in vitro* may not replicate that of the parent plant), it must also supply additional organic supplements such as amino acids and vitamins. Many plant cell cultures, as they are not photosynthetic, also require the

Table 2.1 Some elements important for plant nutrition and their physiological function, supplied by the culture medium to support the growth of healthy cultures *in vitro*

Element	Function
Nitrogen	Component of proteins, nucleic acids, and some coenzymes; element required in the greatest amounts
Potassium	Regulates osmotic potential; principal inorganic cation
Calcium	Cell-wall synthesis, membrane function, cell signalling
Magnesium	Enzyme cofactor, component of chlorophyll
Phosphorus	Component of nucleic acids; energy transfer; component of intermediates in respiration and photosynthesis
Sulphur	Component of some amino acids (methionine, cysteine) and some cofactors
Chlorine	Required for photosynthesis
Iron	Electron transfer as a component of cytochromes
Manganese	Enzyme cofactor
Cobalt	Component of some vitamins
Copper	Enzyme cofactor; electron-transfer reactions
Zinc	Enzyme cofactor; chlorophyll biosynthesis
Molybdenum	Enzyme cofactor; component of nitrate reductase

addition of a fixed carbon source in the form of a sugar (most often sucrose). One other vital component that must also be supplied is water, the principal biological solvent. Physical factors, such as temperature, pH, the gaseous environment, light (quality and duration), and osmotic pressure, also have to be maintained within acceptable limits.

Plant cell culture media

Culture media used for the cultivation of plant cells *in vitro* are composed of three basic components:

- Medium can be complex but stocks can be made in bulk and stored or standard medium can be purchased from chemical suppliers.

1 essential elements, or mineral ions, supplied as a complex mixture of salts;

2 an organic supplement supplying vitamins and/or amino acids; and

3 a source of fixed carbon; usually supplied as the sugar sucrose.

For practical purposes, the essential elements are further divided into the following categories:

1 macroelements (or macronutrients);

2 microelements (or micronutrients); and

3 an iron source.

Complete plant cell culture medium is usually made by combining several different components, as outlined in Table 2.2.

Media components

It is useful to briefly consider some of the individual components of the stock solutions.

Macroelements

As is implied by the name, the stock solution supplies macroelements required in large amounts for plant growth and development. Nitrogen, phosphorus, potassium, magnesium, calcium, and sulphur (and carbon, which is added separately) are usually regarded as macroelements. These elements usually comprise at least 0.1% of the dry weight of plants.

Nitrogen is most commonly supplied as a mixture of nitrate ions (from KNO_3) and ammonium ions (from NH_4NO_3). Theoretically, there is an advantage in supplying nitrogen in the form of ammonium ions, as nitrogen must be in the reduced form to be incorporated into macromolecules. Nitrate ions therefore need to be reduced before incorporation. However, at high concentrations, ammonium ions can be toxic to plant cell cultures and uptake of ammonium ions from the medium causes acidification of the medium. For ammonium ions to be used as the sole nitrogen source, the medium needs to be buffered. High concentrations of ammonium ions can also cause culture problems by increasing the frequency of vitrification (the culture appears pale and 'glassy' and is usually unsuitable for further culture). Using a

Table 2.2 Composition of a typical plant culture medium

Essential element	Concentration in stock solution (mg l^{-1})	Concentration in medium (mg l^{-1})
Macroelements[a]		
NH_4NO_3	33 000	1 650
KNO_3	38 000	1 900
$CaCl_2 \cdot 2H_2O$	8 800	440
$MgSO_4 \cdot 7H_2O$	7 400	370
KH_2PO_4	3 400	170
Microelements[b]		
KI	166	0.83
H_3BO_3	1 240	6.2
$MnSO_4 \cdot 4H_2O$	4 460	22.3
$ZnSO_4 \cdot 7H_2O$	1 720	8.6
$Na_2MoO_4 \cdot 2H_2O$	50	0.25
$CuSO_4 \cdot 5H_2O$	5	0.025
$CoCl_2 \cdot 6H_2O$	5	0.025
Iron source[b]		
$FeSO_4 \cdot 7H_2O$	5 560	27.8
$Na_2EDTA \cdot 2H_2O$	7 460	37.3
Organic supplement[b]		
Myoinositol	20 000	100
Nicotinic acid	100	0.5
Pyridoxine-HCl	100	0.5
Thiamine-HCl	100	0.5
Glycine	400	2
Carbon source		
Sucrose	Added as solid	30 000

The medium described here is that of Murashige and Skoog (MS); many other commonly used plant culture media (such as Gamborg's B5 and Schenk and Hildebrandt (SH) medium) are similar in composition to MS medium and can be thought of as high-salt media. MS is extremely widely used and forms the basis for many other media formulations.

[a]50 ml of stock solution used per litre of medium.

[b]5 ml of stock solution used per litre of medium.

mixture of nitrate and ammonium ions has the advantage of weakly buffering the medium as the uptake of nitrate ions causes OH^- ions to be excreted.

Phosphorus is usually supplied as the phosphate ion of ammonium, sodium, or potassium salts. High concentrations of phosphate can lead to the precipitation of medium elements as insoluble phosphates.

Microelements

Microelements are required in trace amounts for plant growth and development, and have many and diverse roles. Manganese, iodine, copper, cobalt, boron, molybdenum, iron, and zinc usually comprise the microelements, although other elements such as nickel and aluminium are found frequently in some formulations.

Iron is usually added as iron sulphate, although iron citrate can also be used. Ethylenediaminetetra-acetic acid (EDTA) is usually used in conjunction with iron sulphate. The EDTA complexes with the iron to allow the slow and continuous release of iron into the medium. Uncomplexed iron can precipitate out of the medium as ferric oxide.

Organic supplements

Only two vitamins, thiamine (vitamin B_1) and myoinositol (considered a B vitamin), are considered essential for the culture of plant cells *in vitro*. However, other vitamins are often added to plant cell culture media for historical reasons.

Amino acids are also commonly included in the organic supplement. The most frequently used is glycine (arginine, asparagine, aspartic acid, alanine, glutamic acid, glutamine, and proline are also used), but in many cases its inclusion is not essential. Amino acids provide a source of reduced nitrogen and, like ammonium ions, uptake causes acidification of the medium. Casein hydrolysate can be used as a relatively cheap source of a mix of amino acids.

Carbon source

Sucrose is cheap, easily available, readily assimilated, and relatively stable, and is therefore the most commonly used carbon source. Other carbohydrates (such as glucose, maltose, galactose, and sorbitol) can also be used (see Chapter 3), and in specialized circumstances may prove superior to sucrose.

Gelling agents

Media for plant cell culture *in vitro* can be used in either liquid or 'solid' forms, depending on the type of culture being grown. For any culture types that require the plant cells or tissues to be grown on the surface of the medium, it must be solidified (more correctly termed gelled). Agar, produced from seaweed, is the most common type of gelling agent, and is ideal for routine applications. However, because it is a natural product, the agar quality can vary from supplier to supplier and from batch to batch. For more demanding applications (see, for instance, the section on microspore culture below and Chapter 3), a range of purer (and in some cases, considerably more expensive) gelling agents are available. Purified agar or agarose can be used, as can a variety of gellan gums.

Summary

These components, then, are the basic chemical necessities for plant cell culture media. However, other additions are made to manipulate the pattern of growth and development of the plant cell culture.

Plant growth regulators

We have already briefly considered the concepts of plasticity and totipotency. The essential point as far as plant cell culture is concerned is that, due to this plasticity and totipotency, specific media manipulations can be used to direct the development of plant cells in culture.

Plant growth regulators are the critical media components in determining the developmental pathway of the plant cells. The plant growth regulators used most commonly are plant hormones or their synthetic analogues.

Classes of plant growth regulator

● Auxins and cytokinins are the most important hormones for transgenic work.

There are five main classes of plant growth regulator used in plant cell culture, namely:

(1) auxins;

(2) cytokinins;

(3) gibberellins;

(4) abscisic acid;

(5) ethylene.

Each class of plant growth regulator will be looked at briefly below.

Auxins

Auxins promote both cell division and cell growth. The most important naturally occurring auxin is indole-3-acetic acid (IAA), but its use in plant cell culture media is limited because it is unstable to both heat and light. Occasionally, amino acid conjugates of IAA (such as indole-acetyl-L-alanine and indole-acetyl-L-glycine), which are more stable, are used to partially alleviate the problems associated with the use of IAA. It is more common, though, to use stable chemical analogues of IAA as a source of auxin in plant cell culture media. 2,4-Dichlorophenoxyacetic acid (2,4-D) is the most commonly used auxin and is extremely effective in most circumstances. Other auxins are available (see Table 2.3), and some may be more effective or 'potent' than 2,4-D in some instances.

Cytokinins

Cytokinins promote cell division. Naturally occurring cytokinins are a large group of structurally related purine derivatives. Of the naturally occurring cytokinins, two have some use in plant tissue culture media (see Table 2.4), zeatin and

Table 2.3 Commonly used auxins, their abbreviations and chemical names

Abbreviation/name	Chemical name
2,4-D	2,4-Dichlorophenoxyacetic acid
2,4,5-T	2,4,5-Trichlorophenoxyacetic acid
Dicamba	2-Methoxy-3,6-dichlorobenzoic acid
IAA	Indole-3-acetic acid
IBA	Indole-3-butyric acid
MCPA	2-Methyl-4-chlorophenoxyacetic acid
NAA	1-Naphthylacetic acid
NOA	2-Naphthyloxyacetic acid
Picloram	4-Amino-2,5,6-trichloropicolinic acid

Table 2.4 Commonly used cytokinins, their abbreviations and chemical names

Abbreviation/name	Chemical name
BAP[a]	6-Benzylaminopurine
2iP (or IPA)[b]	N^6-(2-Isopentyl)adenine
Kinetin[a]	6-Furfurylaminopurine
Thidiazuron[c]	1-Phenyl-3-(1,2,3-thiadiazol-5-yl)urea
Zeatin[b]	4-Hydroxy-3-methyl-*trans*-2-butenylaminopurine

[a]Synthetic analogues.
[b]Naturally occurring cytokinins.
[c]A substituted phenylurea-type cytokinin.

N^6-(2-isopentyl)adenine (2iP). Their use is not widespread as they are expensive (particularly zeatin) and relatively unstable. The synthetic analogues kinetin and 6-benzylaminopurine (BAP) are therefore used more frequently. Non-purine-based chemicals, such as substituted phenylureas, are also used as cytokinins in plant cell culture media. These substituted phenylureas can also substitute for auxin in some culture systems.

Gibberellins

There are numerous, naturally occurring, structurally related compounds termed gibberellins. They are involved in regulating cell elongation, and are agronomically important in determining plant height and fruit-set. Only a few of the gibberellins are used in plant tissue culture media, GA_3 being the most common.

Abscisic acid

Abscisic acid (ABA) inhibits cell division. It is most commonly used in plant tissue culture to promote distinct developmental pathways such as somatic embryogenesis (see also Box 2.1, below).

Ethylene

Ethylene is a gaseous, naturally occurring, plant growth regulator most commonly associated with controlling fruit ripening in climacteric fruits, and its use in plant tissue culture is not widespread. It does, though, present a particular problem for plant tissue culture. Some plant cell cultures produce ethylene, which, if it builds up sufficiently, can inhibit the growth and development of the culture. The type of culture vessel used and its means of closure affect the gaseous exchange between the culture vessel and the outside atmosphere and thus the levels of ethylene present in the culture.

Plant growth regulators and tissue culture

Generalizations about plant growth regulators and their use in plant cell culture media have been developed from initial observations made in the 1950s. There is, however, some considerable difficulty in predicting the effects of plant growth regulators: this is because of the great differences in culture response among species,

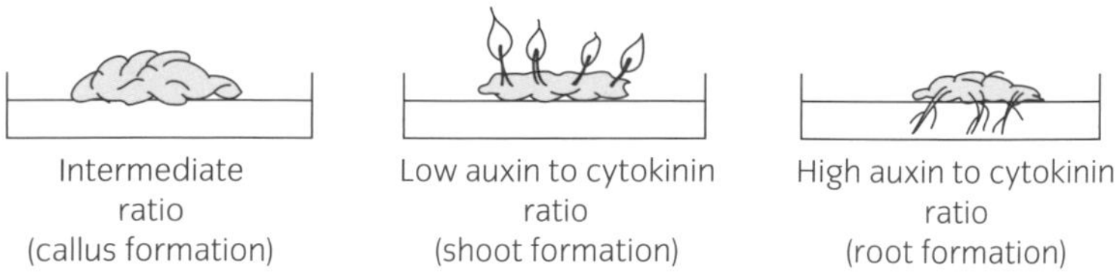

Figure 2.1 The effect of different ratios of auxin to cytokinin on the growth and morphogenesis of callus. High auxin to cytokinin ratios promote root development, while low ratios promote shoot development. Intermediate ratios promote continued growth of the callus without differentiation.

cultivars, and even plants of the same cultivar grown under different conditions. However, some principles do hold true and have become the paradigm on which most plant tissue culture regimes are based.

Auxins and cytokinins are the most widely used plant growth regulators in plant tissue culture and are usually used together, the ratio of the auxin to the cytokinin determining the type of culture established or regenerated (see Figure 2.1). A high auxin to cytokinin ratio generally favours root formation, whereas a high cytokinin to auxin ratio favours shoot formation. An intermediate ratio favours callus production.

Culture types

● Although a range of cultures are potentially available for transformation experiments, some may be more readily available. This is often related to the type of plant being used.

Cultures are generally initiated from sterile pieces of a whole plant. These pieces are termed explants, and may consist of pieces of organs, such as leaves or roots, or may be specific cell types, such as pollen or endosperm. Many features of the explant are known to affect the efficiency of culture initiation. Generally, younger, more rapidly growing tissue (or tissue at an early stage of development) is most effective. Several different culture types most commonly used in plant transformation studies will now be examined in more detail.

Callus

Explants, when cultured on the appropriate medium, usually with both an auxin and a cytokinin, can give rise to an unorganized, growing, and dividing mass of cells. It is thought that any plant tissue can be used as an explant, if the correct conditions are found. In culture, this proliferation can be maintained more or less indefinitely, provided that the callus is subcultured on to fresh medium periodically. During callus formation, there is some degree of dedifferentiation (i.e. the changes that occur during development and specialization are, to some extent, reversed), both in morphology (a callus is usually composed of unspecialized parenchyma cells) and metabolism. One major consequence of this dedifferentiation is that most plant cultures lose the ability to photosynthesize. This has important consequences for the culture of callus tissue, as the metabolic profile will probably not match that of the donor plant. This necessitates the addition of other components—such as vitamins and, most importantly, a carbon source—to the culture medium, in addition to the usual mineral nutrients.

Callus culture is often performed in the dark (the lack of photosynthetic capability being no drawback) as light can encourage differentiation of the callus. During long-term culture, the culture may lose the requirement for auxin and/or cytokinin. This process, known as **habituation**, is common in callus cultures from some plant species (such as sugar beet).

Callus cultures are extremely important in plant biotechnology. Manipulation of the auxin to cytokinin ratio in the medium can lead to the development of shoots, roots, or somatic embryos from which whole plants can subsequently be produced. Callus cultures can also be used to initiate cell suspensions, which are used in a variety of ways in plant transformation studies.

Cell-suspension cultures

Callus cultures, broadly speaking, fall into one of two categories: compact or friable. In compact callus, the cells are densely aggregated, whereas in friable callus, the cells are only loosely associated with each other and the callus becomes soft and breaks apart easily. Friable callus provides the inoculum to form cell-suspension cultures. Explants from some plant species or particular cell types tend not to form friable callus, making it difficult to initiate cell suspension. The friability of the callus can sometimes be improved by manipulating the medium components or by repeated subculturing. The friability of the callus can also sometimes be improved by culturing it on semi-solid medium (medium with a low concentration of gelling agent).

When friable callus is placed into a liquid medium (usually the same composition as the solid medium used for the callus culture) and then agitated, single cells and/or small clumps of cells are released into the medium. Under the correct conditions, these released cells continue to grow and divide, eventually producing a cell-suspension culture. A relatively large inoculum should be used when initiating cell suspensions so that the released cell numbers build up quickly. The inoculum should not be too large though, as toxic products released from damaged or stressed cells can build up to lethal levels. Large cell clumps can be removed during subculture of the cell suspension.

Cell suspensions can be maintained relatively simply as batch cultures in conical flasks. They are continually cultured by repeated subculturing into fresh medium. This results in dilution of the suspension and the initiation of another batch growth cycle. The degree of dilution during subculture should be determined empirically for each culture. Too great a degree of dilution will result in a greatly extended lag period or, in extreme cases, death of the transferred cells.

After subculture, the cells divide and the biomass of the culture increases in a characteristic fashion, until nutrients in the medium are exhausted and/or toxic by-products build up to inhibitory levels: this is called the **stationary phase**. If cells are left in the stationary phase for too long, they will die and the culture will be lost. Therefore, cells should be transferred as they enter the stationary phase. It is therefore important that the batch growth-cycle parameters are determined for each cell-suspension culture.

Protoplasts

Protoplasts are plant cells with the cell wall removed. Protoplasts are most commonly isolated from either leaf mesophyll cells or cell suspensions, although other sources can be used to advantage. Two general approaches to removing the cell wall (a difficult task without damaging the protoplast) can be taken: mechanical or enzymatic isolation. Mechanical isolation, although possible, often results in low yields, poor quality, and poor performance in culture due to substances released from damaged cells. Enzymatic isolation is usually carried out in a simple salt solution with a high osmoticum, plus the cell-wall-degrading enzymes. It is usual to use a mix of both cellulase and pectinase enzymes, which must be of high quality and purity.

Protoplasts are fragile and damaged easily, and therefore must be cultured carefully. Liquid medium is not agitated and a high osmotic potential is maintained, at least in the initial stages. The liquid medium must be shallow enough to allow aeration in the absence of agitation. Protoplasts can be plated out on to solid medium and callus produced. Whole plants can be regenerated by organogenesis or somatic embryogenesis from this callus. Protoplasts are ideal targets for transformation by a variety of means.

Root cultures

Root cultures can be established *in vitro* from explants of the root tip of either primary or lateral roots and can be cultured on fairly simple media. The growth of roots *in vitro* is potentially unlimited, as roots are indeterminate organs. Although the establishment of root cultures was one of the first achievements of modern plant tissue culture, they are not widely used in plant transformation studies.

Shoot tip and meristem culture

The tips of shoots (which contain the shoot apical meristem) can be cultured *in vitro*, producing clumps of shoots from either axillary or adventitious buds. This method can be used for clonal propagation.

Shoot meristem cultures are potential alternatives to the more commonly used methods for cereal regeneration (see Case study 2.1, below) as they are less genotype-dependent and more efficient (seedlings can be used as donor material).

Embryo culture

Embryos can be used as explants to generate callus cultures or somatic embryos. Both immature and mature embryos can be used as explants. Immature, embryo-derived embryogenic callus is the most popular method of monocotyledon plant regeneration.

Microspore culture

Haploid tissue can be cultured *in vitro* by using pollen or anthers as an explant. Pollen contains the male gametophyte, which is termed the microspore. Both callus and embryos can be produced from pollen. Two main approaches can be taken to produce cultures *in vitro* from haploid tissue.

The first method depends on using the anther as the explant. Anthers (somatic tissue that surrounds and contains the pollen) can be cultured on solid medium (agar should not be used to solidify the medium as it contains inhibitory substances). Pollen-derived embryos are subsequently produced via dehiscence of the mature anthers. The dehiscence of the anther depends both on its isolation at the correct stage and on the correct culture conditions. In some species, the reliance on natural dehiscence can be circumvented by cutting the wall of the anther, although this does, of course, take a considerable amount of time. In the second method, anthers can also be cultured in liquid medium, and pollen released from the anthers can be induced to form embryos, although the efficiency of plant regeneration is often very low. Immature pollen can also be extracted from developing anthers and cultured directly, although this is a very time-consuming process.

Both methods have advantages and disadvantages. Some beneficial effects to the culture are observed when anthers are used as the explant material. There is, however, the danger that some of the embryos produced from anther culture will originate from the somatic anther tissue rather than the haploid microspore cells. If isolated pollen is used there is no danger of mixed-embryo formation, but the efficiency is low and the process is time-consuming.

In microspore culture, the condition of the donor plant is of critical importance, as is the timing of isolation. Pretreatments, such as a cold treatment, are often found to increase the efficiency. These pretreatments can be applied before culture, or, in some species, after placing the anthers in culture.

Plant species can be divided into two groups, depending on whether they require the addition of plant growth regulators to the medium for pollen/anther culture; those that do also often require organic supplements, such as amino acids. Many of the cereals (rice, wheat, barley, and maize) require medium supplemented with plant growth regulators for pollen or anther culture.

Regeneration from microspore explants can be obtained by direct embryogenesis, or via a callus stage and subsequent embryogenesis. Haploid tissue cultures can also be initiated from the female gametophyte (the ovule). In some cases, this is a more efficient method than using pollen or anthers.

Plants obtained from haploid cultures may not be haploid. This can be a consequence of chromosome doubling during the culture period. Chromosome doubling (which often has to be induced by treatment with chemicals such as colchicine) may be an advantage, as in many cases haploid plants are not the desired outcome of regeneration from haploid tissues. Such plants are often referred to as di-haploids, because they contain two copies of the same haploid genome.

Plant regeneration

Having looked at the main types of plant culture that can be established *in vitro*, we can now look at how whole plants can be regenerated from these cultures. In broad terms, two methods of plant regeneration are widely used in plant transformation studies: somatic embryogenesis and organogenesis.

Somatic embryogenesis

● Embryogenesis is particularly favoured for transformation work.

In somatic (asexual) embryogenesis, embryo-like structures, which can develop into whole plants in a way analogous to zygotic embryos, are formed from somatic tissues (Figure 2.2). These somatic embryos can be produced either directly or indirectly. In direct somatic embryogenesis, the embryo is formed directly from a cell or small group of cells without the production of an intervening callus. Though common from some tissues (usually reproductive tissues such as the nucellus, styles, or pollen), direct somatic embryogenesis is generally rare in comparison with indirect somatic embryogenesis. An example of direct somatic embryogenesis is given in Box 2.1.

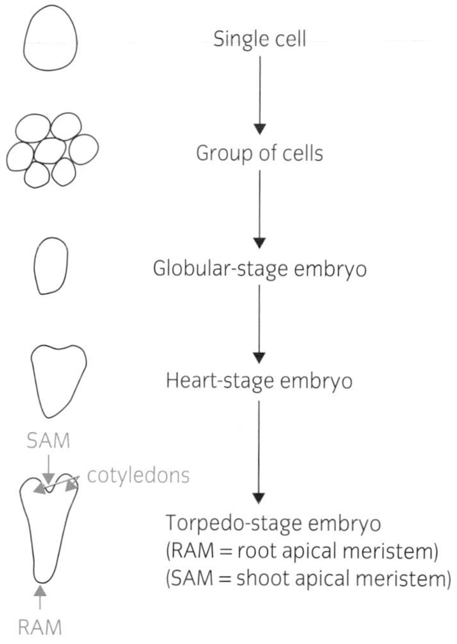

Figure 2.2 A schematic representation of the sequential stages of somatic embryo development. Somatic embryos may develop from single cells or from a small group of cells. Repeated cell divisions lead to the production of a group of cells that develop into an organized structure known as a globular-stage embryo. Further development results in heart- and torpedo-stage embryos, from which plants can be regenerated. Zygotic embryos undergo a fundamentally similar development through the globular (which is formed after the 16-cell stage), heart, and torpedo stages. Polarity is established early in embryo development. Signs of tissue differentiation become apparent at the globular stage and apical meristems are apparent in heart-stage embryos.

BOX 2.1 *Somatic embryogenesis*

Indirect somatic embryogenesis in carrot (*Daucus carota*)

A callus can be established from explants from a wide range of carrot tissues by placing the explant on solid medium (e.g. Murashige and Skoog (MS) medium) containing 2,4-D (1 mg l^{-1}). This callus can be used to produce a cell suspension by placing it in agitated liquid MS medium also containing 2,4-D (1 mg l^{-1}). The cell suspension can be maintained by repeated subculturing into 2,4-D-containing medium. Removal of the old 2,4-D-containing medium and replacement with fresh medium containing ABA (0.025 mg l^{-1}) results in the production of embryos.

Direct somatic embryogenesis from alfalfa (*Medicago falcata*)

Young trifoliate leaves are used as the explant (see figure). These are removed from the plant and chopped into small pieces. The pieces are washed in a plant growth regulator-free medium and placed in liquid medium (B5) supplemented with 2,4-D (4 mg l^{-1}), kinetin (0.2 mg l^{-1}), adenine (1 mg l^{-1}), and glutathione (10 mg l^{-1}). The cultures are maintained in agitated liquid medium for about 10–15 days. Washing the explants and replacing the old medium with B5 medium supplemented with maltose and polyethylene glycol results in the development of the somatic embryos. These somatic embryos can be matured on solid medium containing ABA.

Note that in both cases, although the production of somatic embryos from alfalfa necessitates the use of more complicated media, the production of embryos is fundamentally a two-step process. The initial medium, which contains 2,4-D, is replaced with a medium that does not contain 2,4-D.

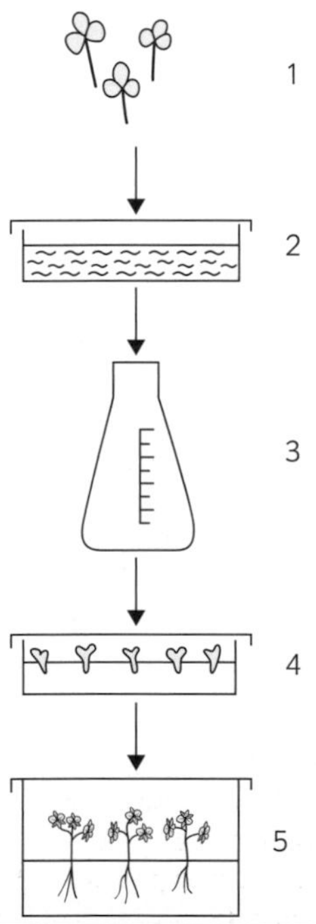

Direct somatic embryogenesis in alfalfa. (1) Explants are removed from plants grown *in vitro*. (2) Explants are placed in liquid medium for embryo induction. (3) Embryos develop to the globular stage in liquid medium supplemented with maltose and polyethylene glycol. (4) Embryos mature on gelled medium containing ABA. (5) Embryos develop into plants on solid medium.

In indirect somatic embryogenesis, callus is first produced from the explant. Embryos can then be produced from the callus tissue or from a cell suspension produced from that callus. Somatic embryogenesis from carrot is the classical example of indirect somatic embryogenesis and is explained in more detail in Box 2.1.

Somatic embryogenesis usually proceeds in two distinct stages. In the initial stage (embryo initiation), a high concentration of 2,4-D is used. In the second stage (embryo production), embryos are produced in a medium with no or very low levels of 2,4-D. In many systems, it has been found that somatic embryogenesis is improved by supplying a source of reduced nitrogen, such as specific amino acids or casein hydrolysate.

CASE STUDY 2.1 **Cereal regeneration via somatic embryogenesis from immature or mature embryos**

The principal method adopted for the tissue culture and regeneration of a wide range of cereal species is somatic embryogenesis, using cultures initiated from immature zygotic embryos. Embryogenic callus is normally initiated by placing the immature embryo on to a medium containing 2,4-D. Shoot regeneration is initiated by placing the embryogenic callus on a medium with BAP (with or without 2,4-D). These shoots can be subsequently rooted. The medium used for induction of embryogenic callus is usually a modified Murashige and Skoog (MS; for Triticeae) or N6 (for rice and maize) medium. Maltose is often used as the carbon source in preference to sucrose, and additional organic supplements (such as specific amino acids, yeast extract, and/or casein hydrolysate) are common.

The isolation and culture of immature embryos is, however, a labour-intensive and relatively expensive procedure. An additional problem is the small target size if the immature embryos are to be used for biolistic transformation (biolistic transformation is explained in Chapter 3). Alternatives are therefore being sought. One alternative is to use mature embryos (or seeds) as the explant to initiate embryogenic callus. This approach has been successfully applied to several cereal species such as rice and oats. The culture techniques and media used for culture establishment and regeneration from mature embryo-/seed-derived cultures are fundamentally the same as those used for cultures initiated from immature embryos.

The importance of genotype

The major influence on tissue culture response appears to be genetic, with culture requirements varying between species and cultivars. Model genotypes that responded well to culture *in vitro* were initially used in plant transformation studies. However, most of the model genotypes used were not elite, commercial cultivars. The commercial cultivars tended to respond poorly to culture *in vitro*. One of the main aims is therefore to identify the components that make up a widely applicable, optimal culture regime.

Many factors have been investigated for their ability to improve the culture response from elite cultivars, including media components (such as alternative carbon sources, macro- and microelement concentrations and composition), media preparation method, and donor plant condition and growth conditions.

BOX 2.2 *Organogenesis in tobacco* (Nicotiana tabacum)

Organogenesis from tobacco pith callus is the classical example of how varying plant growth regulator regimes can be used to manipulate the pattern of regeneration from plant tissue cultures.

When cultured on a medium containing both auxin and cytokinin, callus will proliferate. If the auxin to cytokinin ratio is increased, adventitious roots will form from the callus by organogenesis. If the auxin to cytokinin ratio is decreased, adventitious shoots will be formed. If the explants are cultured on medium containing only a cytokinin, shoots can be produced directly.

Tobacco plants can also be easily regenerated from tobacco leaf pieces. Leaves are cut into approximate 1-cm squares with a sterile scalpel (avoiding large leaf veins and any damaged areas). The leaf pieces are then transferred (right side up) to gelled MS medium supplemented with 1 mg l^{-1} BAP (a cytokinin; see Table 2.4) and 0.1 mg l^{-1} NAA (an auxin; see Table 2.3). Over the next few weeks, callus forms on the explants, particularly around the cut surfaces. After 3–5 weeks, shoots emerge directly from the explants or from callus derived from the explants. When these shoots are about 1 cm long, they can be cut at the base and placed on to solid MS medium without any plant growth regulators. The shoots will form roots and plantlets that will grow in this medium and can subsequently be transferred to soil.

Organogenesis

Somatic embryogenesis relies on plant regeneration through a process analogous to zygotic embryo germination. Organogenesis (Box 2.2) relies on the production of organs, either directly from an explant or from a callus culture. There are three methods of plant regeneration via organogenesis.

The first two methods depend on adventitious organs arising either from a callus culture or directly from an explant. Alternatively, the third method is by axillary bud formation and growth, which can also be used to regenerate whole plants from some types of tissue culture.

Organogenesis relies on the inherent plasticity of plant tissues, and is regulated by altering the components of the medium. In particular, it is the auxin to cytokinin ratio of the medium that determines which developmental pathway the regenerating tissue will take.

It is usual to induce shoot formation by increasing the cytokinin to auxin ratio of the culture medium. These shoots can then be rooted relatively simply.

Integration of plant tissue culture into plant transformation protocols

Various methods of plant regeneration are available to the plant biotechnologist. Some plant species may be amenable to regeneration by a variety of methods, but some may only be regenerated by one method. In Chapter 3, the various methods that

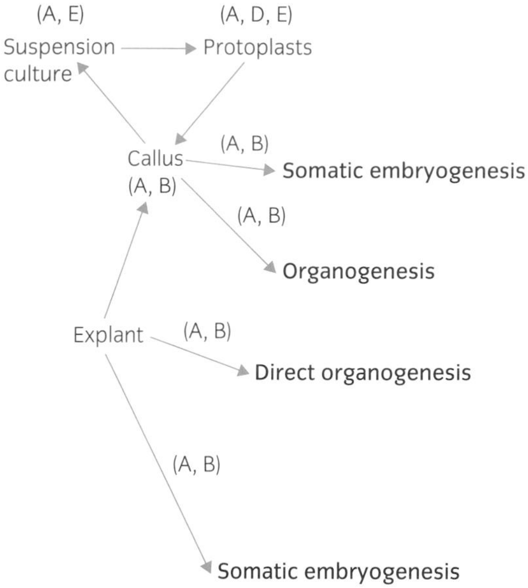

Figure 2.3 A simplified scheme for the integration of plant tissue culture into plant transformation protocols. An explant can be a variety of tissues, depending on the particular plant species being cultured. The explant can be used to initiate a variety of culture types, depending on the explant used. Regeneration by either organogenesis or somatic embryogenesis results in the production of whole plants. Different culture types and regeneration methods are amenable to different transformation protocols. The transformation protocols (see Chapter 3) highlighted in this figure are (A) *Agrobacterium*-mediated, (B) biolistic transformation, (D) direct DNA uptake, and (E) electroporation. Different combinations of culture type and transformation protocol are used depending on the plant species and cultivar being used. In some species, a variety of culture types and regeneration methods can be used, which enables a wide variety of transformation protocols to be utilized. In other species, there is effectively no choice over culture type and/or regeneration method, which can limit the transformation protocols that are applicable. (Redrawn with permission from Walden and Wingender (1995).)

can be used to transform plants will be considered, but it is worthwhile briefly considering the interaction of plant regeneration and transformation methodologies here (Figure 2.3). Not all plant tissue is suited to every plant transformation method, and not all plant species can be regenerated by every method. There is therefore a need to find both a suitable plant tissue culture/regeneration regime and a compatible plant transformation methodology.

SUMMARY

Tissue culture and plant regeneration are an integral part of most plant transformation strategies, and can often prove to be the most challenging aspect of a plant transformation protocol. Key to success in integrating plant tissue culture into plant transformation strategies is the realization that a quick (to avoid too many deleterious effects from somaclonal variation) and efficient regeneration system must be developed. However, this system must also allow high transformation efficiencies from whichever transformation technique is adopted.

Not all regeneration protocols are compatible with all transformation techniques. Some crops may be amenable to a variety of regeneration and transformation strategies, and others may currently only be amenable to one particular protocol. Advances are being made all the time, so it is impossible to say that a particular crop will never be regenerated by a particular protocol. However, some protocols, at least at the moment, are clearly more efficient than others. Regeneration from immature embryo-derived somatic embryos is, for example, the favoured method for regenerating monocotyledon species.

This chapter has only scratched the surface of the problems and potentials of plant tissue culture. General rules are sometimes difficult to make because of the variability of response to particular protocols of different plant species or even cultivars. However, as will be seen in Chapter 3 and subsequent chapters, plant tissue culture has been successfully integrated into plant transformation strategies and the list of plant species that can be transformed routinely continues to grow.

FURTHER READING

Visit www.oxfordtextbooks.co.uk/orc/slaterplants2e/ for hyperlinked references that take you directly to online abstracts for the journal articles listed below:

Barcelo, P., Rasco-Gaunt, S., Thorpe, C., and Lazzeri, P. (2001) Transformation and gene expression. In *Advances in Botanical Research, vol. 34: Biotechnology of Cereals* (P. R. Shewry, P. A. Lazzeri, and K. J. Edwards, eds), pp. 59–126. Academic Press, London.

Dahleen, L. S. (1999) Donor-plant environment effects on regeneration from barley embryo-derived callus. *Crop Science* **39**, 682–685.

Dahleen, L. S. and Bregitzer, P. (2002) An improved media system for high regeneration rates from barley immature embryo-derived callus cultures of commercial cultivars. *Crop Science* **42**, 934–938.

Dodds, J. H. and Roberts, L. W. (1995) *Experiments in Plant Tissue Culture*. Cambridge University Press, Cambridge.

Fowler, M. R. (2000) Plant cell culture, laboratory techniques. In *Encyclopedia of Cell Technology* (R. E. Spier, ed.), pp. 994–1004. Wiley, New York.

Gamborg, O. L. (2002) Plant tissue culture. Biotechnology. Milestones. *In vitro Cellular and Developmental Biology—Plant* **38**, 84–92.

Ramage, C. M. and Williams, R. R. (2002) Mineral nutrition and plant morphogenesis. *In vitro Cellular and Developmental Biology—Plant* **38**, 116–124.

Sugiyama, M. (1999) Organogenesis *in vitro*. *Current Opinion in Plant Biology* **2**, 61–64.

Torbert, K. A., Rines, H. W., and Somers, D. A. (1998) Transformation of oat using mature embryo-derived tissue cultures. *Crop Science* **38**, 226–231.

Torbert, K. A., Rines, H. W., Kaeppler, H. F., Menon, G. K., and Somers, D. A. (1998) Genetically engineering elite oat cultivars. *Crop Science* **38**, 1685–1687.

Walden, R. and Wingender, R. (1995) Gene-transfer and plant-regeneration techniques. *Trends in Biotechnology* **13**, 324–331.

3

Techniques for plant transformation

INTRODUCTION

Plant transformation has become widely adopted as a method both to understand how plants work and to improve crop plant characteristics. Plant transformation depends on the stable introduction of transgene(s) into the genome of the plant. Various methods have been developed to achieve this and many plant species have been transformed successfully. Some transformation methods are based on utilizing *Agrobacterium*, a pathogen of dicotyledonous (broad-leafed) plants that transfers genes into the plant genome. Problems with using *Agrobacterium* to transform monocotyledonous plants (grasses) spurred on the development of other methods, the so-called direct gene-transfer methods. The major direct gene-transfer method, particle bombardment (or biolistics), is the method of choice in many laboratories for the transformation of monocotyledonous plants, despite *Agrobacterium*-based protocols having subsequently been developed for the transformation of monocotyledonous plants.

In this chapter, we will look at both the biology of *Agrobacterium tumefaciens* and how it is used in plant transformation studies. Direct gene-transfer methods for plant transformation, and in particular biolistics, will also be considered and examples of their use for the transformation of monocotyledonous and dicotyledonous crops will be given.

Agrobacterium-mediated gene transfer

The biology of *Agrobacterium*

Agrobacterium tumefaciens is a soil-borne, Gram-negative bacterium, much beloved of plant biologists and biotechnologists alike. The explosion of interest in *A. tumefaciens* has been driven by several observations that have both biotechnological and purely scientific ramifications. *A. tumefaciens* is the causative agent of crown-gall disease, an economically important disease of many plants (particularly grapes, walnuts, apples, and roses). The ability to cause crown galls (tumorous tissue growths) depends on the ability of *Agrobacterium* spp. to transfer bacterial genes into the plant genome. This startling feature is, to date, one of the very few examples of inter-kingdom gene transfer, which has obvious scientific interest. More important to the biotechnologist is the development of plant transformation methods that make use of this unique capability.

Crown-gall disease

In order to fully appreciate and understand how *A. tumefaciens* became such a widely used tool in plant biotechnology it is worth looking at the biology of crown-gall disease. *A. tumefaciens*, the causative agent, is a Gram-negative, rod-shaped, motile bacterium found in the rhizosphere (the region around the roots of plants) where it normally survives on nutrients released from plant roots. However, if a plant is wounded or damaged, *A. tumefaciens* can infect the plant at the wound site and cause disease symptoms. *A. tumefaciens* is attracted to the wound site via chemotaxis, in response to chemicals (sugars and phenolic molecules) released from the damaged plant cells.

● Plasmid DNA is transferred from *Agrobacterium* into the plant cell DNA, causing crown-gall disease.

Crown-gall formation depends on the presence of a plasmid in *A. tumefaciens* known as the Ti plasmid (Ti stands for tumour-inducing). Part of this plasmid (the transfer DNA, or T-DNA region) is actually transferred from the bacterium into the plant cell, where it becomes integrated into the genome of the host plant. The T-DNA carries genes that encode proteins involved in both hormone biosynthesis (of auxin and cytokinin) and the biosynthesis of novel plant metabolites called opines and agropines. The production of auxin and cytokinin causes the plant cells to proliferate and so form the gall. The proliferating cells also produce opines (which are amino acid derivatives) and agropines (sugar derivatives), which are used by *A. tumefaciens* as its sole carbon and energy source (Figure 3.1 shows the structure of the common opines, octopine and nopaline). Different strains of *A. tumefaciens* contain different Ti plasmids that code for the production of different opines. Opines and agropines are not normally part of plant metabolism and are very stable chemicals, which provide a carbon and energy source that only *A. tumefaciens* can use (genes on the Ti

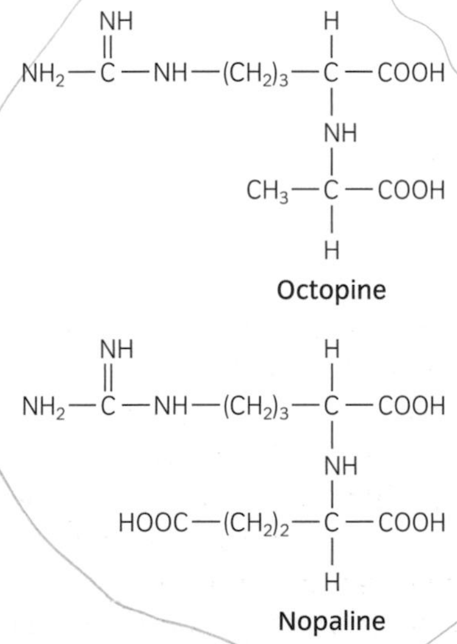

Figure 3.1 The structures of the opines octopine and nopaline are shown. Both octopine and nopaline are amino acid derivatives. Octopine is derived from the amino acid arginine and pyruvate. Nopaline is derived from arginine and α-ketoglutarate.

plasmid that are not transferred to the plant encode proteins involved in opine uptake and catabolism). *A. tumefaciens* has therefore developed the ability to genetically transform plant cells to usurp the plant's biosynthetic machinery and produce nutrients that only it can utilize.

The Ti plasmid

As the Ti plasmid has a central role in crown-gall formation, and it is a portion of the Ti plasmid (the T-DNA region) that is actually integrated into the genome of the host plant, we shall look at the Ti plasmid in more detail.

As we have seen, the ability of *A. tumefaciens* to cause crown-gall disease was found to depend on the presence in the bacterium of a large (≈200 kb) plasmid termed the Ti plasmid. Analysis of the nuclear DNA from plant tumours showed that a portion of the Ti plasmid was integrated into the genome of the host plant. This portion was termed transfer DNA (T-DNA) and was found to be responsible for the tumorous phenotype. Analysis showed that one or more copies of the T-DNA could be integrated into the genome, but, in general, the T-DNA insertions in the plant genome were bordered by small (24 bp), nearly perfect, direct repeats, which also border the T-DNA in the Ti plasmid (Figure 3.2).

Ti-plasmid features

● The Ti plasmid carries genes coding for plant hormones that cause cell proliferation and genes that code for proteins mediating DNA transfer.

Ti plasmids from different strains of *A. tumefaciens* generally have several features in common:

1 they contain one (or more) T-DNA regions;

2 they contain a *vir* region;

3 they contain an origin of replication;

4 they contain a region enabling conjugative transfer;

5 they contain genes for the catabolism of opines (a class of amino acid/sugar conjugates).

The T-DNA

The T-DNA region of any Ti plasmid is defined by the presence of the right- and left-border sequences (see Figure 3.2). These border sequences are 24 bp imperfect

Figure 3.2 The consensus sequences of the border sequences from nopaline- and octopine-strain Ti plasmids. The nucleotide sequence of the enhancer sequence is also shown.

Nopaline strains

TG(G/A)CAGGATATAT(-/T)G(T/G)(G/C)G(T/G)GTAAAC

Octopine strains

(C/T)GGCAGGATATA(T/A)C(A/C)(A/G)TTGTAA(A/T)T

TAAGTCGCTGTGTATGTTTGTTTG (Enhancer or 'overdrive' sequence)

repeats. Any DNA between the borders will be transferred into the genome of the host plant. Although not part of the T-DNA, octopine-strain Ti plasmids contain an overdrive sequence or enhancer sequence (see Figure 3.2) associated with the right-border sequence, which is required for optimal T-DNA transfer.

Nopaline strains of Ti plasmid have one T-DNA of approximately 20 kb, whereas octopine strains have two T-DNA regions, termed T_L and T_R, that are approximately 14 and 7 kb, respectively. However, only the T_L region is oncogenic (the T_R region carries genes for opine biosynthesis).

The T-DNA of nopaline Ti plasmids contains 13 open reading frames (ORFs), whereas the T_L T-DNA of octopine plasmids contains eight. These ORFs have the features of eukaryotic, rather than prokaryotic, genes. The nopaline T-DNA and the T_L octopine T-DNA show extensive similarity in an area known as the core region. This region contains the genes that code for proteins involved in hormone biosynthesis (the oncogenes) and opine synthesis, and for determining tumour size. The organization of the genes in the T_L T-DNA is shown in Figure 3.3.

The oncogenes

Two genes, *auxA* (or *tms*1 or *iaaM*) and *auxB* (or *tms*2 or *iaaH*), encode proteins involved in the production of the auxin indole-3-acetic acid (IAA). *auxA* encodes tryptophan monooxygenase and *aux*B encodes indole acetamidehydrolase. Another gene, *cyt* (or *tmr* or *ipt*), encodes an isopentenyl transferase that catalyses the most important step in cytokinin production. These genes are the prime determinants of tumour phenotype and are therefore often referred to as oncogenes. The T_R region of octopine T-DNA contains no oncogenes, explaining its lack of oncogenicity.

Other genes present on the T-DNA

Genes for the production of opines (either octopine or nopaline) are present on the T-DNA (the octopine T_R T-DNA possesses genes encoding proteins involved in the biosynthesis of other opines such as mannopine, agropine, and fructopine). The *tml* gene, which is involved in determining tumour size in some species, is also found in the T-DNA (see Figure 3.3).

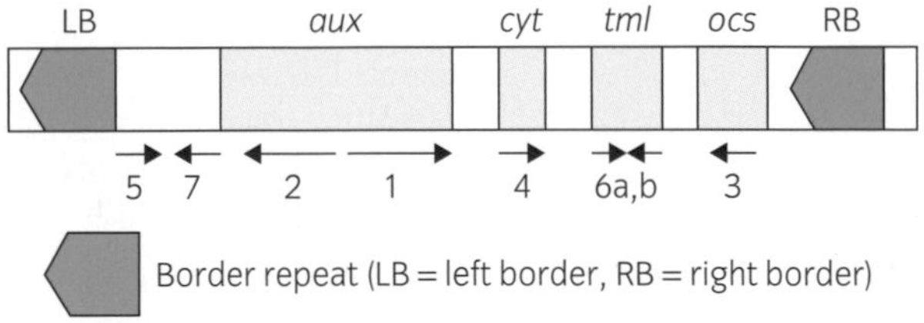

Figure 3.3 The genetic organization of the T_L T-DNA of an octopine-type Ti plasmid. Only the T_L region is shown as this has homology with the T-DNA of nopaline-type Ti plasmids. Eight ORFs are indicated (1–7), although ORFs 5 and 7 are not discussed in this text. Regions of importance are shaded light green and include the *aux* genes (which encode enzymes involved in auxin biosynthesis), *cyt*, which encodes isopentyl transferase (an enzyme involved in cytokinin production), *tml*, which is involved in regulating tumour size, and *ocs* (octopine synthase), which encodes opine synthesis. (Redrawn with permission from Hughes (1996).)

Table 3.1 *Agrobacterium* virulence protein functions

Virulence protein	Function in *Agrobacterium* spp.	Function in plant	Plant proteins that interact
VirA	Phenolic sensor kinase Part of two-component system with VirG; phosphorylates and activates VirG		
VirG	Transcription factor Responsible for induction of *vir* gene expression		
VirB1–B11	Components of membrane structure (transfer apparatus) for transfer of T-DNA		
VirC1	Overdrive-binding protein Enhances efficiency of T-DNA transfer		
VirD1	Required for T-DNA processing Modulates VirD2 activity		
VirD2	Nicks the T-DNA and directs T-DNA through the VirB/VirD4 transfer apparatus	Nuclear targeting of T-DNA Protection of 5′ end of T-DNA from nucleases Possibly involved in integration	Karyopherin-α[a] Cyclophilins[b] pp2C[c]
VirD4	Component of transfer apparatus		
VirE1	VirE2 chaperone Required for VirE2 export from *Agrobacterium* spp.		
VirE2		Single-stranded DNA-binding protein Prevents T-DNA degradation by nucleases Involved in nuclear targeting and passage through NPC	VIP1[d]
VirF		Directs proteins coating T-complex for destruction in proteasome	
VirJ	T-DNA export		

[a]Enables transport through the plant nuclear-pore complex (NPC).
[b]Cyclophilins may cause conformational changes in VirD2 or may be involved in targeting to chromatin and integration of the T-DNA.
[c]Serine/threonine protein phosphatase type 2C (pp2C) negatively affects nuclear import of the T-DNA.
[d]VIP1 facilitates nuclear import and is involved in directing the T-DNA to chromatin.

The vir region

The genes responsible for the transfer of the T-DNA region into the host plant are also situated on the Ti plasmid, in a region of approximately 40 kb outside the T-DNA, known as the *vir* (virulence) region.

There are at least nine *vir* gene operons (Table 3.1 summarizes *vir* gene functions). The functions of the *vir* genes present in these operons have, in many cases, been determined. In the following section, we will look in more detail at *vir* gene functions and the roles they play in T-DNA transfer from *A. tumefaciens* to plant cells.

The process of T-DNA transfer and integration

T-DNA transfer and integration into the plant genome can be divided into the following steps (see Figure 3.4 for a simplified representation of this process).

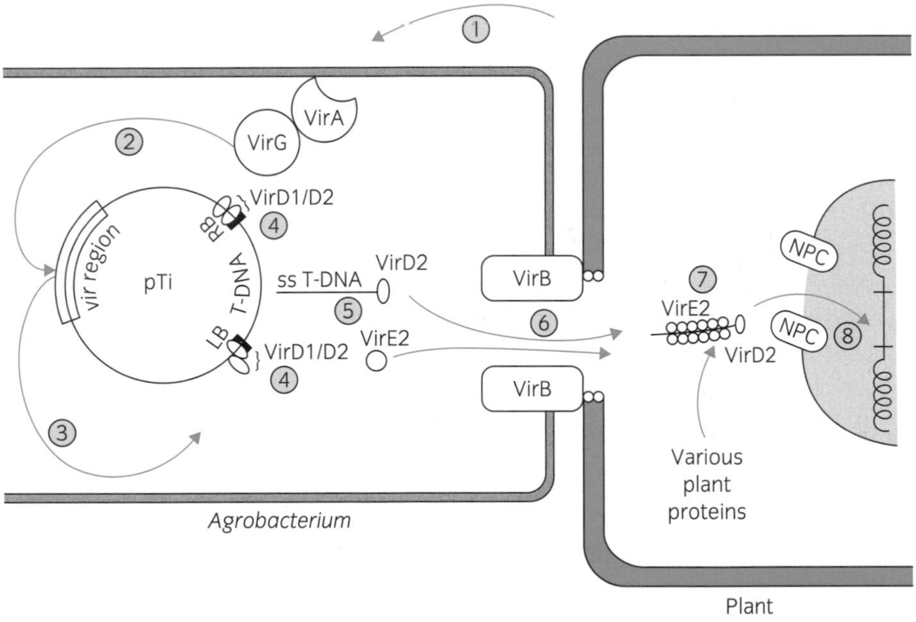

Figure 3.4 A simplified representation of the T-DNA transfer and integration process. Wounded plant cells release phenolic substances and sugars (1) that are sensed by VirA, which activates VirG by phosphorylation. VirG induces the expression of all the genes in the *vir* region of the Ti plasmid (2). Gene products of the *vir* genes (3) are involved in a variety of processes. VirD1 and VirD2 are involved in single-stranded (ss) T-DNA production, protection, and export (4 and 5). VirB products form the transfer apparatus. The single-stranded T-DNA (associated with VirD2) and VirE2 are exported through the transfer apparatus (VirF may also be exported) (6). In the plant cell, the T-DNA becomes coated with VirE2 (7). Various plant proteins interact with either VirD2 or VirE2, which are attached to the T-DNA and influence transport and integration. The T-DNA–VirD2–VirE2–plant protein complex enters the nucleus through the nuclear-pore complex (NPC). Integration into the plant chromosome (8) occurs via illegitimate recombination. LB, left border; pTi, Ti plasmid; RB, right border.

Step 1. Signal recognition by *Agrobacterium*

Agrobacterium perceives signals, such as phenolics and sugars, which are released from wounded plant cells. Normally these substances are probably part of the plant's defence mechanism, being involved in phytoallexin and lignin synthesis. The substances released from wounded cells effectively signal the presence of plant cells that are competent for transformation.

Step 2. Attachment to plant cells

Attachment of *Agrobacterium* to plant cells is a two-step process, involving an initial attachment via a polysaccharide (the product of the *attR* locus). Subsequently, a mesh of cellulose fibres is produced by the bacterium. Several chromosomal virulence genes (*chv* genes) are involved in the attachment of the bacterial cells to the plant cells. The *chvA* and *chvB* genes are involved in the production and secretion of cyclic β-1,2-glycans, and the *pscA* gene (psc, polysaccharide) is involved in secreting succinoglycan. *Agrobacterium* strains carrying mutated *chvA* or *chvB* genes are avirulent or extremely attenuated.

Step 3. Induction of *vir* genes

VirA (a membrane-linked sensor kinase) senses phenolics (such as acetosyringone) and autophosphorylates, subsequently phosphorylating and thereby activating VirG. VirG then induces expression of all the *vir* genes (including *virA* and *virG*). Many sugars, but in particular glucose, galactose, and xylose, enhance *vir* gene induction. This enhancement requires another chromosomal *vir* gene, termed *chvE*, which encodes a glucose/galactose transporter that interacts with VirA.

Step 4. T-strand production

The left and right borders are recognized by a VirD1–VirD2 complex and VirD2 produces single-stranded nicks in the DNA. After nicking, VirD2 becomes covalently attached to the 5′ end of the displaced single-stranded T-DNA strand. Repair synthesis replaces the displaced strand. VirC1 may assist in the process.

Step 5. Transfer of T-DNA out of the bacterial cell

- T-DNA is transferred to the host plant cells as a single strand of DNA covalently linked to VirD2.

The T-DNA–VirD2 complex is exported from the bacterial cell by a T-pilus (effectively a membrane-channel secretory system) composed of proteins encoded by the *virB* operon and VirD4. VirE2 and VirF are also exported from the bacterial cell.

Step 6. Transfer of the T-DNA and Vir proteins into the plant cell and nuclear localization

- Host-plant proteins are heavily implicated in nuclear localization and integration of the T-DNA.

The T-DNA–VirD2 complex and other Vir proteins cross the plant plasma membrane. Once inside the plant cytoplasm, the T-DNA strand becomes covered with VirE2 proteins and is known as a mature T-complex. The VirE2 molecules are postulated to protect the T-DNA from nucleases, facilitate nuclear localization, and confer the correct conformation to the T-DNA–VirD2 complex for passage through the nuclear-pore complex (NPC).

Both VirD2 and VirE2 have been shown to interact with a variety of plant proteins that are important in ensuring nuclear localization of the T-DNA complex. VirD2 contains a nuclear localization signal and interacts with a plant protein. This plant protein belongs to a group of proteins (called the karyopherin-α family) involved in the recognition and transport of proteins through the NPC.

VirE2 possess two nuclear localization signals, but its nuclear localization is mediated by another plant protein, termed VirE2-interacting protein 1 (VIP1), which also has a functional homologue, the bacterial VirE3 protein. VIP1 and VirE3 are thought to facilitate interactions between VirE2 and karyopherin-α, effectively 'piggy-backing' the VirE2 into the nucleus of the plant cell. This contribution of VirE2 to the nuclear localization of T-DNA complexes may be particularly important for large T-DNAs.

The T-DNA and associated proteins pass through the nuclear pore, with the bound VirE2 proteins also giving the correct conformation to the T-DNA strand. VIP1, which interacts with the chromatin protein histone H2A, may be particularly important in targeting the T-complex to the integration site. The T-complex does, however, also interact with various proteins that form part of the transcriptional machinery and these interactions may also be important in directing the T-complex to the site of integration. The proteins bound to the T-complex must also be removed, a process involving VirF, a bacterial F-box protein that can direct VIP1 and VirE2 for degradation by the proteasomal machinery. The T-DNA strand is integrated into the host plant's genome by a process referred to as illegitimate recombination. This process, unlike homologous recombination, does not depend on extensive regions of sequence similarity and is likely to be mediated almost exclusively by plant proteins involved in DNA repair.

Practical applications of *Agrobacterium*-mediated plant transformation

We have looked at the basic biology of *A. tumefaciens* and how the T-DNA is transferred into the plant genome in some detail. We can now look at how *A. tumefaciens* is incorporated into plant transformation experiments. It is important to remember that the successful production of transgenic plants is reliant on combining a suitable transformation protocol with a robust regeneration protocol (see Chapter 2). *Agrobacterium*-mediated transformation may not prove suitable for all types of explant, and may not be the most efficient transformation method for some plant species. Nevertheless, *Agrobacterium*-mediated transformation is a widely used transformation method, partly for historical reasons and partly because of some advantages over other methods. *Agrobacterium*-mediated transformation methods are thought to induce less rearrangement of the transgene and result in a lower transgene copy number than direct DNA-delivery methods.

● *Agrobacterium* was first used to transfer DNA into dicotyledonous plant species. Subsequently it has been developed for some monocotyledon species.

CASE STUDY 3.1 *Agrobacterium*-mediated transformation of tobacco

Tobacco is a relatively easy plant to transform with *Agrobacterium* and provides a good introduction to the use of *Agrobacterium* in plant transformation. In this case study, we will look at the processes used for *Agrobacterium*-mediated plant transformation. Several factors have to be considered in the design and implementation of any plant transformation study.

1 The plant tissue to be transformed. The purpose of most plant transformation experiments in plant biotechnology is to produce whole, transgenic plants. The explant used in transformation experiments must therefore be capable of producing whole plants by regeneration, and should contain a high proportion of cells that are competent for transformation (explants and the regeneration of whole plants from them are discussed in detail in Chapter 2).

2 The vector used to deliver the transgene into the genome of the plant. The construction of vectors for use in *Agrobacterium*-mediated transformation is discussed in detail in Chapter 4. However, there are some key points, relevant to their use for plant transformation, which we can consider here (a simple vector for *Agrobacterium*-mediated plant transformation is shown in Figure 3.5). Vectors used in *Agrobacterium*-mediated transformation are derivatives of, or

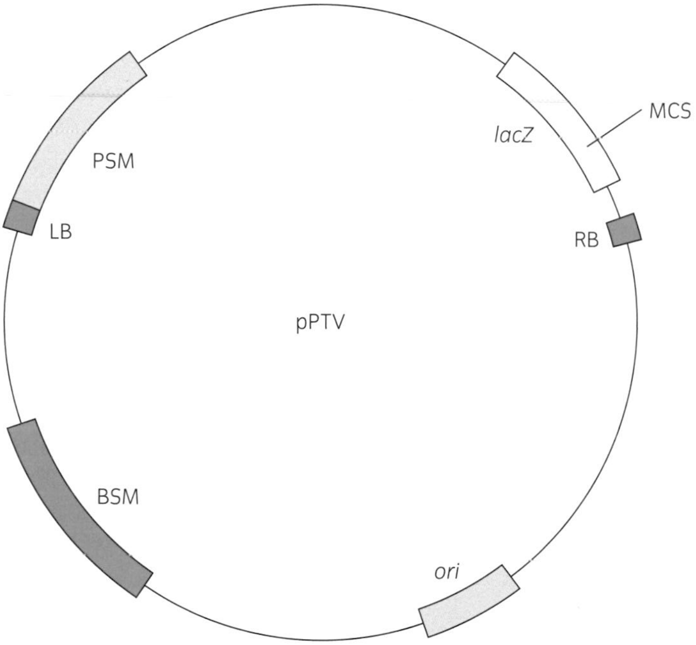

Figure 3.5 A simplified representation of a plasmid plant transformation vector (pPTV) showing the essential features of a binary-type plant transformation vector. The DNA that is transferred to the plant genome (the T-DNA) is situated between, and defined by, the left (LB) and right (RB) borders. It contains a multiple cloning site (MCS) in *lacZα* to facilitate cloning of the transgene into the vector. The T-DNA also contains a selectable marker gene to enable selection of transformed plant cells (PSM). Outside the T-DNA is a bacterial selectable marker gene (BSM) to allow selection of transformed bacteria. There is also an origin of replication (*ori*) to allow replication of the plasmid. Binary vectors like pPTV are described fully in Chapter 4.

are based on, the naturally occurring Ti plasmid. They are, however, modified extensively so that most of the features of a natural Ti plasmid are removed, only the left- and right-border sequences being used to ensure transfer of the T-DNA region between them. The vector also contains a selectable marker on the T-DNA so that transformed plants can be identified, as well as a separate selectable marker (outside the T-DNA) to enable identification of the transformed bacteria. Virulence genes required for transfer of the T-DNA are often located on a separate plasmid in the bacterium.

3 The strain of *Agrobacterium* used. Several widely used strains of *Agrobacterium* are available for plant transformation experiments. For the transformation of crops that, like tobacco, are amenable to *Agrobacterium*-mediated transformation, the choice of strain is not critical to the success of the experiment. However, for more recalcitrant plant species, the choice of strain is a major factor contributing to the success or failure of the experiment.

The basic protocol used for any *Agrobacterium*-mediated transformation experiment can be illustrated with the following example of tobacco transformation.

1 Suitable plant tissue, to be used as a source of explants, is removed from the donor plant and sterilized (if it is not from a sterile plant). For tobacco transformation, leaves are ideal.

2 The leaf tissue is cut into small pieces (using a scalpel or cork-borer) and placed into a culture of *Agrobacterium* (which contains the vector) for about 30 min, a process known as co-cultivation. During this incubation, the bacteria attach to the plant cells. The explants are subsequently removed from the bacterial culture, excess bacterial culture blotted off, and then placed on to solid Murashige and Skoog (MS) medium that contains no selective agent.

3 The incubation of the explants with *Agrobacterium* is allowed to continue for 2 days to allow transfer of the T-DNA to the plant cells.

4 The explants are removed from the medium and washed in an antibiotic solution (such as cefotaxime) that kills *Agrobacterium* cells.

5 The explants are transferred to fresh solid medium. This medium is supplemented with a selective agent (often kanamycin, but this will depend on which selectable marker gene is present in the T-DNA of the vector) to prevent the growth of non-transformed plant cells. It also contains cefotaxime (to prevent the growth of any *Agrobacterium* that were not killed by the initial treatment with cefotaxime). An auxin (1-naphthylacetic acid (NAA), 0.1 mg l^{-1}) and a cytokinin (6-benzylaminopurine (BAP), 1 mg l^{-1}) are also included to encourage regeneration by organogenesis (see Chapter 2). The relatively high cytokinin to auxin ratio promotes shoot formation from the explants. These shoots can be rooted by placing them on solid medium containing a high auxin to cytokinin ratio.

Although this is a specific example, most *Agrobacterium*-mediated transformation protocols (although see *in planta* methods described later in this chapter) follow a similar pattern, which is summarized below:

1 identify a suitable explant;

2 co-cultivate with the *Agrobacterium*;

3 kill the *Agrobacterium* with a suitable antibiotic (which does not harm the plant tissue);

4 select for transformed plant cells;

5 regenerate whole plants.

Experimental details vary depending on the plant species being transformed and the strain of *Agrobacterium* being used. There is a multitude of protocols available for the transformation of a wide range of crops, even ones such as cereals that were previously considered to be incapable of being transformed by *Agrobacterium*-mediated methods. These protocols are constantly being improved, which makes the identification of specific protocols redundant. However, some generalizations can be made.

Dicotyledonous plants, the natural target for *Agrobacterium* transformation, are, in general, easily transformed using standard vectors (see Chapter 4) and standard strains of *Agrobacterium*, such as LBA4404. Some crops, such as cereals (which are not naturally infected by *Agrobacterium*), are more difficult to transform and may require the use of modified vectors and/or so-called supervirulent strains of *Agrobacterium* (such as EHA101 or EHA105). These supervirulence systems rely on extra copies of some of the virulence genes being present either on the vector itself or on a separate plasmid in the *Agrobacterium*. Different selectable markers (Chapter 4) to identify transformed plant cells may also be required for some plant species, particularly cereals. Antibiotics such as kanamycin are commonly used to select for transformed plant cells, but alternatives such as herbicides or more potent antibiotics are often required for cereal transformation.

Transformation

Transformation protocols requiring extensive periods of tissue culture have several drawbacks, including time (typically many months), space, and, importantly, the tendency to induce somaclonal variation (i.e. the variation in plants derived from tissue culture). Hence there is considerable interest in developing transformation methods that obviate the need for tissue culture steps. Recently, such methods have been demonstrated to work in a limited number of plant species, opening up the possibility of widespread application to crop improvement. Methods for transformation *in planta* utilize either *Agrobacterium*-mediated transformation or direct gene-transfer methods (which are discussed later in this chapter). One of the most promising methods, which was developed using a member of the genus *Arabidopsis* as the target plant, is the floral dip. This is an extremely simple method in which plants with young flowers are dipped (with or without a vacuum being applied) into a culture of *Agrobacterium* that also contains a surfactant. The plant is subsequently allowed to set seed, whereupon a small proportion of the seeds produced are transgenic. Although the efficiency of this technique is very low (at present, at least), the vast number of seeds produced results in an acceptable overall transformation efficiency. This and similar techniques have been applied successfully to other plant species, including alfalfa and some brassicas.

The first part of this chapter has dealt with *Agrobacterium*-mediated transformation. However, another species of the same genus, *Agrobacterium rhizogenes*, is also capable of transferring genes to plants, and has been developed into a plant transformation system that is used in some specialized circumstances. Recently other bacterial species capable of transforming plants have also been identified and developed as plant transformation systems (Box 3.1).

● Other bacterial species are now being developed for transformation.

BOX 3.1 *Agrobacterium rhizogenes and alternative organisms to Agrobacterium spp.*

Although *A. rhizogenes* also infects plants, it differs from *A. tumefaciens* in that the resulting pathology is not crown galls but a phenomenon known as hairy roots. At the site of infection there is a proliferation of roots. Hairy roots are important in some areas of plant biotechnology as they can be cultured *in vitro*. For many years they have been used as a source of secondary metabolites, but more recently they have been used as a system for the production of pharmaceutical proteins.

Plasmids (Ri plasmids) in strains of *A. rhizogenes* have been characterized, and it has been shown that there are a number of different types, which can be classified based upon opine usage. Hormone biosynthetic genes are also present on some Ri plasmids. Homologues of the Ti-plasmid auxin-biosynthetic genes *tms*1 and *tms*2 have been identified on the agropine-strain Ri plasmids. However, despite the well-established link between auxin and rooting, the genes are unnecessary for virulence. Rather, a series of other ORFs has been identified within the T-DNA of the plasmids, including the *rolB* and *rolC* genes, which are involved in the metabolism of plant growth regulators and which lead to the plant being sensitized to endogenous auxin. It is this increased sensitivity that leads to the root formation.

As with *A. tumefaciens*, vectors have been constructed that can be used in *A. rhizogenes* as binary systems (binary vectors are described in Chapter 4). *A. rhizogenes* transformation was, at one stage, considered an alternative strategy to *A. tumefaciens* for gene transfer as it led to the production of defined tissues (hairy roots) that could be regenerated into whole plants. This strategy seems to have been discarded, however, as more efficient *A. tumefaciens* systems have been developed.

There has been a move to develop non-*Agrobacterium*-based gene-transfer systems using other microorganisms. This is in part due to the complex landscape of patents and intellectual property rights for many biotechnological processes and in particular for *Agrobacterium*-based transformation. This issue came to a head with the development of Golden rice. This rice was to be provided free to developing countries but the legal issues took a lot of hard work to overcome (see Chapter 10). Several other issues have driven the non-*Agrobacterium* initiative: the main one being the possibility of broadening the target range of microbial-based transformations. The pathogenic nature of *Agrobacterium* has always been a problem; it can cause plant defence responses to be induced, which may lead to a reduction of transformation frequency. Another problem is that many crops are either resistant to the *Agrobacterium* infection or cannot be regenerated from susceptible tissues.

The CAMBIA group (an affiliated research centre of Charles Sturt University in Australia) have tested a number of bacterial species for their ability to transfer DNA sequences into plant tissues

(Continued overleaf)

BOX 3.1 *Continued*

so that they will be integrated, expressed, and inherited in a Mendelian fashion. Initial studies have been with *Rhizobium* sp. NGR234, *Sinorhizobium meliloti*, and *Mesorhizobium loti*. These are not plant pathogens but are either symbionts or benign endo-/epiphytes that interact with plants in different ways (in terms of target cells and growth stages), broadening the potential range of target plants. The transformation methods will need optimization but preliminary experiments have indicated significant transformation frequencies with, for instance, floral dip technologies. The aim of this work is to provide workable transformation systems with open-source licences (www.bioforge.net; Biological Innovation for Open Society, www.bios.net). These licences are without any commercial restrictions other than covenants for sharing of improvements, relevant safety information, and regulatory data, and for preserving the opportunity for others to freely improve and use the technology.

There has also been a lot of work on the development of viral systems of gene transfer; these are considered in Chapter 11.

Direct gene-transfer methods

Non-*Agrobacterium* systems are widely used for recalcitrant species.

The term direct gene transfer (or direct transfer) is used to discriminate between methods of plant transformation that rely on the use of *Agrobacterium* (indirect methods) and those that do not (direct methods). Direct gene-transfer methods all rely on the delivery of large amounts of naked DNA while the plant cell is transiently permeabilized. Several different methods or strategies for direct gene transfer have been developed over the years. Some of these, particularly biolistics or gene bombardment, have become widely adopted by plant biotechnologists.

Direct gene-transfer methods, which have found particularly widespread use in the transformation of cereal crops (which initially proved difficult to transform with *Agrobacterium*), have some advantages and disadvantages (Table 3.2) when compared

Table 3.2 Direct gene-transfer methods

Direct gene-transfer method	Comments
Particle bombardment	Very successful method. Risk of gene rearrangements and high copy number. Useful for transient expression assays.
Electroporation	Transgenic plants obtained from a range of cereal crops. Low efficiency. Requires careful optimization.
DNA uptake into protoplasts	Used for all major cereal crops. Requires optimization with a regenerable cell suspension that may not be available.
Silicon carbide fibres	Requires regenerable cell suspensions. Transgenic plants obtained from a number of species.

BOX 3.2 *Transgene integration and gene rearrangement*

The major complication found with particle-bombardment transformation methods is that the vector DNA is often rearranged and transgene copy number can be very high. Recent work on the mechanism of direct gene transfer to rice by biolistics suggests that the transfer is a two-stage process. During a pre-integration phase, vector molecules (which can be either intact or partial) are spliced together. This gives rise to fragments carrying multiple gene copies, which subsequently integrate into the host plant's genome. This integration event can act as a hot spot for further transgene integration at, or very close to, the initial point of integration. The implications of these observations are that particle bombardment generally results in the DNA being integrated in high copy number at a single locus (or a low number of loci). Despite this high copy number, the single locus may have benefits for subsequent breeding programmes.

It is not just particle-bombardment that has problems. Sequences such as the 35S promoter (Chapter 4) have been identified as recombination hot spots and some problems with *Agrobacterium*-based binary vector systems (Chapter 4) have been identified. Here, between 30 and 80% of transformants from *Agrobacterium*-based transformation have been shown to contain sequences from the plasmid outside the T-DNA region. These observations have led to a rethink in vector design for both transformation approaches. In particular, the use of all-plant-based constructs should be used now and virus or bacterial sequences should, where possible, be avoided.

- Both *Agrobacterium* and direct-transfer systems can lead to abnormal integration.

- Not transgenic but intragenic: plants transformed with 'just native plant DNA'.

with *Agrobacterium*-mediated transformation. One of the major disadvantages with direct gene-transfer methods is that they tend to lead to a higher frequency of transgene rearrangement (see Box 3.2) and a higher transgene copy number. This can lead to high frequencies of gene silencing (see Box 4.4).

The main types of direct gene transfer will be considered in some detail below. Other, less-reproducible methods, which will not be considered further, such as laser-mediated uptake of DNA, microinjection, ultrasound, and *in planta* exogenous application, have mainly been used for the analysis of transient gene expression, although stable transformation has been reported for some of these techniques on rare occasions.

Particle bombardment

- Particle bombardment has been used widely for cereal transformation.

Particle bombardment (biolistics) is the most important and most effective direct gene-transfer method in regular use. In this technique, tungsten or gold particles are coated with the DNA that is to be used to transform the plant tissue. The particles are propelled at high speed into the target plant material, where the DNA is released within the cell and can integrate into the genome. The delivery of DNA using this technology has allowed transient gene expression (which does not depend on integration of the transgene into the plant genome) to be widely studied, but integration of the transgene occurs only infrequently. To generate transgenic plants, the plant material, the tissue culture regime, and the transformation conditions have to be optimized quite carefully.

Particle bombardment systems were first developed in 1987 and used an explosive charge to propel DNA-coated tungsten particles. This technology was the key to cereal transformation. All the major cereals were able to be transformed, and the first commercial genetically modified (GM) crops, such as maize containing the *Bacillus thuringiensis* (*Bt*) toxin gene (see Chapter 6), were produced by this method. Developments to the technology led to the production of a number of systems, such as an electrostatic discharge device, and others based on gas flow. Of the latter type, a commercially produced, helium-driven, particle-bombardment apparatus (PDS-1000/He) has become the most widely used.

Attempts to optimize the system have focused on three aspects of the process: particle type and preparation; particle acceleration; and choice of target material. A balance has to be reached between the number and size of particles fired into the target cells, the damage they do and the amount of DNA they deliver. Too little DNA may lead to low transformation frequencies, but too much DNA may lead to a high copy number and rearrangements of the transgene constructs (Box 3.2). Recently, new methods of preparing the particles for biolistic transformation, which involve the use of aminosiloxanes (instead of calcium chloride and polyamines) to coat the particles with DNA, have led to higher transformation efficiencies at lower DNA concentrations.

Plant tissues used for bombardment are generally of two types: primary explants that are bombarded and then induced to become embryogenic; or proliferating embryogenic cultures that are bombarded and then allowed to proliferate further and subsequently to regenerate. To protect the plant tissue from the damage sustained during the bombardment procedure, treatments to induce limited plasmolysis or culture on high-osmoticum media have been used.

CASE STUDY 3.2 **Biolistic transformation of rice**

In this case study, we will look at how the PDS-1000/He particle-bombardment system can be used to transform rice. The strategy chosen is one where two plasmids (see Figure 3.6) are introduced into the plant cell together. One plasmid (pOZ) carries the transgene of interest and the other (pHAG) carries a selectable marker (*hyg*), which confers resistance to the antibiotic hygromycin (selectable markers are described fully in Chapter 4). This plasmid also carries the *gusA* reporter gene (reporter genes are described in Chapter 4), which can be easily assayed histochemically. Both plasmids are coated on to gold particles (these particles are termed the microcarrier), for bombardment into the plant material. The plant material used for transformation is embryogenic callus derived from mature seeds (see Chapter 2).

After bombardment, cells containing the antibiotic-resistance gene are selected for on culture medium containing the antibiotic hygromycin (the selective agent). There is no direct selection for the other plasmid (pOZ), but it has been shown that in many cells both plasmids will integrate (dual transformation). In this system, then, the aim is to select and regenerate plants

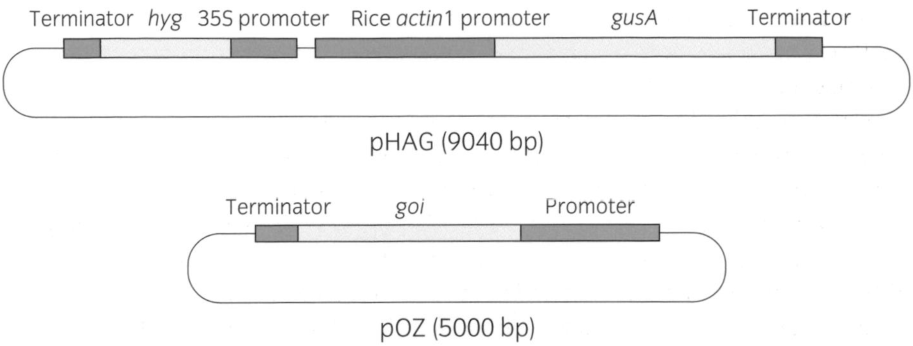

Figure 3.6 Plasmid maps of pHAG and pOZ. The plasmid pHAG contains a selectable marker gene (*hyg*), which is driven by a constitutive promoter (the 35S promoter from cauliflower mosaic virus, which is described in more detail in Chapter 4). The *hyg* gene confers resistance to the antibiotic hygromycin. pHAG also contains a reporter gene (*gusA*), which is driven by the rice *actin*1 promoter, a strong promoter in cereals. The *gusA* gene product can be easily assayed and forms a convenient way of confirming that the hygromycin-resistant plant cells are transformed and not 'escapes'. It is the presence of this plasmid that is selected for after particle bombardment by growing the putatively transformed plant tissue on medium that contains hygromycin. The plasmid pOZ is used to deliver the gene of interest, *goi*. There is no selectable marker on pOZ, and therefore the presence of this plasmid cannot be directly selected for. However, dual transformation with both plasmids is a relatively common event. The *ipt* gene is driven by a suitable promoter, the characteristics of which will depend on the required pattern of expression. The promoter may be inducible, tissue-specific, or constitutive.

in which both plasmids have stably integrated into the genome. This approach ultimately allows the selectable marker to be segregated out in subsequent generations. The resulting plants contain the gene of interest, but do not contain the hygromycin-resistance gene, which is no longer necessary. The desirability of removing antibiotic-resistance marker genes, in particular from crops, is discussed in Chapters 4 and 12.

Preparation of microcarriers

Gold or tungsten particles (beads or balls) are used as the microcarrier. These need to be pre-treated by washing the microcarriers in ethanol and sterile distilled water. Suspensions of the microcarriers can then be stored as aliquots, either at 4°C (gold) or at –20°C (tungsten). Plasmid DNA (isolated from *Escherichia coli*) is attached to the gold microcarriers by mixing the DNA and the microcarriers in 2.5 mol l^{-1} $CaCl_2$ and 0.1 mol l^{-1} spermidine (a polyamine). The microcarriers are then mixed, washed with ethanol, and finally resuspended in ethanol. The microcarriers are then applied to the macrocarrier membrane as an ethanol suspension and are allowed to dry on to the macrocarrier.

Preparation of plant material for transformation: callus induction

Dehulled, mature rice seeds are sterilized in ethanol and bleach, and subsequently washed in sterile distilled water to remove any traces of bleach. The dehulled seeds are then placed on MS1 medium (MS medium supplemented with 2.5 mg l^{-1} 2,4-D/30 g l^{-1} maltose and solidified with 2.5 g l^{-1} Phytagel; medium composition is discussed in detail in Chapter 2). The cultures are

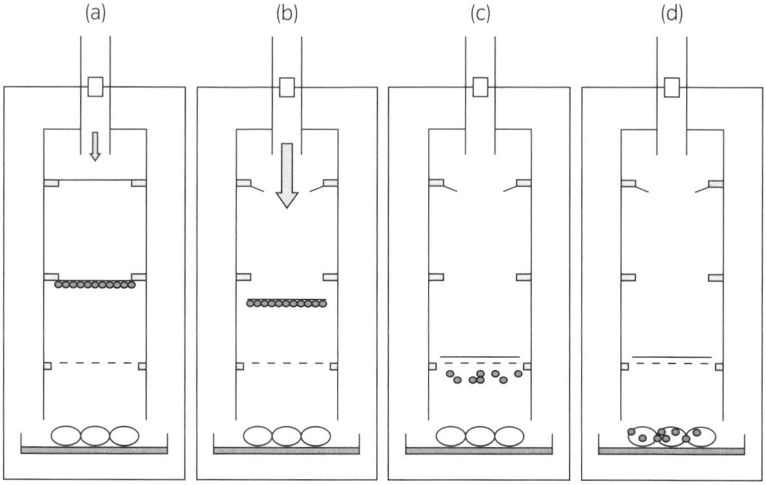

Figure 3.7 The PDS-1000/He particle-bombardment system. The plant tissue is placed into a vacuum chamber (chamber pressure 27 mmHg) 13 cm below the microcarrier stopping plate. The vector DNA-coated particles (the microcarriers) on the macrocarrier membrane are inserted into the apparatus (a). Once the vacuum in the lower part of the apparatus is established, the helium pressure above the rupture disc is increased until at 1100 psi (7584.2 kPa; or whatever pressure the rupture disc is designed to rupture at) the rupture disc bursts (b). This propels fragments of the macrocarrier and the projectiles down the chamber. The macrocarrier is stopped at the stopping plate (c), allowing the microcarriers to pass through and hit the plant material (d). Various parameters can be optimized in this system: the distance between the stopping plate and the plant material can be varied; and the pressure at which the rupture disc bursts can also be selected. Varying these parameters allows the speed and pattern of the microcarriers to be adjusted to suit the needs of the plant material being transformed.

then incubated in the dark at 25°C for 2 weeks. Callus tissue that develops from the seeds is transferred to fresh medium of the same composition and incubated for a further 2 weeks for embryogenic callus to develop. The embryogenic callus is transferred (50–70 calluses are arranged on one 9-cm diameter Petri dish) to a different medium (MS medium supplemented with 1 mg l^{-1} 2,4-D, 14% (w/v) maltose, and 2.5 g l^{-1} Phytagel) and incubated for 4 h. This medium contains a high concentration of maltose to generate a high osmoticum prior to particle bombardment of the plant tissue.

Bombardment conditions

The PDS-1000/He equipment is assembled as shown in Figure 3.7 and used to bombard the rice embryogenic callus. The operation of the PDS-1000/He equipment is also described in the figure.

Plant selection and regeneration

One day after bombardment, the embryogenic callus is transferred to MS1 medium. After 1 week, the embryogenic callus is transferred to selection medium (MS1 medium supplemented with 40 mg l^{-1} hygromycin) and incubated for 2 weeks in the dark. The callus is then transferred to fresh selection medium (of the same composition) and incubated for a further 2 weeks in the dark. Surviving callus is transferred to plant regeneration medium RM1 (MS medium supplemented with 3 mg l^{-1} kinetin, 30 g l^{-1} maltose, 40 mg l^{-1} hygromycin, and 6.25 g l^{-1} Phytagel) and incubated for 3 weeks in the light to encourage shoot formation. Callus is then transferred to RM2 medium (as RM1 but with 2.5 g l^{-1} Phytagel) for a further 3 weeks' incubation in the light. Green shoots that have formed on RM2 medium are rooted by transferring them on to R1 medium (half-strength MS medium supplemented with 1 mg l^{-1} NAA). Young plantlets are then subcultured on to R2 medium (half-strength MS basal medium). The plants are acclimatized in growth chambers and can then be transferred to soil and grown to maturity.

Bombardment has been used mainly for transformation of the nuclear genome, but more recently it has become the main strategy for introducing DNA into plastids (Box 3.3). Chloroplast transformation is a development for the 21st century, and its application to GM crops is discussed in detail in Chapters 11 and 12. Suffice it to say that chloroplast transformation offers opportunities for creating GM crops that are both environmentally friendly and more efficient.

BOX 3.3 *Plastid transformation*

There are a number of important reasons why chloroplast transformation has become an important technique in plant biotechnology. In many, but not all, crop species, chloroplasts display only maternal inheritance so there is no danger of any gene transfer to related weedy species through pollen. Another significant feature of this technology is the very large transgene copy number that may be obtained. Transgene insertion may occur into a single genome contained within an individual chloroplast. However, by using several rounds of selection during the plant regeneration process, it is possible to develop a homoplasmic population of chloroplasts (where all the chloroplasts are transformed). As there may be as many as 100 chloroplasts per cell, each containing up to 100 copies of the genome, there may be as many as 10 000 copies of the transgene per cell. An additional advantage is the absence of reports of gene silencing with chloroplast transformation. This fact, combined with the ability of the chloroplast to correctly fold and crosslink expressed proteins, means there is a tremendous potential in chloroplast transformation for very high level gene expression and synthesis of active proteins.

● The high number of chloroplasts and the multiple copies of the chromosomes means that the cells contain many copies of the transgene, giving high levels of expression without gene silencing.

This has led to the development of vectors designed specifically for chloroplast transformation. The figure shows (in panel a) a basic vector for chloroplast transformation in which selection is based on the antibiotic spectinomycin (the selectable marker gene is referred to as *aadA*). The vector is designed for the expression of a single transgene, which is fused to plastid regulatory sequences, such as the promoter from the *psbA* gene and ribosome-binding regions. The foreign genes are bordered by chloroplast DNA sequences. After introduction into the chloroplast by bombardment, homologous recombination occurs between the plastid sequences on the vector and those on the genome. This site-specific integration event avoids any problems associated with random insertion into the nuclear genome.

(Continued overleaf)

BOX 3.3 *Continued*

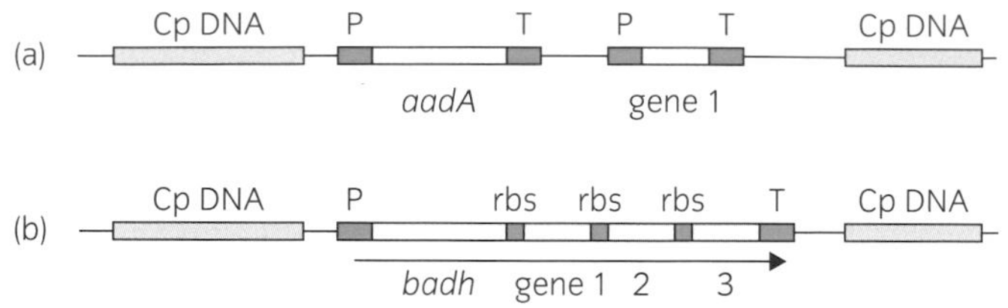

Vectors for chloroplast transformation. (a) A construct designed for the expression of one gene of interest (gene 1). There is also a selectable marker gene (*aadA*), which confers resistance to the antibiotic spectinomycin. Each gene (gene 1 and *aadA*) is driven by a promoter (P) and has a terminator (T). Flanking the two genes are regions of chloroplast DNA (Cp DNA) that direct integration of the DNA at a defined point in the chloroplast genome by homologous recombination. (b) A construct designed for the expression of multiple transgenes. One promoter (P) drives the expression of a polycistronic transcription unit, or operon. In this case the betaine-aldehyde dehydrogenase gene (*badh*) is used as the selectable marker. Each transgene is preceded by a ribosome-binding site (rbs) to ensure efficient translation. Other features are as in (a). (Redrawn with permission from Daniell *et al.* (2002).)

More-advanced vectors have been designed that take advantage of the prokaryotic nature of some of the plastid genome (see Chapter 1). Multiple ORFs can be introduced as a single transcription unit that is expressed as a polycistronic message (panel b in the figure). This offers a significant advantage over nuclear transformation, where individual genes are introduced separately and may have to be brought together by breeding strategies after plants with similar expression levels have been identified. A similar approach has been the development of a two-vector system, each vector carrying separate subunits that can combine in the organelle to form a functional protein. Another recombination-based approach has been to design promoterless vectors. These vectors are targeted to integrate adjacent to existing promoters.

Further development is also being carried out on the selection systems used. Strategies have been developed to remove the selectable markers, by techniques such as reversal of the homologous recombination events; some vectors incorporate the use of the betaine-aldehyde selection system, which generates much higher transformation efficiencies (see Chapters 4 and 9).

Polyethylene glycol-mediated transformation

Plant protoplasts (plant cells that have had their cell walls removed; see Chapter 2) can be transformed with naked DNA by treatment with polyethylene glycol (PEG) in the presence of divalent cations (usually calcium). The PEG and the divalent cations destabilize the plasma membrane of the plant protoplast and render it permeable to naked DNA. Once inside the protoplast the DNA enters the nucleus and integrates into the genome.

Plant protoplasts are not easy to work with, and the regeneration of fertile plants from protoplasts is problematical for some species, limiting the usefulness of the technique. The DNA used for transformation is also susceptible to degradation and rearrangement. Despite these limitations, the technique does have the advantages that protoplasts can be isolated and transformed in large numbers from a wide range of plant species.

Electroporation

- Electroporation has been used successfully to transform cereals.

The electroporation of cells can be used to deliver DNA into plant cells and protoplasts. The vectors used can be simple plasmids; the genes of interest require plant regulatory sequences, but no specific sequences are required for integration. Material is incubated in a buffer solution containing DNA and subjected to high-voltage electrical pulses. The DNA then migrates through high-voltage-induced pores in the plasma membrane and integrates into the genome. Electroporation has been successfully used to transform all the major cereals, particularly rice, wheat, and maize. Initially, protoplasts were used for transformation, but one of the advantages of the system is that both intact cells and tissues (such as callus cultures, immature embryos, and inflorescence material) can be used. This reduces some, but not all, of the tissue culture problems. However, the plant material used for electroporation may require specific treatments, such as pre- and post-electroporation incubations in buffers of high osmotic pressure. The efficiency of electroporation is also questionable: it is very dependent on the condition of the plant material used and the electroporation and tissue treatment conditions chosen.

Some studies have indicated that linear, rather than supercoiled, plasmid DNA and the addition of spermidine to the incubation buffer (which can induce condensation of the DNA) improves the efficiency of electroporation. Optimization of what is known as the field strength (the voltage applied across the electrodes) is also important.

Even under optimal conditions the amount of DNA delivered to each cell is low, which has the benefit of producing transformants with a low transgene copy number. The delivery rate (the proportion of electroporated cells that actually receive DNA) can be high: in some experiments it has been shown that between 40 and 60% of the incubated cells receive DNA, and that as many as 50% of the cells survive the treatment. Electroporation also has the advantage that all the cells are in the same physiological state after transformation, unlike the situation with particle bombardment where transformed cells may be at a disadvantage due to damage from the transformation procedure.

Silicon carbide fibres: WHISKERS™

A simple technique using silicon carbide fibres requires no specialized equipment. Plant material (such as cells in suspension culture, embryos, and embryo-derived calluses) is introduced into a buffer containing DNA and silicon carbide fibres, which is then vortexed. The fibres, which are about 0.3–0.6 μm in diameter and 10–100 μm long, penetrate the cell wall and plasma membrane, allowing the DNA to gain access to the inside of the cell. The drawbacks of this technique relate to the availability of suitable plant material and the inherent dangers of the fibres, which require careful handling.

Although the procedure has been utilized with friable callus from maize, this type of friable callus is limited only to a few genotypes of maize and oats. Many cereals, the natural targets for the procedure, produce embryogenic callus that is hard and compact and not easily transformed, at present, with this technique. Recently though, some progress has been made in transforming such material, and procedures are being developed to allow transformation of cereals such as rice, wheat, barley, and maize without the need to initiate cell suspensions.

SUMMARY

A variety of techniques for plant transformation are available to the plant biotechnologist. These techniques can be split into two groups: *Agrobacterium*-mediated transformation and direct gene-transfer methods, of which the biolistics approach is probably the most widely used. These two groups of techniques are fundamentally different in mechanism, and are, in general, applied to different crops. *Agrobacterium*-mediated transformation is most widely used with dicotyledonous crops, which reflects the natural host range of members of the genus *Agrobacterium*. Direct gene-transfer methods are most commonly used to transform monocotyledonous crops, such as cereals. In part, this reflects the initial difficulties with using *Agrobacterium* to transform monocotyledonous plants. Direct gene-transfer methods and *Agrobacterium*-mediated methods have their own advantages and disadvantages. However, all plant transformation methods can suffer from a problem known as gene silencing (see Box 4.4), where transgene (and homologous endogenous gene) expression is actually repressed. Gene silencing is impossible to predict precisely, although some precautions in vector design and transformation protocol can be taken to reduce its frequency. Gene silencing can prove a major hurdle to the commercialization of plant transformation products.

Despite any problems, improvements in plant transformation technologies, especially when coupled with an efficient plant regeneration protocol, have seen the list of crop species that can be routinely transformed grow. Crops that were once considered impossible to transform (cereals may well fall into this category) are now transformed routinely in many laboratories around the world.

These plant transformation technologies provide the basis for the advances in plant biotechnology discussed in the remainder of this book.

FURTHER READING

Visit www.oxfordtextbooks.co.uk/orc/slaterplants2e/ for hyperlinked references that take you directly to online abstracts for the journal articles listed below:

General

Barcelo, P., Rasco-Gaunt, S., Thorpe, C., and Lazzeri, P. (2001) Transformation and gene expression. In *Advances in Botanical Research, vol. 34: Biotechnology of Cereals* (P. R. Shewry, P. A. Lazzeri, and K. J. Edwards, eds), pp. 59–126. Academic Press, London.

Birch, R. G. (1997) Plant transformation: problems and strategies for practical application. *Annual Reviews in Plant Physiology and Plant Molecular Biology* **48**, 297–326.

Hansen, G. and Wright, M. S. (1999) Recent advances in the transformation of plants. *Trends in Plant Science* **4**, 226–231.

Hughes, M. A. (1996) *Plant Molecular Genetics*. Addison Wesley Longman, Harlow.

Walden, R. and Wingender, R. (1995) Gene transfer and plant-regeneration techniques. *Trends in Biotechnology* **13**, 324–331.

Agrobacterium-mediated transformation

Bent, A. F. (2000) *Arabidopsis in planta* transformation. Uses, mechanisms, and prospects for transformation of other species. *Plant Physiology* **124**, 1540–1547.

Broothaerts, W., Mitchell, H. J., Weir, B., Kaines, S., Smith, L. M. A., Yang, W., Mayer, J. E., Roa-Rodríguez, C., and Jefferson, R. A. (2005) Gene transfer to plants by diverse species of bacteria. *Nature* **433**, 629–633.

Chug, S.-M., Vaidya, M., and Tzfira, T. (2006) *Agrobacterium* is not alone: gene transfer to plants by viruses and other bacteria. *Trends in Plant Science* **11**, 1–4.

De la Riva, G. A., Gonzalez-Cabrera, J., Vazquez-Padron, R., and Ayra-Paedo, C. (1998) *A. tumefaciens*: a natural tool for plant transformation. *Electronic Journal of Biotechnology* **1**, 1–16.

Ditt, R. F., Nester, E. W., and Comai, L. (2001) Plant gene expression response to *A. tumefaciens*. *Proceedings of the National Academy of Sciences USA* **98**, 10954–10959.

Dumas, F., Duckely, M., Pelczar, P., Van Gelder, P., and Hohn, B. (2001) An *Agrobacterium* VirE2 channel for transferred-DNA transport into plant cells. *Proceedings of the National Academy of Sciences USA* **98**, 485–490.

Gelvin, S. B. (2000) *Agrobacterium* and plant genes involved in T-DNA transfer and integration. *Annual Reviews of Plant Physiology and Plant Molecular Biology* **51**, 223–256.

Ingram, J. (1998) Plant cyclophilins and *Agrobacterium*. *Trends in Plant Science* **3**, 292.

Rommens, C. M. (2004) All-native DNA transformation: a new approach to plant genetic engineering. *Trends in Plant Sciences* **9**, 457–464.

Schrammeijer, B., Risseeuw, E., Pansegrau, W., Regensburg-Tuink, T. J. G., Crosby, W. L., and Hooykaas, P. J. J. (2001) Interaction of the virulence protein VirF of *A. tumefaciens* with plant homologs of the yeast Skp1 protein. *Current Biology* **11**, 258–262.

Tzfira, T. and Citovsky, V. (2006) *Agrobacterium*-mediated genetic transformation of plants: biology and biotechnology. *Current Opinion in Biotechnology* **17**, 147–154.

Ward, D. V., Zupan, J. R., and Zambryski, P. C. (2002) *Agrobacterium* VirE2 gets the VIP1 treatment in plant nuclear import. *Trends in Plant Science* **7**, 1–3.

Zupan, J., Muth, T. R., Draper, O., and Zambryski, P. (2000) The transfer of DNA from *A. tumefaciens* into plants: a feast of fundamental insights. *Plant Journal* **23**, 11–28.

Direct gene transfer

Kohli, A., Leech, M., Vain, P., Laurie, D. A., and Christou, P. (1998) Transgene organisation in rice engineered through direct DNA transfer supports a two-phase integration mechanism mediated by the establishment of integration hot spots. *Proceedings of the National Academy of Sciences USA* **95**, 7203–7208.

Sorokin, A. P., Ke, X.-Y., Chen, D. F., and Elliott, M. C. (2000) Production of fertile transgenic wheat plants via tissue electroporation. *Plant Science* **156**, 227–233.

Plastid transformation

Daniell, H., Khan, M. S., and Allison, L. (2002) Milestones in chloroplast genetic engineering: an environmentally friendly era in biotechnology. *Trends in Plant Science* **7**, 84–91.

Maliga, P. (2004) Plastid transformation in higher plants. *Annual Reviews of Plant Biology* **55**, 289–313.

Web-links

Plant transformation, web-link 3.1: www.ppge.ucdavis.edu/Transformation/transform1.htm

Plant transformation, web-link 3.2: www.uoguelph.ca/~jdberg/plantran.htm

The microbial world. Biology and control of crown gall (*A. tumefaciens*), web-link 3.3: http://helios.bto.ed.ac.uk/bto/microbes/crown.htm

Transgenic crops: an introduction and resource guide, web-link 3.4: www.colostate.edu/programs/lifesciences/TransgenicCrops/how.html

4

Vectors for plant transformation

INTRODUCTION

Vectors for plant transformation have to be created in the same way as any other plasmid-derived cloning vector, by using molecular biology techniques. It is therefore useful to look at some of the basic features that are necessary or desirable in any vector. Such features are described here because the usefulness of vectors often equates with the ease with which they can be manipulated, both in standard laboratory hosts such as *Escherichia coli* and *in vitro*, before being used to transform plants. Later in this chapter we will see that many of the recent advances in the design of plant transformation vectors are aimed at improving the ease with which they can be manipulated in the laboratory, rather than their ease of use for plant transformation per se. In addition, this chapter will describe the basic features of vectors used for plant transformation, their development into systems allowing the routine transformation of a range of plant species, and some of the problems associated with vector design and transgene expression.

Desirable features of any plasmid vector

Ideally, as well as the ability to replicate independently from the chromosome (which depends on the presence of an origin of replication), cloning vectors should have the following properties.

- Plasmids need certain features to function efficiently as vectors.

1 Be of a small size (low molecular mass). Small size has several important advantages. Small plasmids are easier to manipulate *in vitro* as they are less liable to damage by shearing. Small plasmids are also usually present in higher copy number and therefore plasmid yields are higher. There is also less chance of the vector having other sites for restriction endonucleases, making the design and integration of a multiple cloning site (MCS) simpler.

2 Confer a selectable phenotype on the host cells so that transformed cells can be selected for. Most plasmid cloning vectors therefore carry genes that confer resistance to an antibiotic, very often ampicillin (a penicillin derivative).

3 Contain single sites for a large number of restriction enzymes to enable the efficient production of recombinant vectors.

4 Enable the identification of bacterial colonies containing recombinant plasmids. This is usually achieved by clustering the unique restriction sites in one small area, the MCS, which is situated in the *lacZ*α sequence that encodes the N-terminal part of *E. coli* β-galactosidase. In suitable *E. coli* hosts, complementation results in the production of active β-galactosidase, which is detected by the bacterial colony turning blue due to hydrolysis of a chromogenic substrate. Cloning DNA into the MCS effectively results in the insertional mutagenesis of *lacZ*α. After transformation with these recombinant plasmids, *E. coli* will produce white colonies as no complementation can occur. Although not 100% reliable (both false-positives and false-negatives can be generated), such a system drastically reduces the number of colonies that need to be screened in order to identify those harbouring recombinant plasmids.

These features are illustrated in Figure 4.1, which shows a typical cloning vector widely used in molecular biology.

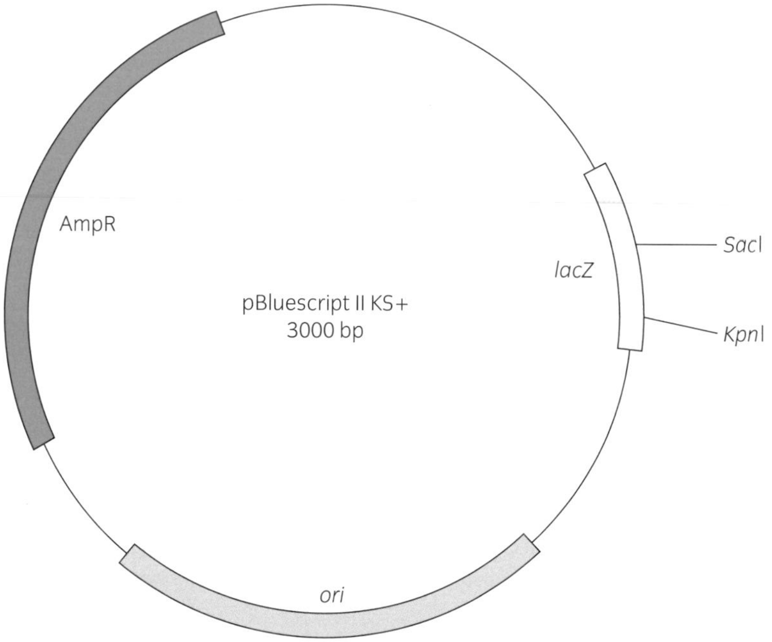

Figure 4.1 Diagram of a typical 'standard' cloning vector commonly used in molecular biology, pBluescript II KS+ in this example, although many fundamentally similar products exist. The following features are worth noting.

1 It is of a small size (3 kb) to facilitate ease of handling and high transformation efficiencies.
2 It has an origin of replication (a pUC *ori*) to enable replication of the plasmid to high copy number in *E. coli*.
3 It has an antibiotic-resistance marker to facilitate the recovery of transformed bacteria.
4 It has a multiple cloning site (MCS) in the *lacZ*α fragment to ease cloning and facilitate the identification of recombinant plasmid-containing colonies by blue/white screening. The MCS is between the *Sac*I and *Kpn*I sites (in this particular plasmid there are restriction sites for over 20 other restriction enzymes in the MCS). (Redrawn with permission from the pBluescript II KS+ figure on the Stratagene website, www.stratagene.com.)

Development of plant transformation vectors

As the process of *Agrobacterium*-mediated plant transformation was elucidated (see Chapter 3), important information came to light that made the development of efficient plant transformation vectors possible. Chief among these were the following discoveries: that the only features of the T-DNA necessary for integration into the host plant genome were the short border sequences; that the removal of the oncogene sequences enabled plants to be regenerated from transformed plant tissue by manipulating the plant hormone composition of the medium; and that the *vir* genes function in *trans*.

Basic features of vectors for plant transformation

Some of the basic factors that contribute to the design of a successful vector have already been considered. However, when considering the design of plant transformation vectors, several additional degrees of complication become apparent, as follows.

● Like all cloning vectors, plant transformation vectors need a minimum set of features to function efficiently.

1 The plasmid (if it is to be used in *Agrobacterium*-mediated transformation) needs to be able to replicate not only in *E. coli* (so that routine manipulations can be carried out) but also in *Agrobacterium*.
2 Additional selectable markers need to be included so that the successfully transformed plants can be identified.
3 Border sequences need to be incorporated into the design of plasmid vectors for *Agrobacterium*-mediated transformation to ensure integration of the genes of interest into the host plant's genome.
4 The genes (particularly if they are from prokaryotes or non-plant eukaryotes) that are to be integrated into the genome of the host plant may need to be made 'plant-like'. This includes the use of appropriate promoters and terminators to ensure that expression of the genes occurs. These features are often incorporated into the basic vector.

Promoters and terminators

An obvious requirement for any genes that are to be expressed as transgenes in plants is that they are expressed correctly (or at least in a predictable fashion). It is known that the major determinant of gene expression (level, location, and timing) is the region upstream of the coding region. This region, termed the promoter, is therefore of vital importance (see Chapter 1). Any genes that are to be expressed in the transformed plant have to possess a promoter that will function in plants. This is an important consideration as many of the genes that are to be expressed in plants, particularly reporter genes and selectable marker genes, are bacterial in origin. They therefore have to be supplied with a promoter that will drive their expression in plants.

● A variety of promoters can be used to direct transgene expression.

Transgenes also need to have suitable terminator sequences at their 3′ terminus to ensure that transcription ceases at the correct position. Failure to stop transcription can lead to the production of aberrant transcripts and can result in a range of deleterious effects, including inactivation of gene products and increased gene silencing.

In addition to the basic need for the promoter to be capable of driving expression of the gene in plants, there are other considerations that need to be taken into account, such as promoter strength, tissue specificity, and developmental regulation.

Agrobacterium-derived promoter and terminator sequences

The genes from the Ti plasmid of *Agrobacterium* that code for opine synthesis, and in particular the nopaline synthase (*nos*) gene, are widely used as a source of both promoters and terminators in plant transformation vectors. Although derived from bacterial genes, their presence on the T-DNA means they are adapted to function in plants. The *nos* promoter is usually considered to be constitutive.

The 35S promoter

The most widely used promoter used to drive expression of genes in plant transformation vectors is the promoter of the cauliflower mosaic virus 35S RNA gene (35S promoter). This promoter is considered to be expressed in all tissues of transgenic plants (although not necessarily in all cell types). In dicotyledons it drives expression at high levels, although in monocotyledons the level of expression is not as high. This makes the 35S promoter ideal for driving the expression of selectable marker genes, and in some cases of reporter genes, as expression is more or less guaranteed. The activity of the 35S promoter can be further increased by the inclusion of one or more copies of the enhancer region (Figure 4.2 shows the structure of the 35S promoter). High-level constitutive expression is not always a desirable feature though, particularly if the gene product is toxic in large amounts.

In monocotyledons, alternatives, such as the maize ubiquitin I promoter or the rice actin promoter/first intron sequence, are often used to drive the high-level expression of transgenes.

Tissue-specific promoters

Considerable effort has been put into isolating promoters that can be used to drive expression in a tissue-specific manner, although the expression is, in reality, tissue-enhanced, as expression is not completely restricted to a given tissue. Such promoter systems are potentially useful in a number of situations. The expression of any potentially harmful substances can be limited to tissues that are not consumed by animals or humans, and genes involved in specific processes can be limited to tissues in which that process occurs. It is beyond the scope of this chapter to look at all the different tissue-specific promoters that have been isolated. However, it should be noted that several have been used to successfully drive transgene expression in the predicted pattern (some examples will be seen in later chapters). It should also be noted, however, that there are examples where promoters have been found not to function, or not to drive expression in the predicted pattern, in heterologous systems. Considerable care has therefore to be taken with the use of promoters. Unless the promoter is well

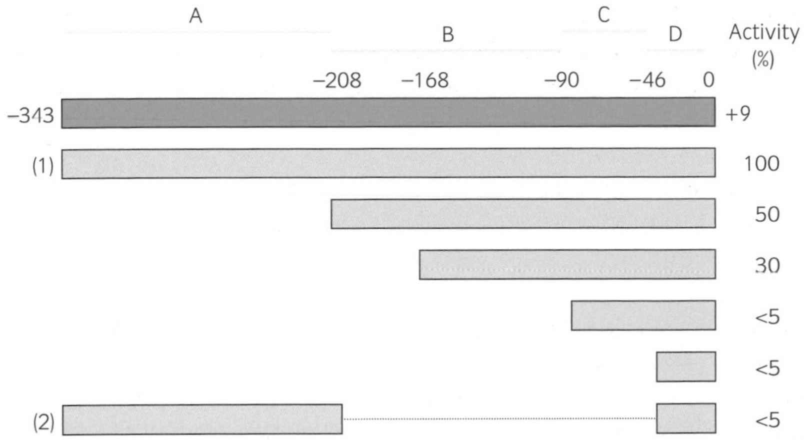

Figure 4.2 Structure/activity relationships in the 35S promoter as elucidated by deletion analysis. The uppermost box represents the 35S promoter region as defined in Fang *et al.* (1989), including 343 bp of upstream 'promoter'. The letters above designate four functional regions that can be identified in the 35S promoter. The numbers above the uppermost box give the position of the end of the promoter fragments (5′ deletions) used in the deletion analysis. (1) The boxes below represent the promoter fragments used to assess the chloramphenicol acetyltransferase (CAT) activity of various promoter fragments. (2) The broken line represents a piece of the promoter removed from a fragment, the two flanking pieces (−343 to −208 and −46 to 0) being ligated together to form a single promoter fragment. Functional regions: (A) (−343 to −208) This has no activity in itself (activity depends on the presence of the −90 to −46 region). It is responsible for about 50% of the promoter activity. (B) (−208 to −90) This functions as an enhancer and is responsible for most of the rest of the activity of the promoter. (C) (−90 to −46) This region has an accessory role in increasing the transcriptional activity of regions A and B. (D) (−46 to 0) The core promoter region, containing the TATA box. The −208 to −46 region of the promoter is often used in transformation vectors as an enhancer to increase the expression of the 35S promoter. Increasing the number of copies of the enhancer increases the transcriptional activity linearly, until a maximum stimulation of activity is reached with four copies. The enhancer also functions with heterologous promoters to increase transcriptional activity. The −343 to −90 region is responsible primarily for promoter activity in leaves, while the −90 to −46 region is required for expression in roots. (Data taken with permission from Fang *et al.* (1989).)

characterized and widely used in the plant species you intend to transform, characterization of the expression pattern conferred by the promoter (promoter::reporter gene fusions could be used) should be carried out. This is particularly so as it becomes apparent that many *cis*-acting regulatory sequences are, in fact, found in regions other than the 'promoter'.

Inducible promoters

Considerable effort has also been put into isolating and developing inducible promoter systems for use in plants. Inducible expression systems can be divided into three main categories: (1) non-plant-derived systems; (2) plant-derived systems that respond to environmental signals; and (3) plant-derived systems based on developmental control of gene expression (examples are given in Box 4.1). Theoretically, such systems allow the timing of transgene expression to be controlled carefully.

● Promoters that confer inducible expression can be used to direct expression of transgenes.

BOX 4.1 *Systems for inducible gene expression in plants*

Inducible expression gives finer control over transgene expression than does a constitutive promoter such as the 35S promoter. Several systems have been developed with the aim of giving inducible transgene expression in plants. Some examples will be considered here.

Non-plant-derived systems

Some features are considered to be desirable in any inducible expression system that depends on the application of an exogenous chemical. These are:

1. there should be no expression of the transgene in the absence of the inducer;
2. the system should respond to only one inducer, or one class of inducer, so that the system is specific to the inducer;
3. induction of gene expression should be rapid following the application of the inducer;
4. gene expression should cease rapidly following withdrawal of the inducer;
5. the inducer should be non-toxic;
6. the inducer should not cause non-specific changes in gene expression.

Some examples of inducible expression systems are given below.

Tetracycline

The antibiotic tetracycline can be used to either de-repress or inactivate gene expression. In bacteria, the tetracycline repressor (TetR) binds to the *tet* operator and negatively regulates its expression. Upon binding to tetracycline, the TetR is modified and released from the operator. This system has been adapted to function in plants. The TetR is constitutively overexpressed from a 35S promoter. The transgene is under the control of a chimaeric promoter consisting of a core 35S promoter and several copies of the *tet* operator (both upstream and downstream of the TATA box). The TetR normally represses expression of the transgene, but with the application of tetracycline it is released from the operators in the promoter of the transgene. This then results in transgene expression. This system was one of the first to be developed, but it does have some drawbacks. It does not appear to work in all plant species (such as *Arabidopsis*) as the plant cells must be able to tolerate the required levels of TetR, and tetracycline has to be applied constantly to ensure transgene expression because it has a short half-life.

Tetracycline can also be used to inactivate transgene expression. A similar promoter is used to control the expression of the transgene, but the TetR is modified to convert it to an activator (tetracycline transactivator or tTA). The tTA binds to the operators and induces gene expression in the absence of tetracycline. When tetracycline is added to the system, the tTA is released and transgene expression ceases.

Systems based on promoter repression are proposed to be less efficient at controlling gene expression than systems based on promoter activation.

Alcohol (ethanol)-inducible

In this system a chimaeric promoter is again used to drive transgene expression. In addition to a core 35S promoter region, the promoter contains binding sites for the AlcR transcription factor

BOX 4.1 *Continued*

(from *Aspergillus nidulans*), which is expressed constitutively. Upon application of ethanol to the system, AlcR binds to the AlcR-binding sites and activates transcription. The ethanol (it is actually ethanal that causes binding of the AlcR to its cognate binding sites) can be applied as a drench or vapour and there is no endogenous activation of the transgene in plants in the absence of AlcR. Activation of the promoter in tissue culture due to the induction of AlcR by endogenous metabolites (proposed to be produced due to anoxia and tissue culture conditions) precludes widespread use of this system in tissue and cell culture. There are also some doubts as to its general applicability, with some researchers showing that ethanol, at levels that would be used for inducing gene expression, can have profound effects in non-transformed plants. The system has therefore to be used with great care after carefully controlled experimentation to determine specific and non-specific ethanol effects.

The original alcohol-inducible expression system utilized a modified 35S promoter, thereby conferring expression wherever the inducing ethanol reached. Although not a problem in some applications, this lack of specificity in spatial expression pattern limited the usefulness of the system. It was known from experiments using physical confinement of the ethanol (using plastic bags for instance) that transgene expression could be limited only to areas exposed to ethanol. This was not a particularly elegant solution to the problem of achieving some spatial specificity to transgene expression and recently modifications to the system have produced ethanol-inducible promoters that do confer some spatial specificity in expression. The simplest approach has been to replace the constitutive 35S promoter elements with tissue-specific promoters (which were discussed previously in this chapter).

Steroid-inducible

A variety of systems conferring steroid inducibility on transgene expression exist. Most consist of a modified transcriptional activator capable of binding a steroid hormone or its analogue. Binding sites for the modified transcription factor are included in a chimaeric promoter, which controls transgene expression. In the absence of an inducer the transcription factor is inactive, but on application of the inducer it binds to the promoter and activates expression. Systems that respond to different steroidal inducers have been developed. Among these are ones that respond to glucocorticoids (the synthetic analogue dexamethasone is used), oestrogen, and ecdysone (an insect hormone). Although the dexamethasone system is the one most widely adopted for inducible expression in *Arabidopsis*, it appears to have some serious drawbacks that may limit its usefulness. The development of the ecdysone-based system is, however, particularly interesting as it also responds to the non-steroidal ecdysone agonists such as RH5992 (tebufenozide) and methoxyfenozide. Tebufenozide and methoxyfenozide are widely used as agrochemicals, which opens up the possibility of using this system in greenhouse and field situations.

Copper-inducible

This system is based on the yeast metallothionein regulatory system. A transcription factor is expressed constitutively. Upon binding copper, the transcription factor binds to elements in a chimaeric promoter and activates expression.

(Continued overleaf)

BOX 4.1 *Continued*

Plant-derived systems based on response to environmental signals

Although not inducible in the same sense as the non-plant-derived systems, some systems have been developed that respond to a variety of environmental signals. These systems depend on either the construction of chimaeric promoters or on the use of complete promoters from wound-inducible genes to control the expression of transgenes. The chimaeric promoters usually consist of a core promoter element (usually from the 35S promoter) and sequence elements that bind transactivating factors that respond to particular environmental signals. Several systems are described briefly here.

Wound-inducible

Many wound-inducible genes have been identified, and the promoters of some of these, particularly some proteinase inhibitors, characterized in some detail. The promoters of these genes have been found to confer wound inducibility on transgene expression. The wound inducibility may be mediated by other factors such as methyl jasmonate (see also Chapter 1), which can be used to mimic the wound response in some cases. Wound-inducible promoters can be used to drive the expression of pest-resistance genes after insect damage.

Pathogen-induced

The promoters from a range of genes induced by pathogen attack have been cloned and characterized. Many pathogen-induced promoters also show wound inducibility, and may not therefore confer the optimum expression pattern on transgenes. However, several pathogen-induced promoters have been identified that at least have a greater degree of specificity in response, being insensitive to wounding, but responding to signals such as salicylic acid.

Heat-shock-inducible

Sequence elements, termed heat-shock elements or HSEs, mediate the heat-shock inducibility of genes (see Chapter 9). If included in chimaeric promoters, they confer heat-shock inducibility on transgene expression.

Plant-derived systems based on developmental control of gene expression

Again, although not inducible in the sense of the non-plant-derived systems, many genes have been identified that are expressed at particular stages in plant development. The promoters from some of these genes have been studied in detail. As in the previous cases, either chimaeric promoters or complete promoters are used to drive transgene expression.

Senescence-specific gene expression

The promoters from two *Arabidopsis* genes, termed *SAG*12 and *SAG*13 (for senescence-activated gene), which are induced during senescence, have been shown to confer senescence-inducible expression on transgenes. For example, expressing the *ipt* gene from *Agrobacterium* (see Chapter 3), which is involved in cytokinin production, under the control of the *SAG*12 promoter delays senescence in tobacco.

BOX 4.1 *Continued*

Abscisic acid-inducible gene expression

Sequence elements responsible for abscisic acid (ABA) (see Chapter 2) induction of gene expression have been identified (see Chapter 1) and if combined with the core 35S promoter in a chimaeric promoter confer ABA inducibility on transgene expression.

Auxin-inducible gene expression

Similarly, several sequence elements responsible for auxin (see Chapter 2)-induced gene expression have been identified. These auxin-response elements (AuxREs) when combined with a core 35S promoter confer auxin (and inactive auxin analogue) inducibility on transgene expression.

Synthetic promoters for inducible gene expression

In some of the previous examples, we have seen how sequence elements from genes that, for example, confer heat and ABA inducibility, can be combined with core promoters to effectively build synthetic promoters that confer the desired expression characteristics on the transgene. As more and more *cis*-elements are identified and characterized, the possibility of building synthetic promoters combining one or more of these elements with a minimal promoter to confer the desired expression characteristics becomes more realistic. Multiple sequence elements responsible for pathogen and wound induction, tissue-enhanced expression, cell-specific expression, and transcriptional up- and down-regulation have all been identified. Combinations of several sequence elements involved in pathogen-induced expression have been combined with a minimal promoter to create synthetic promoters that have been tested and shown to function effectively as pathogen-induced promoters, with the added advantage that multiple elements involved in the same response may act synergistically. It may also be possible to construct synthetic promoters that confer the characteristics of several of these elements in some cases. Construction of these synthetic promoters is not without its difficulties, as the spatial arrangement of the various *cis*-elements has to be controlled carefully and can be difficult to predict. However, the potential benefits of synthetic promoters suggests an important role in driving transgene expression, at least in demanding situations requiring precise regulation of expression.

The advantage of non-plant-derived systems is that they are independent of normal plant processes, requiring the application of a specific substance to induce expression. This is, however, also their weakness, as the application of (in some cases expensive) inducers on an agricultural scale is, in most but not all cases, not feasible or economically viable (see Box 4.1 for an example where it may prove to be both feasible and economically viable). Experimentally, however, some of these have proved to be quite useful.

Systems based on plant-derived components do not have the advantage of independence from normal plant processes. This does, however, make their use in agriculture potentially simpler as the application of an inducer is not required. It should be noted that no one system is suited to all situations, and potentially more damaging is the fact that some of the systems developed have proved to be difficult to use or have proved to be 'leaky' (i.e. they direct expression even in the absence of the inducing agent).

Selectable markers

Selectable marker genes introduced during transformation allow transformed tissue to be selected, normally by conferring resistance to a toxic substance.

Plant transformation is, in many cases, a very-low-frequency event. It is therefore vital that some means for selecting the transformed plant tissue is provided by the plant transformation vector. In most cases this selection is based on the inclusion into the culture medium of a substance that is toxic to plants. The selectable marker on the vector confers resistance to the toxic substance when expressed in transformed plant tissue. In many cases, as in the situation when using *E. coli*, antibiotic-resistance genes can be used as selectable markers in plants. Although plants are eukaryotic, antibiotics efficiently inhibit protein synthesis in the organelles, particularly the chloroplasts. Perhaps the most widely used selectable marker gene is the *nptII* gene, which confers resistance to the antibiotic kanamycin. Other antibiotic-resistance genes have also been used with some considerable success in plant transformation vectors. In part, this use of alternative selectable marker genes was driven by the observations that some plant species exhibited a very high degree of natural resistance to kanamycin (such as cereals), and that some species (some soft fruits for example) were too sensitive to kanamycin for it to be used successfully. This made the selection of transformed tissue very difficult, leading to a high number of false-positive results or the inability to recover transformed plants. Thus, the use of any selectable marker has to be controlled closely, with appropriate concentrations being determined by kill-curves, on a case-by-case basis. The development of alternative selectable marker genes also allows for the re-transformation of plant tissue that already expresses one or more different selectable markers. Thus, genes conferring resistance to antibiotics such as bleomycin, spectinomycin, and hygromycin are used quite widely. These selective agents can be used at lower concentrations than kanamycin, and therefore usually result in a cleaner selection of transformed tissue. Other resistance genes can also be used as selectable markers in plants. Among those used widely are genes that confer resistance to herbicides such as chlorsulphuron and bialaphos (see Chapter 5).

Public concern has, in recent years, questioned the general desirability of growing transformed crops, and, more specifically, the merits of using antibiotic- or herbicide-resistance genes as selectable markers during plant transformation. There is obvious concern about the creation of so-called super-weeds and the transfer of antibiotic-resistance genes. Although in the long term it is undoubtedly better to remove these selectable marker genes from the transformed plants once their job has been done (this issue will be considered in more detail later in this chapter), other, more acceptable, selectable marker genes are also being introduced.

Some of the most interesting are based on the principal of facilitating alternative carbon-source utilization. Thus, genes from bacteria that allow the use of mannose or xylose (and, in fact, other carbohydrates) as carbon sources have been successfully used as selectable markers in plant transformation. These genes have also generally proved to be superior to standard antibiotic-selection genes in some transformation protocols. These are examples of positive-selection methods, so-called because untransformed tissue does not die (due to the presence of low concentrations of sucrose in the medium used for regeneration) but does not proliferate.

Other potential selectable markers are also being investigated; these include: the reporter gene *gfp*, which can be used for the visual screening of transformants; the *Agrobacterium* isopentenyl transferase (*ipt*) gene, which determines growth and morphology during regeneration; and the use of cytokinin glucuronides in combination with β-glucuronidase (GUS). Examples of various selectable markers are shown in Table 4.1.

Reporter genes

Reporter genes (see Table 4.2) are widely used in plant transformation vectors, both as a means of assessing gene expression by promoter analysis (see Chapter 1) and as easily scored indicators of transformation (indeed, they have been used in place of selectable marker genes in some cases). Ideally, reporter genes should be easy to assay, preferably with a non-destructive assay system, and there should be little or no endogenous activity in the plant to be transformed. At present, only a small number of reporter genes are in widespread use in plant transformation vectors, these being β-glucuronidase (*uidA* or *gus*), green fluorescent protein (*gfp*), luciferase genes (*lux* and *luc*), and, to a lesser degree (although it is widely used in animal systems), the chloramphenicol acetyltransferase gene (*cat*). Each of these reporter genes will be looked at in more detail.

● Reporter genes confer an easily screened phenotype on transformed plant tissues.

β-Glucuronidase

This is perhaps the most widely used reporter gene in plant transformation vectors. Its widespread acceptance is due to its many advantages over the previously existing marker genes. GUS can be assayed extremely sensitively using quick, easy, and non-radioactive methods. It can be used to obtain both quantitative (i.e. the level of gene expression) and qualitative (i.e. localization of gene expression) data. There is also little or no endogenous activity in most plant tissues (with the possible exception of reproductive tissues).

Quantitative data is obtained by assays utilizing fluorogenic substrates such as 4-methylumbelliferryl-β-D-glucuronide (4-MUG), which is hydrolysed to 4-MU (4-methylumbelliferone). Standard enzyme assay protocols can be used and results compared with a standard curve of 4-MU fluorescence. Qualitative data can be obtained from histochemical assays that allow tissue- and cell-specific localization of GUS activity. These histochemical assays are conducted *in situ* with the chromogenic substrate 5-bromo-4-chloro-3-indolyl β-D-glucuronide (X-gluc). GUS activity results in the deposition of an insoluble blue precipitate, effectively identifying the precise location of expression.

In order to avoid problems associated with GUS expression in *Agrobacterium*, a *gus* gene containing an intron (which can therefore only be expressed in eukaryotes) is often used in plant transformation vectors.

Table 4.1 Selectable marker genes used in plant transformation, their source, and mode of action

Selectable marker gene	Abbreviation	Source of gene	Selection mechanism	Selective agent
Antibiotic resistance				
Aminoglycoside adenyltransferase	*aadA*	*Shigella flexneri*	Antibiotic resistance	Streptomycin, spectinomycin
Bleomycin resistance	*ble*	*E. coli*	Antibiotic resistance	Bleomycin
Dihydropteroate synthase	*sul/dhps*	*E. coli*	Antibiotic resistance	Sulphonamides
Dihydrofolate reductase	*dhfr*	Mouse	Antibiotic resistance	Methotrexate
Hygromycin phosphotransferase	*hpt/aphIV/hyg*	*E. coli*	Antibiotic resistance	Hygromycin
Neomycin phosphotransferase II	*nptII/neo*	*E. coli*	Antibiotic resistance	Kanamycin, geneticin (G418)
Neomycin phosphotransferase III	*nptIII*	*Streptococcus faecalis*	Antibiotic resistance	Kanamycin, geneticin (G418)
Herbicide resistance				
Acetolactate synthase	*als*	*Arabidopsis* spp./ maize/tobacco	Herbicide resistance	Sulphonylureas
Enolpyruvylshikimate phosphate synthase	*epsps/aroA*	*Petunia hybrida/ Agrobacterium* spp.	Herbicide resistance	Glyphosate
Glyphosate oxidoreductase	*gox*	*Achromobacter LBAA*	Herbicide resistance	Glyphosate
Phosphinothricin acetyltransferase	*bar/pat*	*Streptomyces hygroscopicus/S. viridochromogenes*	Herbicide resistance	Bialophos, glufosinate, L-phosphinothricin
Cyanamide hydratase	*cah*	*Myrothecium verrucaria*	Herbicide resistance	Cyanamide
Bromoxynil nitrilase	*bxn*	*Klebsiella pneumoniae*	Herbicide resistance	Bromoxynil

Others				
Lysine–threonine aspartokinase	*lysC*	*E. coli*	Resistance to high levels of lysine or threonine	Lysine and threonine
D-Serine ammonia lyase	*dsdA*	*E. coli*	Resistance to D-serine	D-Serine
Mannose-6-phosphate isomerase[a]	*pmi/manA*	*E. coli*	Alternative carbon source	Mannose
Xylose isomerase	*xylA*	*Thermoanaerobacterium thermosulfurogenes/ Streptomyces rubiginosus*	Alternative carbon source	Xylose
Arabitol dehydrogenase	*atl*D	*E. coli*	Alternative carbon source	Arabitol
Trehalose-6-phosphate synthase	*AtTPS1*	*Arabidopsis*	Resistance to high levels of glucose	Glucose
β-Glucuronidase	*gus/uidA*	*E. coli*	Complementation of missing medium component	Cytokinin glucuronide
Betaine aldehyde dehydrogenase[b]	*badh*	Spinach	Detoxification	Betaine aldehyde
Green fluorescent protein	*gfp*	*Aequorea victoria* (jellyfish)	Visual screening	None
Isopentenyl transferase[c]	*ipt*	*Agrobacterium* spp.	Growth in absence of exogenous cytokinin	None
Leafy cotyledon[d]	*lec*	Maize	Growth	None
Knotted1	*Kn*1	Maize	Growth	None

[a]Also known as phosphomannose isomerase.
[b]See Chapter 9. Used as a selectable marker in chloroplast transformation.
[c]If used with a constitutive promoter, plants have poor root formation and are often sterile, thus limiting the use of the *ipt* gene as a selectable marker. However, use with an inducible promoter has resulted in the development of an antibiotic-free selectable-marker system.
[d]Leafy cotyledon gives the transformed cells a competitive growth advantage, and when coupled with a visual marker to identify transformed cells can be used as an antibiotic-free selectable-marker system.

Table 4.2 Reporter genes used in plant transformation

Reporter gene	Abbreviation	Source of gene	Detection/assay
β-Glucuronidase	*gus/uidA*	*E. coli*	Fluorimetric (quantitative) or histochemical (*in situ*), non-radioactive
Green fluorescent protein	*gfp*	*Aequorea victoria* (jellyfish)	Fluorescence, non-destructive
Chloramphenicol acetyltransferase	*cat*	*E. coli*	Radioactive assay of plant extract, sensitive, semi-quantitative
Luciferase	*luc*	*Photinus pyralis* (firefly)	Luminescence
Luciferase	*luxA, luxB*	*Vibrio harveyi*	Luminescence

Green fluorescent protein (GFP)

Nowadays, the green fluorescent protein gene (*gfp*) is rapidly becoming a very widely used reporter gene. GFP has the advantages that it is even easier to assay than GUS (given the correct equipment) and the assay is non-destructive (although non-destructive assays for GUS activity have been reported, they have not been adopted widely). This means that GFP can be used in situations where GUS cannot, for example in screening primary transformants, in time-course experiments, or for analysing segregation in small seedlings.

The gene was isolated from the jellyfish *Aequorea victoria*, which are brightly luminescent organisms. To work efficiently in plants it was found that the *gfp* gene had to be modified significantly in order to: (1) remove a cryptic intron (i.e. *gfp* mRNA is efficiently *mis*-spliced in some plants, resulting in the removal of 84 nucleotides); (2) make the codon usage more 'plant-like'; and (3) prevent accumulation in the nucleoplasm.

Luciferases

The firefly luciferase gene (*luc*) encodes an enzyme that catalyses the oxidation of D-luciferin in an ATP-dependent fashion. This oxidation results in the emission of light. Highly sensitive assays (based on photomultipliers, luminometers, or film exposure) have been developed to detect the extremely rapid emission of light. Other assay systems are also available commercially. The firefly luciferase gene is not a widely used marker gene as assaying the gene product is difficult, but it is useful for the detection of low-level or highly localized expression.

The bacterial luciferase genes (*luxA* and *luxB* from *Vibrio harveyi*) are found in a few plant transformation vectors. They catalyse the oxidation of long-chain fatty aldehydes resulting in the emission of light.

Chloramphenicol acetyltransferase

Although widely used as a reporter gene in mammalian cells, the availability of the GUS and GFP reporter systems has generally limited the use of the chloramphenicol acetyltransferase (CAT) system in plants, although it was the first bacterial gene to be expressed in plants. Some plant transformation vectors do, however, carry this reporter gene as it can be assayed very sensitively, but it requires a radioactive assay procedure.

Origins of replication

As both *E. coli* and *Agrobacterium* are Gram-negative bacteria, a single origin (*ori*) with a broad host range can be used to allow vector replication in both *E. coli* and *Agrobacterium*. However, the broad-host-range origins used in many plant transformation vectors (e.g. the RK2 origin used in pBIN19 and derivatives) often results in a low plasmid copy number in *E. coli*, making manipulation of the DNA more difficult or time-consuming. An alternative approach is to use two separate origins of replication, one to enable replication in *Agrobacterium* and a second to direct replication in *E. coli*. This second origin is usually derived from a high-copy-number cloning vector (such as the *Col*E1 *ori* from pBR322 or the pUC *ori*). This results in a higher copy number in *E. coli*, therefore facilitating manipulation, but it also results in a larger vector (as two separate origins are present) with the associated difficulties.

Co-integrative and binary vectors

Initially, vectors used for *Agrobacterium*-mediated plant transformation were deletion derivatives of Ti plasmids that were termed co-integrative vectors. The DNA to be introduced into the plant transformation vector is cloned into an intermediate vector (which is small and easily manipulated *in vitro*) and recombination within the bacterium is used to clone the sequences. In most systems of this type, the modified T-DNA (containing the gene of interest) is on the same molecule as the transfer apparatus as part of one Ti plasmid. Vectors of this type have become less common as binary vector systems are more easily manipulated.

In binary vectors, the transfer apparatus (i.e. the *vir* genes) and the T-DNA are located on separate plasmids (Figure 4.3). As only the border sequences are needed to define the T-DNA region and the *vir* region is absent, binary vectors are relatively small (and therefore easily manipulated). Box 4.2 describes the development of binary vectors.

Families of binary vectors

It soon became apparent that, although vectors such as pBIN19 (which offer one selectable marker) were sufficient for routine transformation studies, the application of transgenic technology to an increasing number of plant species required a more flexible set of binary vectors.

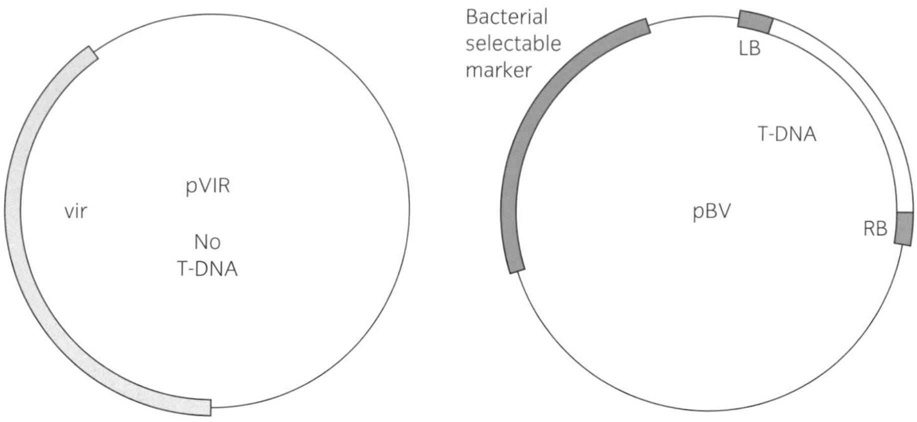

Figure 4.3 The binary-vector strategy illustrated by a pair of hypothetical plasmids. One plasmid (pVIR in this case, a derivative of a Ti plasmid) carries the *vir* genes. This plasmid has no T-DNA region, including border sequences, and therefore no T-DNA can be transferred to the plant genome. A smaller binary vector (pBV, which is over 20 times smaller than pVIR) has a T-DNA region defined by the presence of the left and right borders (LB, RB). As there are no *vir* genes on this plasmid (making the binary vector pBV relatively small), their function is supplied *in trans* by pVIR. pVIR would normally be maintained in suitable strains of *Agrobacterium* and is not generally isolated or manipulated. Binary vectors are designed to be easily manipulated using standard *in vitro* techniques and contain selectable markers for selection in bacteria (*E. coli* and *Agrobacterium*) outside the T-DNA region. The T-DNA region will, in all probability, contain a selectable marker for use in plants. pBV is normally maintained in *E. coli*, the gene of interest (the transgene) being cloned into it. Recombinant *E. coli* colonies are identified after bacterial transformation, and the pBV vector is checked to ensure that the transgene is correctly inserted and has not been mutated. pBV can then be transferred, by electroporation, into the *Agrobacterium* strain used for plant transformation (which also harbours pVIR).

Thus, modern binary vectors tend to come in families. These offer a choice of selectable marker (both antibiotic and herbicide resistance in many cases) and, in some cases, the opportunity for dual selection by incorporating two selectable markers on the one vector. This dual selection approach allows a much more rigorous selection pressure to be applied to plants, in which it has traditionally proved difficult to avoid false positives during the selection process.

Optimization

● Many features of transformation-vector design influence transgene expression.

Considerable advances have been made in the design of plant transformation vectors (see Box 4.2) and the ease with which these vectors can be produced (see Box 4.3 for an example of how plant transformation vector construction has been simplified). However, despite the improvements made in vector design and advances in our understanding of both the mechanisms of transgene integration and plant gene expression, plant transformation is still in many ways an imprecise art. Several aspects of vector design are known to influence the efficiency of transgene expression.

BOX 4.2 *The evolution of binary plant transformation vectors*

Having looked at the basic requirements for plant transformation vectors, we can look at how the design of plant transformation vectors has improved over the years. The binary plant transformation vectors pBIN19 and pGreen can be compared (see figure).

pBIN19 was one of the first binary plant transformation vectors (developed in the early 1980s) and has been one of the most widely used. pGreen is a new binary vector (one of a family available with different promoters, selectable markers, and reporter genes). As can be seen, pBIN19 (11 777 bp) is much larger than pGreen (4632 bp) and is therefore that much more difficult to work with. This reduction in size is, at least in part, due to the unusual origin of replication present in pGreen. The pSa origin is in itself small, but in pGreen it has been further divided so that only the replication origin is present on the plasmid. The replicase gene is present on another plasmid and functions *in trans*. The restriction sites in pBIN19 available for use in cloning are limited in comparison with pGreen, although in both cases blue/white screening is possible. The selectable marker, which in both cases is *nptII*, is situated close to the right border in pBIN19, but in pGreen it is situated close to the left border. The advantage of having the selectable marker next to the left border becomes apparent if one considers that there is a polarity in T-DNA transfer (Chapter 3). Transfer originates from the right border and therefore the selectable marker will be transferred before the gene of interest if the selectable marker is next to the right border. If T-DNA transfer is interrupted, this could result in plants expressing the selectable marker (and therefore being isolated as genuine positive transformants) but not containing, or containing a truncated version of, the gene of interest.

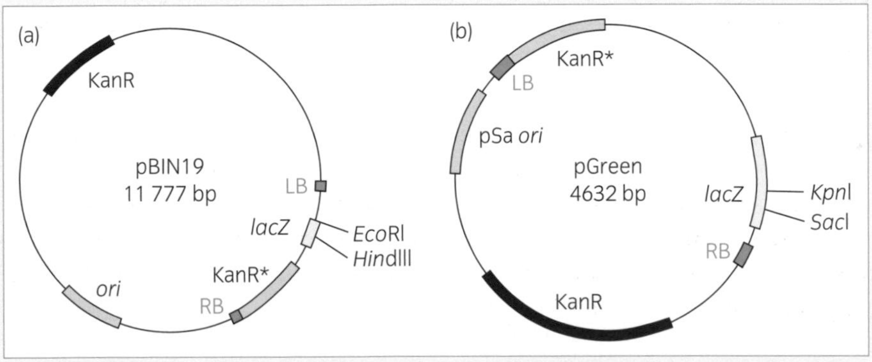

Simplified representation of the binary plant transformation vectors (a) pBIN19 and (b) pGreen (in this case pGreen 0029), showing the pertinent features. pBIN19 is a large plasmid, making manipulations using this plasmid difficult. The selectable marker for selecting transformed plants (KanR*) is located close to the right border (RB), leading to the possibility of false positives (due to truncation of the T-DNA during transfer, kanamycin-resistant plants that do not express the transgene might be isolated). The MCS, situated between the *Eco*RI and *Hin*dIII sites, has eight restriction enzyme sites. The MCS is in the *lacZ*α fragment so blue/white screening for recombinant colonies is possible. pGreen 0029 is one of a large family of plant transformation vectors. It is much smaller than pBIN19 and has a much larger MCS (with over 15 restriction enzyme sites) situated in *lacZ*α (between the *Kpn*I and *Sac*I sites), so blue/white screening is possible. The pSa origin is small and contributes to the small overall size of pGreen 0029. (Redrawn with permission from figures on the pGreen website, www.pgreen.ac.uk/a_hom.htm.)

BOX 4.3 Gateway™ vectors

In the introduction to this chapter we discussed how many of the developments in plant transformation vector design were aimed at making manipulations involving the vector simpler. Binary vectors, even modern ones such as pGreen (shown in Box 4.2), tend on the whole to be quite large and the cloning of sequences into these vectors is time-consuming and laborious. Recently though, a new technology developed by Invitrogen in the USA, termed Gateway™ conversion technology, has opened up the possibility of rapid, facile cloning of sequences into suitably designed binary vectors. Key advantages of Gateway technology are that the cloning of DNA into plant transformation vectors is not dependent on the constraints of sequence that are normally imposed by the need to use restriction enzymes in conventional cloning strategies, and so manipulations are eased and less laborious. Once a piece of DNA is flanked by suitable sites that allow Gateway technology to be used, it can be easily cloned into a large and expanding range of suitable Gateway-compatible plant transformation vectors with the minimum of manipulation.

Gateway cloning

Gateway technology depends in essence on recombination technology, as opposed to the more standard cloning technology based on restriction enzymes and DNA ligase. Later in this chapter

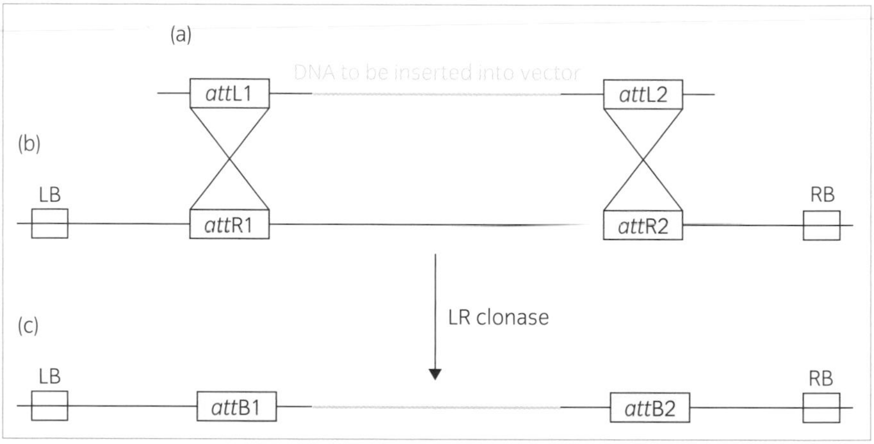

Gateway™ cloning. The DNA to be introduced into the plant transformation vector (the destination vector) is flanked by *att* (in this case *att*L) recombination sites, which are small (25-bp) sequence elements. The destination vector contains similar *att*R sites between the left (LB) and right (RB) borders that define the T-DNA. The LR clonase™ enzyme causes exchange of the DNA fragments between the recombination sites as indicated. The destination vector now contains the DNA fragment flanked by two *att*B recombination sites, which means that it can be further mobilized by recombination into a vector containing *att*P sites (the reaction being catalysed by another clonase enzyme). The efficiency of the system is further improved by enabling positive selection of destination vectors that have the DNA of interest inserted. This positive selection comes from the inclusion of the *ccd*B gene, which prevents the growth of *E. coli*, between the *att*R sites of the destination vector. Only *E. coli* harbouring destination vectors in which the *ccd*B gene has been replaced by the DNA of interest through recombination will grow.

BOX 4.3 *Continued*

the role of site-specific recombinases in facilitating clean-gene technologies is explained; here though a different site-specific recombinase system is exploited. In Gateway cloning, the site-specific recombinase system used by bacteriophage lambda (phage λ) is utilized.

Any DNA fragments that need to be cloned into a plant transformation vector that are flanked by recombination sites will, in the presence of an enzyme, recombine with recombination sites in the vector, leading to incorporation of the fragment into the plant transformation vector. The precise way in which the overall cloning scheme works varies slightly depending on which group has developed the vectors, but all are based on this simple principle.

The recombinase sites that flank the DNA that is to be cloned into the plant's transformation vector can be introduced simply in a variety of ways. One common method is to use topoisomerase-mediated cloning, developed by Invitrogen. In this approach, the polymerase chain reaction (PCR) is used to amplify the target sequence using a forward primer that has a CACC motif at its 5′ end. This 5′ CACC motif mediates directional incorporation of the PCR product into a suitable vector that results in the PCR-amplified target DNA being flanked by *att*L recombination sites. The technology can be used with more traditional approaches as well. Recombination sequences can be introduced into vectors so as to flank MCSs and can be included in PCR primers. These different variations require subtle modification of the technology but have all been used successfully.

Gateway-compatible plant transformation vectors

A variety of laboratories have developed destination vectors for use in plants, many of which are available, through either a material transfer agreement or freely. These destination vectors have been developed with a range of applications in mind, from simple constitutive expression of transgenes, through inducible expression, to protein interactions, protein localization, and protein purification.

The key advantages of Gateway technology in facilitating the quick and easy cloning of DNA into plant transformation vectors means that such vectors are ideally suited to high-throughput applications associated with genomics and proteomics.

Arrangement of genes in the vector

Analysis of the efficiency of transgene expression allows a few simple 'rules' about vector design to be drawn up. If more than one gene is in the vector, different promoters and terminators should be used for each. The use of the same promoter and/or terminator can lead to an increase in gene silencing (Box 4.4 discusses how vectors designed specifically to induce gene silencing can be produced). Multiple genes on one vector should not be immediately adjacent to each other, and should be in the same orientation. This avoids adjacent inverted repeats that cause plasmid instability in bacteria and increased gene silencing in plants.

BOX 4.4 Gene silencing

In Chapter 1 we saw how early experiments with transgenic plants resulted in the discovery of gene silencing, and instability of transgene expression is still a problem encountered in experiments involving transgenic plants. There are a number of practical steps in vector design that can be taken to reduce the risk of gene silencing. Reducing the number of transgenes inserted (the copy number), avoiding the use of multiple copies of the same promoter or terminator (particularly in repeat structures), and avoiding using promoters and transgenes with a high degree of similarity to endogenous ones can help to reduce the probability of gene silencing.

However, despite the problems presented by unintended gene silencing, no discussion of plant transformation vectors would be complete without looking at how vectors deliberately designed for gene silencing can be designed.

A full explanation of gene silencing is given in Chapter 1, but a short introduction to the pertinent points as far as vector design is concerned is given here. Two observations were key in leading to the development of gene-silencing strategies for the down-regulation of gene expression in transgenic plants: attempts to achieve high-level transgene expression could paradoxically result in very low or no transgene expression and the expression of sense-orientation transgenes as controls for antisense RNA experiments resulted in gene silencing. In both cases, the silencing was post-transcriptional, and dependent on the production of double-stranded (ds) RNA that is subsequently processed by nucleases into small interfering RNAs (siRNAs), which direct the degradation of homologous RNA by a different nuclease. This post-transcriptional dsRNA-mediated gene silencing is known as RNA interference (RNAi). By constructing plant transformation vectors that indirectly or directly lead to the production of dsRNA, RNAi can be deliberately induced to silence the expression of specific genes in transgenic plants.

RNAi vectors

Normally, when expression of a transgene is desired, the arrangement of promoter and sequence to be expressed mimics the usual arrangement of the gene, so that the promoter directs expression of sense RNA that is in the correct orientation to be translated (panel a in the figure). The simplest RNAi-related technology is to produce so-called antisense RNA. In vectors designed to produce antisense RNA, the sequence to be expressed (which can be all or a part of the coding sequence of an endogenous gene) is placed in the reverse orientation with respect to the promoter (panel b in the figure). The promoter therefore directs the production of antisense RNA, which is complementary to the endogenous mRNA. The complementarity between antisense RNA and mRNA results in the production of dsRNA.

However, antisense RNA production is not the most effective way of directing RNAi; far more efficient is the production of hairpin RNA (hpRNA). In a vector designed to produce hpRNA, the transgene sequence is cloned as an inverted repeat, with the repeat units separated by a spacer (panel c in the figure). The expression of the inverted repeat is driven by a strong promoter such as the 35S promoter for dicotyledonous plants or the ubiquitin 1 promoter for monocotyledonous plants. The inverted-repeat structure allows the two complementary regions to double-back on each other, pivoting at the spacer region. This double-stranded hpRNA is extremely efficient in RNAi-induced destruction of any homologous mRNA.

BOX 4.4 *Continued*

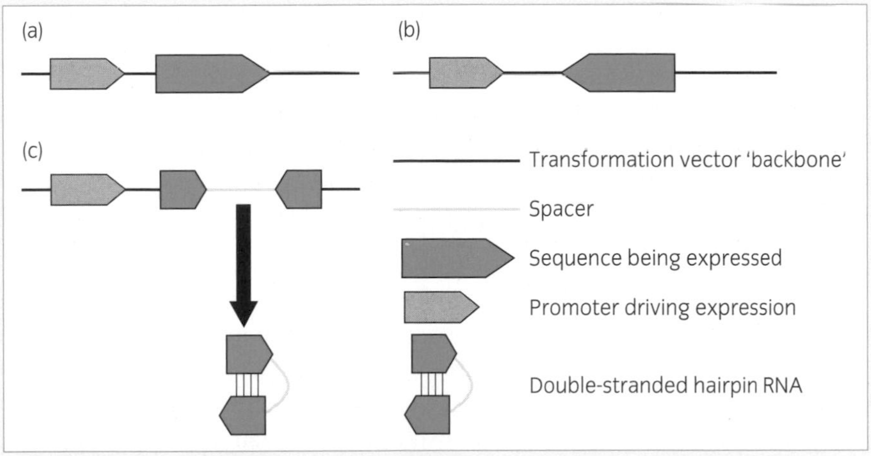

A simplified representation of constructs used for RNAi. When expression of a transgene is required, the arrangement of promoter and sequence to be expressed mimics the usual arrangement of the gene, so that the promoter directs expression of sense RNA (a). In antisense constructs, the orientation of the sequence to be expressed is reversed relative to the promoter, so that antisense RNA is produced (b). More efficient down-regulation is achieved by the expression of hpRNA. The expressed sequence is composed of an inverted repeat, with the repeat units separated by a spacer (c). The spacer is usually composed of an intron sequence, which is necessary for construct stability during cloning and the promoter is often one driving high-level constitutive expression. When the sequence is expressed from the promoter, the RNA can form a hairpin structure with a loop formed from the intron spacer region. This hpRNA directs RNAi-mediated destruction of any homologous mRNA.

hpRNA-induced RNAi has many advantages as a technology for down-regulation of gene expression.

- hpRNA-induced RNAi may be stably inherited, unlike some other mechanisms that induce post-transcriptional gene silencing. The stability of any introduced phenotypes in transgenic plants is obviously crucial, particularly if applied to crop plants. hpRNA-induced RNAi is known to be inherited through multiple generations in *Arabidopsis*.
- Down-regulation of gene expression by hpRNA-induced RNAi is also effective when the target RNA is encoded by a multigene family. Down-regulation of multigene families is difficult to achieve with other technologies used for down-regulation.
- The degree of down-regulation can also be fine-tuned by either moderating the expression level by the choice of an appropriate promoter (although very weak promoters are not suitable) or by using sequences with varying degrees of identity to the target RNA.
- The expression of the hpRNA can be restricted to given organs by the use of organ-specific promoters or can be made to respond to various stimuli by the use of inducible promoters.
- hpRNA constructs are relatively simple to make, and can be combined with Gateway™ technology to facilitate rapid vector construction.

Transgene copy number

In any plant transformation protocol, multiple copies of the transgene can be incorporated into the target plant genome. Often high levels of transgene expression are associated with multiple copies of the transgene, but this is not always the case. Multiple T-DNA insertions into the genome may lead to erratic transgene expression.

Transgene position

Transgene integration into the plant genome is basically a random event, and it has been demonstrated that the position of the transgene in the genome can have a marked effect on transgene expression levels. Thus, independent transformants with single copies of the transgene can exhibit large differences in the level of transgene expression. This positional effect on transgene expression remains a problem in plant transformation, despite efforts to find ways of targeting gene integration. Unfortunately, higher plants do not possess an endogenous system enabling homologous recombination (genetic recombination involving the exchange of homologous loci) at a high efficiency (but see Chapter 5) that allows transgenes to be targeted to a particular region of the genome. However, alternatives utilizing bacteriophage or yeast site-specific recombinases have been used with some success.

One approach taken to assuage position effects is the inclusion of matrix-attachment regions (MARs). These AT-rich sequences are thought to be involved in maintaining the chromatin in an open structure allowing gene expression. The inclusion of MARs flanking transgenes has been shown to provide position-independent expression in animal systems, and some, albeit limited, success has been achieved with their use in plants. MARs may also help to stabilize transgene expression.

Transgene features

It is well documented that heterologous genes, particularly those from non-plant species, tend to be expressed poorly in plants, even when driven by a strong promoter. This can be due to a variety of factors, many of which are associated with the structure of the transgene.

Genes from different organisms tend to have different G+C contents, with bacterial genes having a particularly low G+C content. It has been found that a high A+T content in transgenes interferes with mRNA processing in plants, leading to little or no expression of the transgene. Differences in A+T content between the transgene and the isochore (long stretches of DNA with a homogeneous base composition) into which it is integrated may also contribute to the transgene being recognized as foreign. A high A+T content also often results in the presence of sequences (AUUUA) that can destabilize mRNA and may also form potential plant polyadenylation signals. So-called cryptic introns may also be found. These can be efficiently mis-spliced from transgenes (i.e. part of the coding sequence is recognized by the plant as an intron and spliced out) by the plant's mRNA-processing machinery, effectively resulting in a deletion mutation of the transgene.

Even if the transgene is transcribed efficiently it may not be translated efficiently, probably due to the architecture of the translation initiation region and the presence of codons that are used infrequently by plants. Consensus sequences flanking the ATG initiation codon are known to differ between plants and other species. The inclusion of plant-specific sequences upstream of the translation initiation codon is known to improve translatability. It has also been demonstrated recently that modification (by the insertion, after the translation initiation codon, of codons that are found in a number of highly expressed plant genes) of the region downstream of the translation initiation codon may also improve transgene translatability. The inclusion of the tobacco mosaic virus Ω leader between the promoter and transgene has also been shown to improve translatability.

Mutagenesis can be used to alter the codon usage of a particular sequence, although if the sequence is large and requires a large number of changes, this may prove difficult and/or expensive. Some examples of transgene modification to optimize expression are given in Box 4.5.

BOX 4.5 *Examples of optimization of vector components*

Modification of the *Bacillus thuringiensis cry1*A(b) gene

The *cry1*A(b) gene encodes a protein that can be used to control insect pests. The wild-type gene was found to be expressed at relatively low levels in transgenic plants. The effect of modifying the sequence of the gene on expression was investigated. Codon usage and overall G+C content was altered and several potential plant polyadenylation signals and ATTTA sequences were removed. These changes were seen to result in large increases (up to 100-fold) in the levels of the *cry1*A(b) gene product.

	Wild-type gene	Modified gene
No. of altered bases (%)	0	390 (21%)
No. of modified codons (%)	0	356 (60%)
G+C content (%)	37	49
No. of potential polyadenylation sites	18	1
No. of ATTTA sequences	13	0

(Data taken with permission from Perlak *et al.* (1991).)

Modification of the *gfp* gene

The *Aequorea victoria gfp* gene is used as a reporter gene in plant transformation experiments. To improve its use in plants, and in particular in *Arabidopsis*, it has been modified extensively. A cryptic intron that is efficiently mis-spliced in some plant species was removed by altering codon usage (altering the codon usage has also been demonstrated to increase the expression level). Translatability of the gene was improved by introducing an AACA sequence upstream of the translation initiation codon. In order to prevent build-up of GFP in the nucleoplasm (where it may prove to be mildly toxic), a 5′ terminal signal peptide and a 3′ endoplasmic reticulum retention signal (HDEL, His–Asp–Glu–Leu) were introduced to target the GFP to the endoplasmic reticulum. (Data taken with permission from Haselhoff *et al.* (1997).)

Clean-gene technology

● New technologies are being developed to reduce the possible environmental impact and increase the acceptability of transgenic plants.

Since the introduction of transgenic crop plants into agriculture, public concern about the safety of such technology, particularly about the safety of antibiotic- and herbicide-resistance genes, has increased to the point where governments have been forced to take action (see Chapter 12). Whatever the scientific merits of the argument, it is a fact that the perceived dangers of transgenic crops are preventing their widespread adoption, particularly in Europe. Even in countries where the growth of such crops is not seen as a major issue (like the USA), the effect is still felt as export markets for foodstuffs derived from genetically engineered crops are affected. There has therefore been a move to so-called clean-gene technologies where the selectable marker is no longer present in the field-grown crop, or, as described earlier, a more acceptable marker gene is used. To develop crops that contain no selectable marker genes, several approaches are available, as follows.

1 Theoretically, the use of selectable marker genes could be avoided altogether by simply inserting the transgene of interest. Molecular biology techniques, such as the polymerase chain reaction (PCR), could be used to screen populations of putative transformed plants for those containing the transgene. In practice, this approach would simply be too expensive and labour-intensive in most situations.

2 A second approach is to introduce the gene of interest and the selectable marker genes on separate (unlinked) T-DNA molecules. Subsequent segregation analysis of the offspring can be used to identify plants that contain the gene of interest but not the selectable marker. Various strategies to deliver separate T-DNAs have been developed, the most effective appearing to be those in which one binary vector contains several T-DNA regions that are integrated at unlinked sites in the plant genome.

3 An alternative approach is to excise the selectable marker from the plant genome. This can be done by utilizing site-specific recombinase systems. Several recombinase systems (phage P1 Cre-*lox*, *Saccharomyces cerevisiae* FLP-*frt*, and the R-*RS* system of *Zygosaccharomyces rouxii*) are available. Vectors in which the selectable marker gene is placed between recognition sites (*lox*, *frt*, or *RS*) for a site-specific recombinase can be used to transform plants. Subsequently, the recombinase (Cre, FLP, R) is expressed (by transient expression, crossing of plants with recombinase-expressing plants, or by using an inducible expression system), resulting in the selectable marker gene being excised from the plant genome.

4 Recently, vectors have been demonstrated in which the selectable marker gene was flanked by sequences that apparently considerably increase the frequency of intrachromosomal recombination leading to excision of the marker gene.

SUMMARY

This chapter has looked at the construction of binary vectors for use in *Agrobacterium*-mediated transformation strategies. These binary vectors are, in effect, highly specialized derivatives of the *Agrobacterium tumefaciens* Ti plasmid looked at in Chapter 3. Key to their successful use in plant transformation protocols are the modifications made, resulting in removal of the oncogenes (thereby preventing abnormal plant development) and a general improvement in their ease of use. *Agrobacterium*-mediated transformation with binary vectors has been successfully applied to a wide range of crop plants, but improvements to many aspects of vector design are still being made. Some of these improvements are based on purely scientific grounds, to prevent or reduce gene silencing for instance, whereas others are based on widening the applicability of binary vectors (for example, alternative selectable markers). Other modifications are being made in response, at least in part, to concerns (public and scientific) about the safety of such vectors. These concerns, mainly over selectable marker genes, are particularly evident in Europe. In the UK, the Royal Society's 1998 report, *Genetically Modified Plants for Food Use*, questioned the future use of antibiotic-resistance genes and recommended that if they were used they should be removed at an early stage of development of the genetically modified plant. The Advisory Committee on Releases to the Environment (ACRE), in its *Guidance on Principles of Best Practice in the Design of Genetically Modified Plants*, similarly noted the desirability of reducing the amount of 'extraneous' DNA (such as marker genes) in transgenic plants. Clean-gene technologies and alternative selectable markers are therefore becoming increasingly attractive.

Binary vectors and *Agrobacterium*-mediated transformation have contributed to the considerable successes that plant biotechnology has already achieved, and will no doubt continue to contribute well into the future. In the chapters that follow, we will look at specific examples of plant biotechnology in practice, including ones that have relied on the use of *Agrobacterium*-mediated transformation and binary vectors.

FURTHER READING

Visit www.oxfordtextbooks.co.uk/orc/slaterplants2e/ for hyperlinked references that take you directly to online abstracts for the journal articles listed below:

Vector construction

Croy, R. R. D. (1993) Plant selectable genes, reporter genes and promoters. In *Plant Molecular Biology Labfax* (R. R. D. Croy, ed.), pp. 149–182. BIOS, Oxford.

Early, K., Haag, J. R., Pontes, O., Opper, K., Juehne, T., Song, K., and Pikaard, C. S. (2006) Gateway-compatible vectors for plant functional genomics and proteomics. *The Plant Journal* **45**, 616–629.

Guerineau, F. and Mullineaux, P. (1993) Plant transformation and expression vectors. In *Plant Molecular Biology Labfax* (R. R. D. Croy, ed.), pp. 121–147. BIOS, Oxford.

Hellens, R., Mullineaux, P., and Klee, H. (2000) A guide to *Agrobacterium* binary Ti vectors. *Trends in Plant Science* **5**, 446–451.

Hull, G. A. and Devic, M. (1995) The β-glucuronidase (*gus*) reporter gene system. In *Methods in Molecular Biology* (H. Jones, ed.), pp. 125–141. Humana Press, Totowa.

Jefferson, R. A., Kavanagh, T. A., and Bevan, M. A. (1987) GUS fusions: β-glucuronidase as a versatile gene fusion marker in higher plants. *EMBO Journal* **6**, 3901–3907.

Karimi, M., Inze, D., and Depicker, A. (2002) GATEWAY vectors for *Agrobacterium*-mediated plant transformation. *Trends in Plant Science* **7**, 193–195.

Karimi, M., De Meyer, B., and Hilson, P. (2005) Modular cloning in plant cells. *Trends in Plant Science* **10**, 103–105.

Selectable markers

Erikson, O., Hertzberg, M., and Nasholm, T. (2005) The *dsdA* gene from *Escherichia coli* provides a novel selectable marker for plant transformation. *Plant Molecular Biology* **57**, 425–433.

Haldrup, A., Noerremark, M., and Okkels, F. T. (2001) Plant selection principle based on xylose isomerase. *In vitro Cellular and Developmental Biology—Plant* **37**, 114–119.

Haselhoff, J., Siemering, K. R., Prasher, D. C., and Hodge, S. (1997) Removal of a cryptic intron and subcellular localisation of green fluorescent protein are required to mark transgenic *Arabidopsis* plants brightly. *Proceedings of the National Academy of Sciences USA* **94**, 2122–2127.

Joersbo, M. (2001) Advances in the selection of transgenic plants using non-antibiotic marker genes. *Physiologia Plantarum* **111**, 269–271.

Kaeppler, H. F., Carlson, A. R., and Menon, G. K. (2001) Routine utilization of green fluorescent protein as a visual selectable marker for cereal transformation. *In vitro Cellular and Developmental Biology—Plant* **37**, 120–126.

LaFayette, R. R., Kane, P. M., Phan, B. H., and Parrott, W. A. (2005) Arabitol dehydrogenase as a selectable marker for rice. *Plant Cell Reports* **24**, 596–602.

Leyman, B., Avonce, N., Ramon, M., Van Dijck, P., Iturriaga, G., and Thevelein, J. M. (2006) Trehalose-6-phosphate synthase as an intrinsic selection marker for plant transformation. *Journal of Biotechnology* **121**, 309–317.

Luo, K., Zheng, X., Chen, Y., Xiao, Y., Zhoa, D., McAvoy, R., Pei, Y., and Li, Y. (2006) The maize *knotted1* gene is an effective positive selectable marker gene for *Agrobacterium*-mediated tobacco transformation. *Plant Cell Reports* **25**, 403–409.

Penna, S., Sagi, L., and Swennen, R. (2002) Positive selectable marker genes for routine plant transformation. *In vitro Cellular and Developmental Biology—Plant* **38**, 125–128.

Reed, J., Privalle, L., Powell, M. L., Meghji, M., Dawson, J., Dunder, E., Suttie, J., Wenck, A., Launis, K., Kramer, C., Chang, Y.-F., Hansen, G., and Wright, M. (2001) Phosphomannose isomerase: an efficient selectable marker for plant transformation. *In vitro Cellular and Developmental Biology—Plant* **37**, 127–132.

Inducible gene expression

Gurr, S. J. and Rushton, P. J. (2005) Engineering plants with increased disease resistance: how are we going to express it? *Trends in Biotechnology* **23**, 283–290.

Kong, H.-G., Fang, Y., and Singh, K. B. (1999) A glucocorticoid-inducible transcription system causes severe growth defects in *Arabidopsis* and induces defence-related genes. *Plant Journal* **20**, 127–133.

Malnoy, M., Reynoird, J., Borejsza-Wysocka, E., and Aldwinckle, H. (2006) Activation of the pathogen-inducible *Gst1* promoter of potato after elicitation by *Venturia inaequalis* and *Erwinia amylovora* in transgenic apple (*Malus* x *domestica*). *Transgenic Research* **15**, 83–93.

Martinez, A., Sparks, C., Hart, C. A., Thompson, J., and Jepson, I. (1999) Ecdysone agonist inducible transcription in transgenic tobacco plants. *Plant Journal* **19**, 97–106.

Moore, I., Samalova, M., and Kurup, S. (2006) Transactivated and chemically inducible gene expression in plants. *The Plant Journal* **45**, 651–683.

Padidam, M. (2003) Chemically regulated gene expression in plants. *Current Opinion in Plant Biology* **6**, 169–177.

Rushton, P. J., Reinstädler, A., Lipka, V., Lippok, B., and Somssich, I. E. (2002) Synthetic plant promoters containing defined regulatory elements provide novel insights into pathogen- and wound-induced signaling. *The Plant Cell* **14**, 749–762.

Vreugdenhill, D., Claassens, M. M. J., Verhees, J., van der Krol, A. R, and van der Plas L. H. W. (2005) Ethanol-inducible gene expression: non-transformed plants also respond to ethanol. *Trends in Plant Science* **11**, 9–11.

Zuo, J. and Chua, N.-H. (2000) Chemical-inducible systems for regulated expression of plant genes. *Current Opinion in Biotechnology* **11**, 146–151.

Transgene integration and expression

Butaye, K. M. J., Goderis, I. J. W. M., Wouters, P. F. J., Pues, J.-T. G., Delaure, S. L., Broekaert, W. F., Depicker, A., Cammue, B. P. A., and De Bolle, M. F. C. (2004) Stable high-level transgene expression in *Arabidopsis thaliana* using gene silencing mutants and matrix attachment regions. *The Plant Journal* **39**, 440–449.

Fang, R.-X., Nagy, F., Sivasubramaniam, S., and Chua, N.-H. (1989) Multiple *cis* regulatory elements for maximal expression of the cauliflower mosaic virus 35S promoter in transgenic plants. *Plant Cell* **1**, 141–150.

Gallie, D. R. (1998) Controlling gene expression in transgenics. *Current Opinion in Plant Biology* **1**, 166–172.

Kumar, S. and Fladung, M. (2001) Controlling transgene integration in plants. *Trends in Plant Science* **6**, 155–159.

Miki, B. (2002) Transgene expression and control. *In vitro Cellular and Developmental Biology—Plant* **38**, 139–145.

Perlak, F. J., Fuchs, R. L., Dean, D. A., McPherson, S. L., and Fischoff, D. A. (1991) Modification of the coding sequence enhances plant expression of insect control protein genes. *Proceedings of the National Academy of Sciences USA* **88**, 3324–3328.

RNAi

Byzova, M., Verduyn, C., De Brouwer, D., and De Block, M. (2004) Transforming petals in sepaloid organs in *Arabidopsis* and oilseed rape: implementation of the hairpin RNA-mediated gene silencing technology in an organ-specific manner. *Planta* **218**, 379–387.

Chuang, C.-F. and Meyerowitz, E. M. (2000) Specific and heritable genetic interference by double-stranded RNA in *Arabidopsis thaliana*. *Proceedings of the National Academy of Sciences USA* **97**, 4985–4990.

Dugas, D. V. and Bartel, B. (2004) MicroRNA regulation of gene expression in plants. *Current Opinion in Plant Biology* **7**, 512–520.

Guo, H.-S., Fei, J.-F., Xie, Q., and Chua, N.-H. (2003) A chemical-regulated inducible RNAi system in plants. *The Plant Journal* **34**, 383–392.

Kusaba, M. (2004) RNA interference in crop plants. *Current Opinion in Plant Biotechnology* **15**, 139–143.

Miki, D., Itoh, R., and Shimamoto, K. (2005) RNA silencing of single and multiple members in a gene family of rice. *Plant Physiology* **138**, 1903–1913.

Clean-gene technology

Ebinuma, H. and Komamoine, A. (2001) MAT (multi-auto-transformation) vector system. The oncogenes of *Agrobacterium* as positive markers for regeneration and selection of marker free transgenic plants. *In vitro Cellular and Developmental Biology—Plant* **37**, 103–113.

Jia, H., Pang, Y., Chen, X., and Fang, R. (2006) Removal of the selectable marker gene from transgenic tobacco plants by expression of Cre recombinase from a tobacco mosaic virus vector through agroinfection. *Transgenic Research* **15**, 375–384.

McCormac, A. C., Elliott, M. C., and Chen, D. F. (1999) pBECKS2000: a novel plasmid series for the facile creation of complex binary vectors which incorporates 'clean-gene' facilities. *Molecular and General Genetics* **261**, 226–235.

McCormac, A. C., Fowler, M. R., Chen, D. F., and Elliott, M. C. (2001) Efficient co-transformation of *Nicotiana tabacum* by two independent T-DNAs, the effect of T-DNA size and implications for genetic separation. *Transgenic Research* **10**, 143–155.

Puchta, H. (2000) Removing selectable marker genes: taking the short cut. *Trends in Plant Science* **5**, 273–274.

Wang, Y., Chen, B., Hu, Y., Li, J., and Lin, Z. (2005) Inducible excision of selectable marker gene from transgenic plants by the Cre/*lox* site-specific recombination system. *Transgenic Research* **14**, 605–614.

Web-links

Best practice in vector design (Advisory Committee on Releases to the Environment (ACRE) report), web-link 4.1: www.defra.gov.uk/environment/acre/bestprac/consult/guidance/bp/index.htm

Gateway™ vectors, Mark Curtis website, web-link 4.2: www.unizh.ch/botinst/Devo_Website/curtisvector/index_2.html

Gateway™ vectors, Plant Systems Biology, University of Ghent, web-link 4.3: www.psb.rug.ac.be/gateway/

pGreen and pBIN19, web-link 4.4: www.pgreen.ac.uk/a_hom.htm

The Royal Society, web-link 4.5: www.royalsoc.ac.uk/

5

The genetic manipulation of herbicide tolerance

INTRODUCTION

The preceding four chapters have given an overview of the tools of plant biotechnology that can be used for genetic manipulation. The remaining section of the book will look in detail at the major targets for the genetic manipulation of crops and how they are being approached or have been achieved. Thus, each chapter will describe the scientific strategies that could be used to attain a particular goal for crop improvement: why certain genes might be useful, how they might be transformed into plants, and how their expression might be regulated. In addition, some of the wider issues surrounding genetically modified (GM) crops will be explored: why have certain targets received so much commercial attention, which GM strategies have actually been successful in the field, and are the concerns about the safety of GM crops justified?

- Herbicide tolerance is the predominant GM trait.

The first chapter in this sequence of different GM applications deals with the genetic manipulation of herbicide-tolerant plants. Herbicide tolerance was one of the first GM traits to be tested in the field, and subsequently for commercial production. Table 5.1 shows that the greatest number of trial sites set up during the period 1987–1997 were to test crops carrying this trait, and that herbicide-tolerant crops form more than three-quarters of the total released GM crops in terms of area of commercial planting. (Indeed, note that of all the traits described in this book, only

Table 5.1 Most frequent transgenic traits in commercial plantings

Trait	Area of commercial planting (million ha)	
	2000	2005
Herbicide tolerance	32.7 (74%)	63.7 (71%)
Insect resistance	8.3 (19%)	16.2 (18%)
Herbicide tolerance + insect resistance	3.1 (7%)	10.1 (11%)

(Data from James, 2001, 2005, web-link 5.12.)

herbicide tolerance and insect resistance (see Chapter 6) are widely grown, as yet.) One of the aims of this chapter is to explain why there has been such rapid development of this particular trait, and why crops genetically modified to be herbicide tolerant remain the most widely grown.

● There are a number of scientific reasons for the rapid development of herbicide-tolerant crops.

There are scientific and commercial reasons why herbicide-tolerant crops have been developed so rapidly compared with other GM traits. The scientific reasons include the following.

1 Considerable information was already available about the modes of action of certain herbicides, and about the affected biochemical pathways.
2 Biological resources were available to gain an understanding of the mechanisms of resistance, and provide genes for genetic manipulation, including:
   - resistant bacteria via the laboratory or from the wild;
   - tolerant plants selected in tissue culture;
   - field-selected resistant crops and weeds.
3 Single-gene mechanisms to obtain tolerance were relatively simple to devise. (This is not the case with some of the traits discussed in subsequent chapters.)

● Herbicide-tolerant crops have been developed and commercialized by agrochemical companies.

The major commercial impetus for their development flowed from the advantage to the agrochemical/seed industries of producing crops tolerant to specific herbicides, particularly those manufactured/owned by the same company. The engineering of a crop to be tolerant to a particular herbicide allows it to be treated at the optimal times for the reduction of weeds, hence greatly extending the applications and effectiveness of that herbicide. The role of the agrochemical industry in driving forward this technology (and GM crop science in general) will be returned to in this and subsequent chapters.

The use of herbicides in modern agriculture

● Weeds, pests, and diseases have a major impact on global crop yield.

Weeds have a significant effect on the yield and quality of crops, as a result of competition for light and nutrients, contamination of the harvested crop and because weed populations harbour pests and diseases. Thus, weeds are one of the three classes of biotic stress that have a major impact on the proportion of world crop yield available for human consumption. Figure 5.1 indicates that the reduction in world crops caused by weeds, pests, and diseases amounts to over one-third of the potential yield, with a roughly equal contribution made by each. Modern agriculture has developed a range of effective herbicides (weedkillers) to tackle the effects of weeds on crop yield. Some of the most useful of these are broad-spectrum herbicides because they are active against a wide range of weeds. However, these can only be used at times when the crop is not itself vulnerable to herbicide action. The development of herbicide-tolerant crops therefore offers the opportunity to spray crops at the most effective time to kill weed species without damaging the crop plants. However, it should be noted that some crop plants are naturally resistant to certain herbicides, and that tolerant strains may appear through the normal processes of mutation and natural

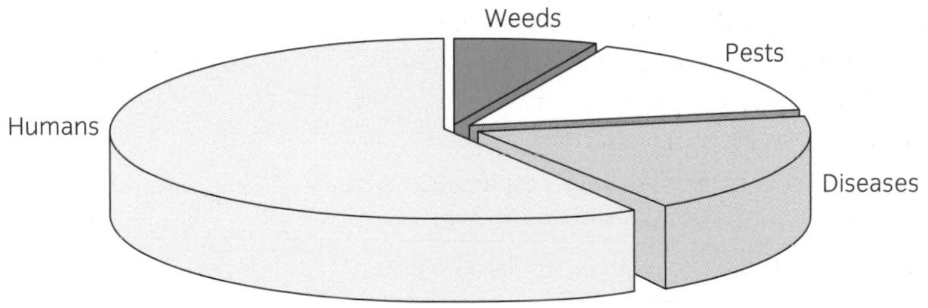

Figure 5.1 The relative proportions of total world crop production lost to weeds, pests, and diseases. The figure shows that of the total potential yield, only 63% is available for human consumption. The remainder is lost in more or less equal proportions to weeds, pests, and diseases.

selection. Thus, the concept of herbicide-tolerant crops is neither novel nor unique to GM technology.

What types of compounds are herbicides?

By definition, herbicides are much more toxic to plants than to animals. Therefore, it is not surprising that they generally affect plant-specific biological processes. Consider for a moment the type of biochemical pathways that are likely to occur in plants but not in animals. Although one's first thought is likely to be of photosynthesis, many other processes are potential targets. Remember that plants are autotrophic; that is, they can synthesize all their macromolecular components *de novo*. For example, the compounds that are essential in the human diet (vitamins, essential amino acids, etc.) are synthesized by plants, and these biosynthetic pathways are therefore possible targets for herbicides. Many of these pathways are located in the chloroplast (Box 5.1),

● Herbicides are inhibitors of plant-specific processes.

BOX 5.1 *Plastids as major biosynthetic factories of plant cells*

The plastids are major contributors to the biosynthetic activity in plant cells. The chloroplast is the sole site of the photoreduction of carbon dioxide to phosphoglyceric acid, the photo-oxidation of water, and photophosphorylation. In addition, all an animal's essential dietary amino acids are synthesized in chloroplasts. The vast majority of lipid biosynthesis occurs in this organelle, as does the synthesis of purines and pyrimidines. The plastids also play a major role in the assimilation of nitrogen, reducing nitrite (produced by reduction of nitrate in the cytosol) to ammonia, and in the assimilation of sulphur, reducing sulphate to sulphide. The chlorophyll pigments of the chloroplast are also produced *in situ*, as are the carotenoid pigments of chromoplasts. Starch biosynthesis is also an important pathway in chloroplasts and is the predominant function of amyloplasts.

The chloroplast is also responsible for a proportion of plant cell protein synthesis. Note that the chloroplast genome encodes all the components required for the plastid translational machinery, as well as for a number of the chloroplast proteins (see Chapter 1).

● Many plant-specific pathways are located in the plastids.

Table 5.2 Classification of herbicides according to their mode of action

HRAC group	Mode of action	Chemical family	Example
A	Inhibition of ACCase	Aryloxyphenoxy-propionates ('FOPs'), cyclohexanediones ('DIMs')	
B	Inhibition of acetolactate synthase (ALS)	Sulphonylureas, imidazolinones, triazolopyrimidines, pyrimidinyl(thio)benzoate, sulphonylaminocarbonyl-triazolinones	Chlorsulphuron, imazapyr
C1	Inhibition of photosynthesis at photosystem II	Triazines, triazinones, triazolinone, uracils, pyridazinones, phenylcarbamates,	Atrazine
C2	Inhibition of photosynthesis at photosystem II	Ureas, amides	
C3	Inhibition of photosynthesis at photosystem II	Nitriles, benzothiadiazinone, phenylpyridazines	Bromoxynil
D	Photosystem I electron diversion	Bipyridyliums	Paraquat
E	Inhibition of protoporphyrinogen oxidase (PPO)	Diphenylethers, phenylpyrazoles, *N*-phenylphthalimides, thiadiazoles, oxadiazoles, triazolinones, oxazolidinediones, pyrimidindiones, others	
F	Bleaching: inhibition of carotenoid biosynthesis at the phytoene desaturase step	Pyridazinones, pyridinecarboxamides, others	
F2	Bleaching: inhibition of 4-hydroxyphenylpyruvate dioxygenase (4-HPPD)	Triketones, isoxazoles, pyrazoles, others	
F3	Bleaching: inhibition of carotenoid biosynthesis (unknown target)	Triazoles, isoxazolidinones, ureas, diphenylethers	
G	Inhibition of EPSPS	Glycines	Glyphosate
H	Inhibition of glutamine synthetase	Phosphinic acids	Glufosinate ammonium, bialaphos
I	Inhibition of dihydropteroate synthase	Carbamates	Asulam
K1	Microtubule assembly inhibition	Dinitroanilines, phosphoroamidates, pyridines, benzamides, benzenedicarboxylic acids	Oryzalin
K2	Inhibition of mitosis/microtubule organization	Carbamates	
K3	Inhibition of cell division	Chloroacetamides, acetamides, oxyacetamides, tetrazolinones, others	
L	Inhibition of cell-wall (cellulose) synthesis	Nitriles, benzamides, triazolocarboxamides	

Table 5.2 Continued

HRAC group	Mode of action	Chemical family	Example
M	Uncoupling (membrane disruption)	Dinitrophenols	
N	Inhibition of lipid synthesis—not ACCase inhibition	Thiocarbamates, phosphorodithioates, benzofuranes, chlorocarbonic acids	Trichloroacetic acid
O	Action like indole-3-acetic acid (synthetic auxins)	Phenoxycarboxylic acids, benzoic acids, pyridine carboxylic acids, quinoline carboxylic acids, others	2,4-D, Dicamba, Picloram
P	Inhibition of auxin transport	Phthalamates, semicarbazones	
Z	Unknown	Arylaminopropionic acids, pyrazolium, organoarsenicals	

ACCase, acetyl-CoA carboxylase; 2,4-D, 2,4-dichlorophenoxyacetic acid; EPSPS, 5-enolpyruvylshikimate-3-phosphate synthase; HRAC, Herbicide Resistance Action Committee.
(Data from the Herbicide Resistance Action Committee and the Weed Science Society of America.)

and this has implications for the targeting of many of the transgenic proteins designed to enhance herbicide tolerance.

The herbicidal activity of many herbicides has been found to result from the specific inhibition of a single enzyme/protein. Table 5.2 shows a classification of the major chemical groups of herbicides according to their modes of action. It can be seen that herbicides belong to a wide range of different chemical families, with about 15 broad classes of mode of activity. It is also worth noting that most herbicides only have one mode of action, and that certain enzymes seem to be relatively vulnerable to herbicide activity. For example, five chemical families of herbicide target the enzyme acetolactate synthase (ALS; or more correctly, acetohydroxy-acid synthase or AHAS). ALS catalyses the first reaction in the biosynthetic pathways of branched-chain amino acids (see Figure 5.6, below), and Table 5.3 shows two common classes of herbicide with widely differing chemical structures (sulphonylureas and imidazolinones) that target this enzyme. As will be seen, knowledge of the herbicide's site of action is essential for some strategies aimed at engineering tolerance, but less important for others.

The wide range of chemical families shown in Table 5.2 means that herbicides differ greatly in other properties besides their mode of action. For example, they can also be classified according to their:

- site of uptake into the plant (root or shoot);
- degree of translocation within the plant (systemic or contact);
- time of application (preplanting, pre-emergence, postemergence, or preharvesting).

The widely varied chemical properties of these compounds also means they will differ greatly with regard to toxicity, environmental persistence, and biodegradability. The environmental impact of herbicide-tolerant crops will be considered at the end

● Herbicides differ considerably in structure, properties and mode of action.

Table 5.3 The structures of common herbicides affecting biochemical targets

Class of herbicide	Compound/ herbicide	Chemical formula	Inhibited pathway	Target protein
Glycine	Glyphosate (Roundup)	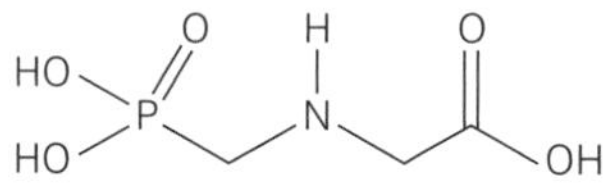	Aromatic amino acids	EPSPS
Phosphinic acid	Phosphinothricin (Basta)	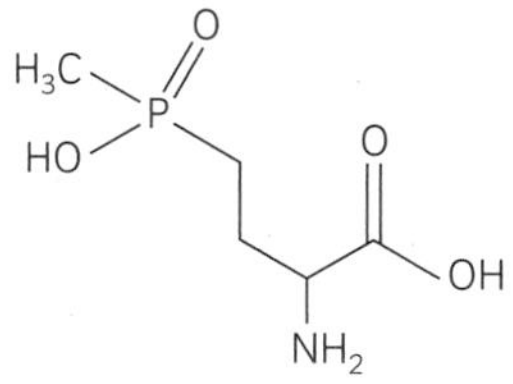	Nitrogen assimilation	Glutamine synthase
Sulphonylurea	Chlorsulphuron (Glean)	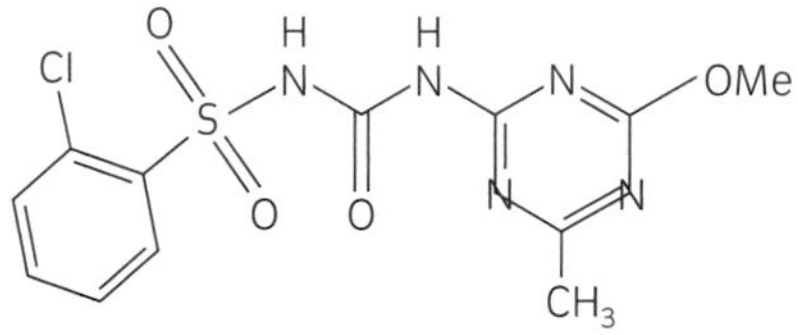	Branched-chain amino acids	Acetolactate synthase
Imidazolinone	Imazathapyr	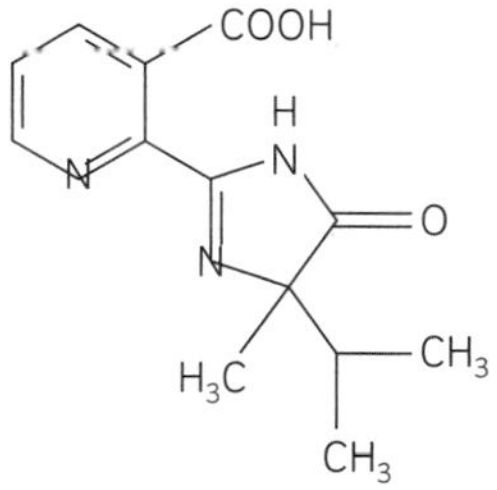	Branched-chain amino acids	Acetolactate synthase
S-Triazine	Atrazine	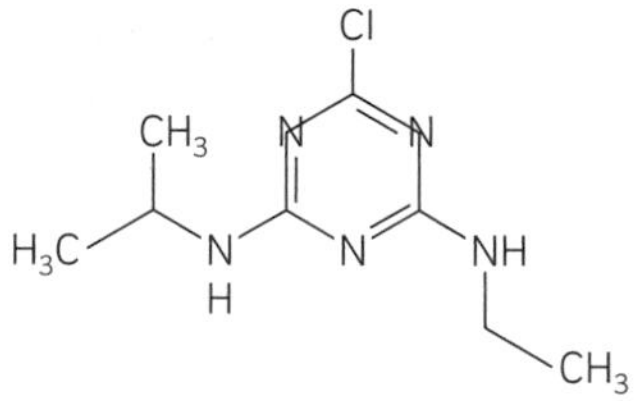	Photosynthesis	33 kDa protein

EPSPS, 5-enolpyruvylshikimate-3-phosphate synthase.

of this chapter. However, this is a useful place to make the point that an assessment of the risks and benefits of growing a particular herbicide-tolerant crop must take account of the properties of that particular herbicide.

Table 5.3 shows the chemical structure of a few common herbicides that affect well-characterized biochemical targets. This again emphasizes the point that herbicides are a heterogeneous group of compounds with differing modes of action. In consequence, most transgenic strategies for tolerance to a particular herbicide have to be designed specifically for that class of herbicide. Rather than describe in detail all the herbicides against which tolerance has been engineered, this chapter will adopt a case-study approach to highlight general approaches and strategies.

Strategies for engineering herbicide tolerance

The engineering of herbicide tolerance demonstrates the way in which quite different strategies can be used to achieve the same objective. Mullineaux (1992) identifies four distinct strategies for engineering herbicide tolerance, detailed below.

- There are four distinct strategies for engineering herbicide tolerance.

1 *Overexpression of the target protein.* This strategy effectively involves titrating the herbicide out by overproduction of the target protein. For example, if the herbicide is a specific inhibitor of one particular enzyme, production of sufficient excess enzyme will partially overcome the inhibition. Overexpression can be achieved by the integration of multiple copies of the gene and/or the use of a strong promoter plus translational enhancer to drive expression of the gene.

2 *Mutation of the target protein.* The logic behind this approach is to find a modified target protein that substitutes functionally for the native protein and which is resistant to inhibition by the herbicide, and to incorporate the resistant target protein gene into the plant genome. Several sources of resistant proteins can be exploited. Note that both the overexpression and mutated target protein strategies require knowledge of the mode of action of the herbicide.

3 *Detoxification of the herbicide, using a single gene from a foreign source.* Detoxification is a means of converting the herbicide to a less toxic form and/or removing it from the system. This strategy can be contrasted with the previous two because it does not require a detailed knowledge of the site of action. Table 5.4 shows several examples of specific detoxification reactions for common herbicides.

4 *Enhanced plant detoxification.* The aim here is to improve the natural plant defences against toxic compounds. This requires detailed information about endogenous plant detoxification pathways and the mechanisms by which compounds are recognized and targeted for detoxification by the plant.

CASE STUDY 5.1 **Glyphosate tolerance**

Glyphosate is a broad-spectrum herbicide that is reputedly effective against 76 of the world's worst 78 weeds, and is marketed as Roundup by the North American chemical company Monsanto. It is a simple glycine derivative (see Table 5.2) that acts as a competitive inhibitor of the enzyme 5-enolpyruvylshikimate-3-phosphate synthase (EPSPS). Figure 5.2 shows the similarity in structure between glyphosate and one of the substrates of the enzyme, phosphoenolpyruvate (PEP). Glyphosate binds more tightly to the EPSPS–shikimate 3-phosphate complex than does PEP: its dissociation rate from the complex is 2300 times slower than PEP. Consequently, EPSPS is effectively inactivated once glyphosate binds to the enzyme–substrate complex. EPSPS is a key enzyme in the biosynthetic pathways of the aromatic amino acids phenylalanine, tyrosine, and

- Glyphosate is a competitive inhibitor of the enzyme EPSPS.

Table 5.4 Herbicide detoxification reactions

Compound/herbicide		Enzyme	Metabolite	Source organism
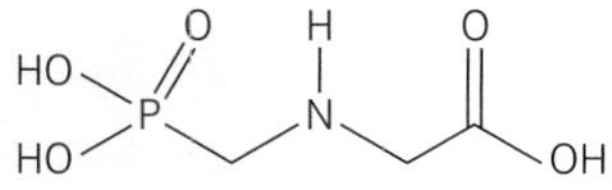	Glyphosate (Roundup)	Oxidoreductase	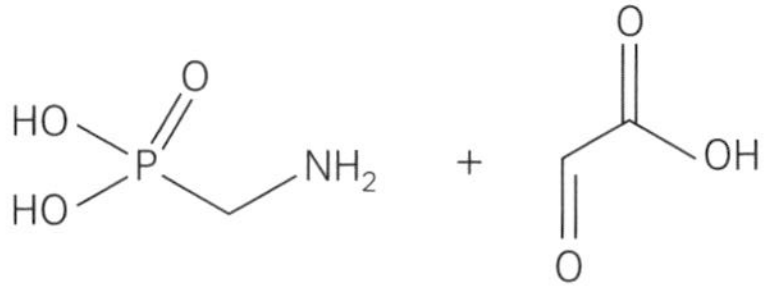	*Ochrabactrum anthropi*
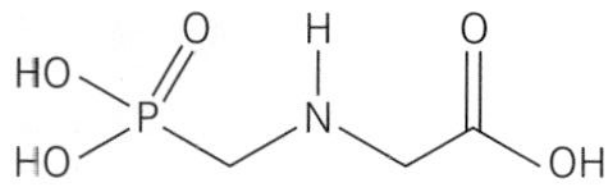	Glyphosate (Roundup)	N-Acetyltransferase	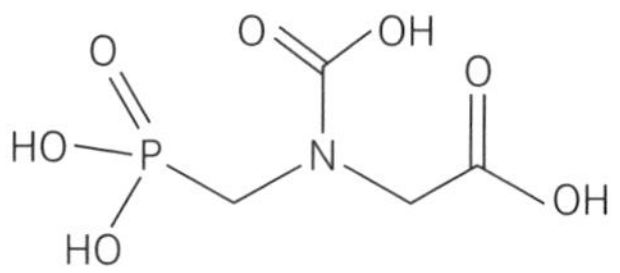	*Bacillus licheniformis*
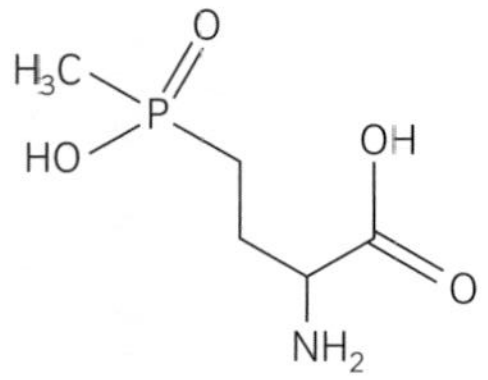	Phosphinothricin (Basta)	Acetyltransferase	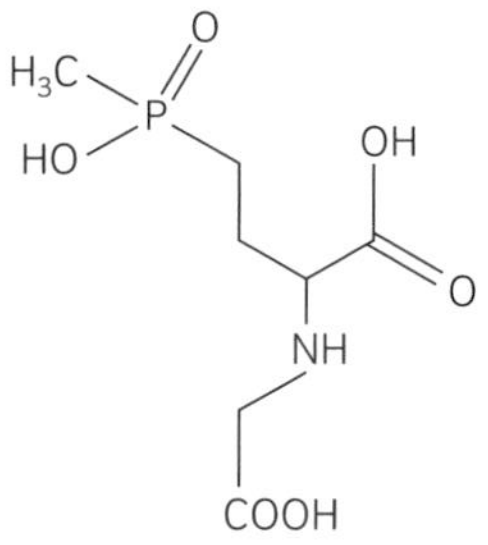	*Streptomyces hygroscopicus*

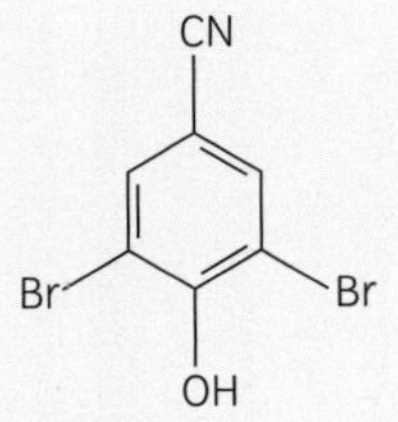 	Bromoxynil (Buctril)	Nitrilase	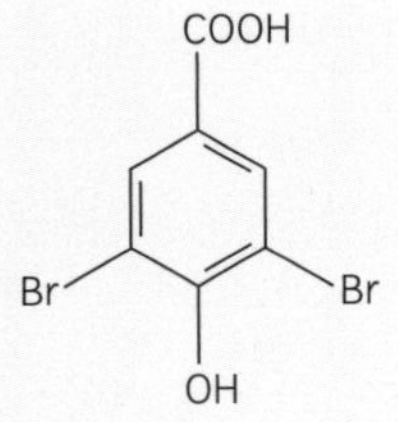 	*Klebsiella ozaenae*
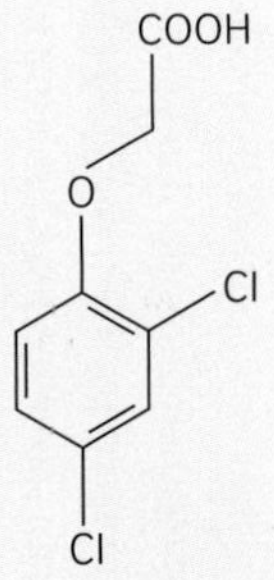 	2,4-D	Monooxygenase	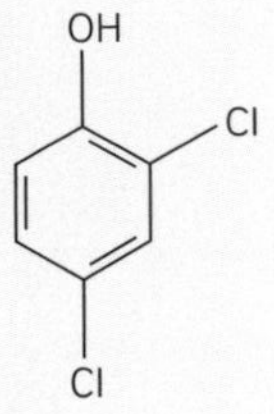 	*Alcaligenes eutrophus*

2,4-D, 2,4-dichlorophenoxyacetic acid.

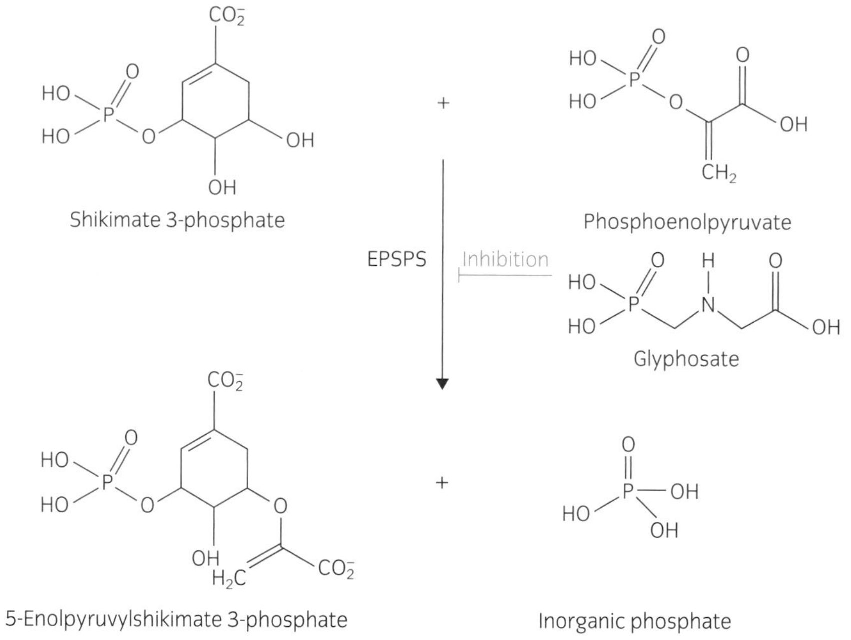

Figure 5.2 Glyphosate and the reaction catalysed by EPSPS. The biochemical reaction catalysed by EPSPS involves the addition of phosphoenolpyruvate to shikimate 3-phosphate. The structural similarity between glyphosate and phosphoenolpyruvate explains the competitive inhibition of EPSPS by the herbicide.

tryptophan (Box 5.2). Thus, the herbicidal activity of glyphosate results from its inhibition of the biosynthesis of aromatic amino acids and other products of the shikimate pathway.

From this knowledge of the mode of action of glyphosate, is it possible to predict what effects the herbicide is likely to have on the growth of a treated plant? The first deduction is that protein synthesis will be blocked due to the insufficient supply of aromatic amino acids. The most immediate effects of this inhibition would be expected in regions of the plant involved in rapid growth and division, including the meristems. Certain specialized organs, such as the developing

BOX 5.2 *Aromatic amino acid biosynthesis pathways*

The biosynthesis of the aromatic amino acids and related compounds shares a common biochemical pathway up to the formation of chorismate. The pathway to chorismate starts with the condensation of phosphoenolpyruvate (PEP) (from glycolysis) and erythrose 4-phosphate (from the pentose phosphate pathway) to form 3-deoxy-D-arabino-heptulosonate 7-phosphate (DAHP). The next step involves a complex redox/cyclization reaction producing 3-dehydroquinate. The next two reactions (to form 3-dehydro-shikimate and then shikimate) are catalysed by a bifunctional enzyme in plants. Shikimate kinase then phosphorylates shikimate to produce shikimate 3-phosphate, one of the substrates of 5-enolpyruvylshikimate-3-phosphate synthase (EPSPS). EPSPS adds the enolpyruvyl side chain to shikimate 3-phosphate to form 5-enolpyruvylshikimate 3-phosphate (EPSP). Finally, chorismate is formed by the elimination of phosphate from EPSP.

BOX 5.2 *Continued*

Chorismate is the precursor of the phenolic and indole rings of the aromatic amino acids, and of other aromatic compounds. Chorismate mutase is the committing enzyme for phenylalanine and tyrosine synthesis, forming prephenate. Prephenate aminotransferase utilizes glutamate or aspartate as amino donors to form arogenate, the immediate precursor of phenylalanine (via arogenate dehydratase) and tyrosine (via arogenate dehydrogenase). These amino acids are themselves the precursors of a wide range of secondary products, including lignins, flavonoids, hydroxycinnamic acids, and alkaloids.

Alternatively, chorismate may be converted to anthranilate by anthranilate synthase, an enzyme complex with two catalytic subunits: the α-subunit catalyses the amination of chorismate and the removal of the enolpyruvyl side chain, while the β-subunit has glutamine aminotransferase activity. Four subsequent steps are involved in the biosynthesis of tryptophan. Tryptophan itself is the precursor of several important secondary products, including IAA, indole alkaloids, indole glucosinolates, phytoalexins, and acridone alkaloids.

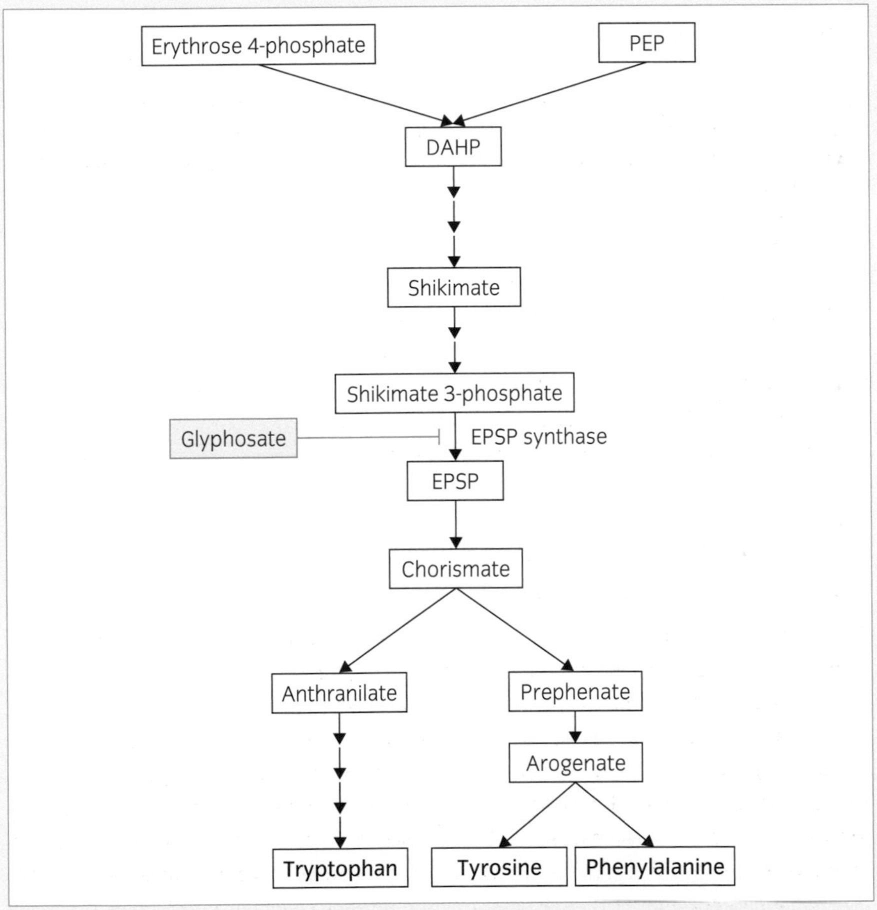

Biosynthetic pathways of the aromatic amino acids tryptophan, tyrosine, and phenylalanine.

endosperm, also vigorously accumulate proteins. In addition, other pathways will be affected by the depletion of aromatic amino acids. The shikimate pathway supplies aromatic precursors for a range of phenolic compounds, including lignins, alkaloids, and flavonoids (Box 5.2). Indeed, 20% of the carbon fixed by plants flows through this pathway, primarily for lignin biosynthesis. Indole compounds other than tryptophan are produced by the same pathway, so biosynthesis of auxin (indole-3-acetic acid, IAA) will also be affected by this herbicide. It is therefore not surprising that glyphosate has such a profound effect on plants.

Strategy 1 for glyphosate tolerance: overexpression of a plant EPSPS gene

● Three different strategies for the production of glyphosate-tolerant crops have been tested.

Of the four strategies for engineering herbicide tolerance described above, three have actually been tested in the laboratory, two of which form the basis of the current commercial plantings of glyphosate-tolerant crops. It is therefore instructive to compare the three strategies. One of the earliest approaches to engineering glyphosate tolerance involved the overexpression of a plant EPSPS gene. This was facilitated by the isolation of petunia complementary DNA (cDNA) from glyphosate-tolerant tissue cultures. The stepwise selection of petunia cells capable of growing in the presence of increasing amounts of glyphosate led to the isolation of cultures in which the levels of EPSPS enzyme were much higher than normal. This was found not to be due to increased expression of the EPSPS gene, but rather the result of gene amplification, such that there were multiple (up to 20) copies of the EPSPS gene in an otherwise normal petunia genome (see Box 5.3). The EPSPS enzyme was not itself mutated: the tolerance was simply due to the increased amount of enzyme. However, the high levels of EPSPS mRNA made it simpler to isolate the cDNA for this gene and use it for re-introduction into plants.

The observation that the endogenous mechanism of resistance that had been selected for in the petunia cells was one of excess normal EPSPS (as a result of gene amplification) indicates that Strategy 1—overexpression of the target protein—should be feasible. The effect of overexpressing the EPSPS gene in the transgenic petunia was tested in one of the earliest

BOX 5.3 *Gene amplification*

In prokaryotes, the isolation of a strain tolerant to a particular toxic compound would usually result from the selective advantage derived from mutations in the gene coding for the affected protein itself, or in related genes involved in transport, detoxification, or gene regulation. In contrast, the selection of eukaryotic cells in culture resistant to a particular toxic compound often results from the amplification of the gene encoding the target enzyme, rather than from mutations in that gene. This process is particularly observed during a stepwise selection procedure, in which the level of toxin is increased gradually in each round of selection. One of the best-characterized examples is that of methotrexate resistance in cultured mammalian cells. The anticancer drug methotrexate is an inhibitor of the enzyme dihydrofolate reductase (DHFR), which supplies single carbon units for, among other things, thymidine synthesis. It is therefore particularly important in cycling cells for the replication of DNA. A stepwise selection of methotrexate-resistant cells led to the predominance of cells in which the DHFR gene had been specifically amplified in tandem 40–400-fold. This indicates that gene duplication occurs more frequently than mutation in eukaryotes. The stepwise selection applies a quantitative selection pressure for repeated gene duplications, resulting in a process of accelerated evolution.

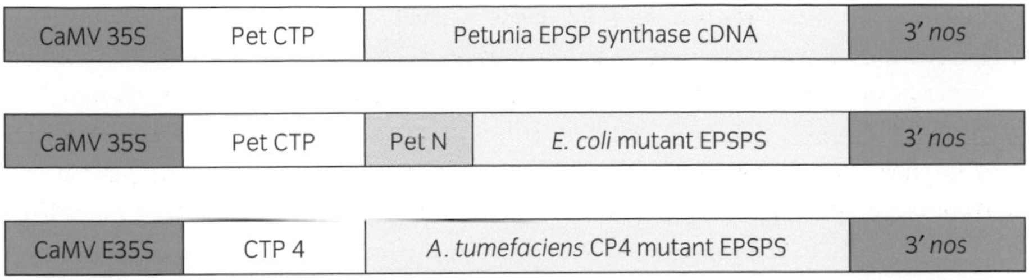

Figure 5.3 Constructs for engineering glyphosate tolerance. Maps of some of the gene constructs that have been used to engineer glyphosate tolerance are shown. The upper two constructs are early experimental designs that were used to demonstrate the feasibility of engineering herbicide tolerance. The first construct directed overexpression of a wild-type petunia (Pet) EPSPS gene, whereas the second shows a resistant *Escherichia coli* gene fused to the N-terminal region of the petunia gene, including the chloroplast transit peptide (CTP). The third construct is one currently used to generate Roundup Ready crops such as soybean and cotton using a resistant EPSPS gene from *A. tumefaciens* with an enhanced CaMV 35S promoter (E35S).

experiments on the engineering of herbicide tolerance. The EPSPS cDNA was fused to the cauliflower mosaic virus 35S promoter and a *nos* terminator sequence in the vector pMON 546 (Figure 5.3) and transformed into petunia using *Agrobacterium*. The use of the plant gene enabled the researchers to avoid one of the major obstacles to transgene expression: protein targeting. As noted above, many of the potential target pathways of herbicides are located in the plastids. The targeting of proteins to the plastid is discussed in Box 5.4. Note that no further

BOX 5.4 *Chloroplast transit peptides and protein targeting*

Some chloroplast proteins are encoded by the plastid genome and translated on 70S ribosomes in the organelle. However, most chloroplast proteins are encoded in the nuclear genome and translated on 80S ribosomes in the cytoplasm. There must therefore be mechanisms for the recognition and transport of proteins destined for the chloroplast. Note that the chloroplast envelope comprises an outer and inner membrane, and that the thylakoids comprise an additional internal membrane system. Thus, there are three distinct compartments in the chloroplast: the intermembrane space of the envelope, the stroma, and the thylakoid lumen. Depending upon the precise destination, a chloroplast protein might be transported across three membrane systems. The means by which all plastid proteins are transported into the chloroplast is by recognition of a sequence of about 40–50 amino acids at the N-terminal end of the protein (the transit peptide). This peptide directs translocation into the stroma, where a specific peptidase removes the transit peptide. Note that the process is post-translational; that is, the protein is transported after synthesis, unlike the co-translational transfer of membrane-bound and secreted proteins into the endoplasmic reticulum during synthesis on rough endoplasmic reticulum-bound ribosomes. The transport process into the stroma appears to involve a complex protein-import apparatus that spans the inner and outer membranes.

Targeting into the thylakoids requires a bipartite transit peptide. Removal of the stromal targeting sequence exposes a second transit peptide that acts as a lumenal targeting peptide.

(Continued overleaf)

BOX 5.4 *Continued*

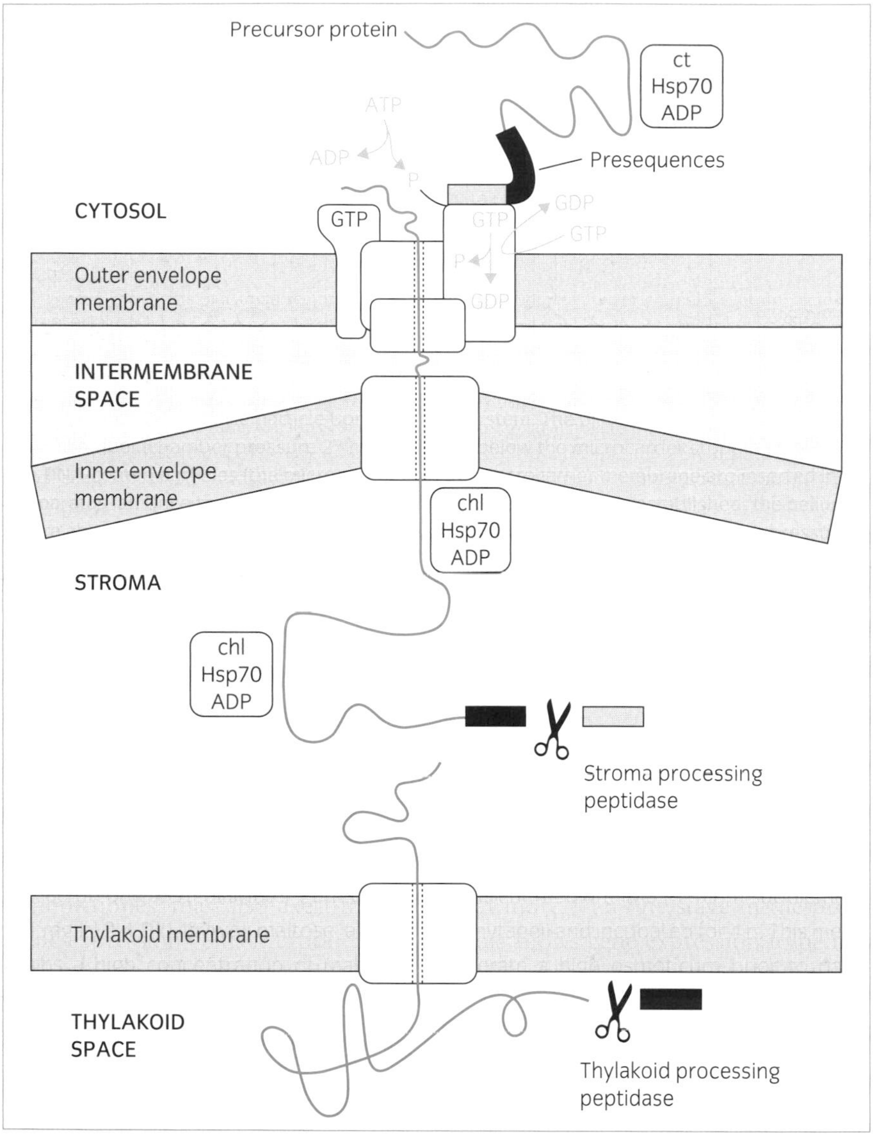

Protein transport into the chloroplast. chl, chloroplastic; ct, cytoplastic; Hsp70, 70 kDa heat-shock protein. (Adapted with permission from Heldt (1997).)

This directs the protein across the thylakoid membrane into the thylakoid lumen, where it is also removed by a specific protease.

Transit peptides will direct the transfer of a chimaeric non-plastid protein into the chloroplast. Hybrid transgenes containing an N-terminal transit peptide can therefore be constructed so as to target the protein to a specific chloroplast compartment.

manipulation of the cDNA was required to target this protein to the plastid site of activity, since the plant EPSPS cDNA sequence contains its own transit peptide. The result of this experiment was a 40-fold increase in EPSPS activity in the transgenic plants and a tolerance to glyphosate sprayed in the field at a dose two to four times higher than that required to kill wild-type plants.

Strategy 2 for glyphosate tolerance: mutant EPSPS genes

Mutated EPSPS genes have been isolated from a number of glyphosate-resistant bacteria. It is instructive to compare two of the early experiments using glyphosate-resistant genes from bacteria. In one, a mutated *aroA* gene from *Salmonella typhimurium* was inserted between the promoter and terminator sequences of the *ocs* gene of the *Agrobacterium tumefaciens* Ti plasmid. Only a moderate increase in herbicide tolerance was obtained. With reference to Box 5.4, you should be able to explain why tolerance was limited: the prokaryote gene did not have a plastid transit peptide sequence, so the resistant enzyme was not transported into the chloroplast.

The requirement for a functional plastid transit peptide was demonstrated by the construction of a hybrid EPSPS gene by fusion of the C-terminal end of a mutated *aroA* gene from *E. coli* to the N-terminal end of the petunia EPSPS cDNA sequence containing the transit peptide sequence (Figure 5.3). Expression of the chimaeric enzyme increased glyphosate tolerance in transgenic tobacco from 0.01 to 1.2 mmol l^{-1} glyphosate. This experiment was an important validation of the strategy, and demonstrated the feasibility of expressing prokaryotic genes incorporated into the plant nuclear genome, given appropriate promoter and termination signals. However, note that the expression of prokaryotic genes is often not optimal in transgenic plants and extensive modifications may be required to obtain high levels of expression (see Chapter 4).

These early experiments provided useful evidence for the feasibility of Strategy 2 (using resistant target proteins), but also revealed problems associated with it. One is that a mutant enzyme with reduced affinity for a competitive inhibitor may also have a lower affinity (increased K_M, the Michaelis–Menten constant) for the substrate. This proved to be the case with the *E. coli* and *S. typhimurium* genes. For this reason, glyphosate-resistant genes from other sources have been tested for their effectiveness in different plants. A gene from the herbicide-resistant *A. tumefaciens* strain CP4 encodes an EPSPS that is resistant to glyphosate, but retains a low K_M for PEP. This gene, in conjunction with an enhanced CaMV 35S promoter (see Chapter 4, Figure 4.2) and a chloroplast transit peptide sequence from *Arabidopsis* or petunia (Figure 5.3), is incorporated into the current range of Monsanto's major dicotyledonous Roundup Ready crops (soybean, cotton, and oilseed rape). On the other hand, Roundup Ready maize contains a construct optimized for monocotyledonous crops, with a resistant EPSPS gene from maize (isolated after mutagenesis and selection in tissue culture), fused to a rice promoter and maize chloroplast transit peptide sequence.

Strategy 3 for glyphosate tolerance: detoxification by heterologous genes

Alternative strategies have been developed for engineering glyphosate tolerance based upon a specific detoxification mechanism. In soil microorganisms, glyphosate can be degraded by cleavage of the C–N bond, catalysed by an oxidoreductase, to form aminomethylphosphonic acid (AMPA) and glyoxylate (Table 5.4). A gene encoding the enzyme glyphosate oxidase (GOX) has been isolated from a soil organism, *Ochrobactrum anthropi* strain LBAA, and modified by addition of a transit peptide. Transgenic crops such as oilseed rape transformed with this gene show very good glyphosate tolerance in the field. However, this strategy is not generally used in isolation. Monsanto now employ a dual strategy for canola (a variety of oilseed rape grown in the USA), in which both the resistant *Agrobacterium* CP4 EPSPS gene and the GOX gene are expressed. In addition to enhanced glyphosate tolerance, this approach avoids the accumulation of the herbicide in the tolerant plant, because the glyphosate is broken down into relatively harmless products (glyoxylate is a normal plant metabolite and AMPA can be converted to glycine). This highlights one significant difference between Strategy 2 and Strategy 3. Strategy 2 enables the plant to function in the presence of the herbicide and therefore the herbicide may

BOX 5.5 *Directed evolution of glyphosate acetyltransferase*

Glyphosate is detoxified by acetylation because *N*-acetylglyphosate is not an effective inhibitor of EPSPS and is consequently not herbicidal. The research team associated with Pioneer Hi-Bred (a subsidiary of DuPont) tested a variety of microbial N-acetytransferases for their ability to modify glyphosate, but none was effective. They therefore screened a microbial collection for glyphosate-acetylating activity, by incubating permeabilized cells with acetyl-CoA and glyphosate, and conducting a sensitive assay for *N*-acetylglyphosate. *Bacillus licheniformis* showed the most promising activity, and a glyphosate acetyltransferase (GAT) gene was isolated from two different strains by shotgun cloning of genomic fragments into *E. coli* and assaying for GAT activity. The enzyme belongs to the GNAT superfamily of N-acetyltransferases, and a third gene was subsequently isolated from a different strain of *B. licheniformis* by sequence homology. The enzymes encoded by the three genes were characterized in terms of k_{cat} (the rate constant of the catalytic reaction) and the K_M (a measure of enzyme affinity for substrate). The enzymes had a relatively low k_{cat} (1.0–1.7 min^{-1}) and showed high affinity (i.e. a low K_M) for acetyl-CoA (K_M = 1.0–1.2 μM) but a low affinity for glyphosate (K_M = 1.2–1.6 mM). Given this low efficiency, none of the three genes was capable of conferring significant glyphosate resistance in transgenic plants.

To improve the efficiency of enzyme activity (as measured by a high k_{cat}, a low K_M for glyphosate, or an overall high k_{cat}/K_M ratio), a process of molecular evolution or directed evolution was employed. This involved gene shuffling by random fragmentation and reassembly of fragments derived from the three genes. These were cloned into *E. coli* to create libraries of shuffled gene variants, and the permeabilized cells were screened for GAT activity. Enzymes from the most promising clones were assayed, and shuffled genes from the best 3–12 variants were chosen for the next round of DNA shuffling. Even after the third iteration of shuffling, several enzymes showed approximately 1000-fold improvement in k_{cat}/K_M. At this stage it was possible to start selecting the shuffled libraries on glyphosate, but further improvement of enzyme efficiency was restricted to high k_{cat} activity, with K_M reaching a plateau. Therefore, a synthetic library of sequences was created using the best GAT variant from the fourth iteration as a template, and incorporating sequence diversity from related sequences found in *Bacillus subtilis* and *Bacillus cereus*. At the eighth iteration, further sequence diversity was incorporated based on *Listeria inocua* and *Zymomonas mobilis* hypothetical protein sequences. After 11 iterations, the most efficient GAT variant had a high k_{cat} of 416 min^{-1} and a K_M of 0.05 M glyphosate, giving a k_{cat}/K_M of 8320 min^{-1}, representing an approximately 10 000-fold improvement over the original enzymes! Improved *gat* genes have been introduced into *Arabidopsis*, tobacco, and maize, and are currently undergoing trials to evaluate their glyphosate tolerance in the field.

accumulate to higher levels than those normally found in that crop. In contrast, the detoxification strategy should result either in the destruction of the herbicide, or in the accumulation of a conjugate less harmful than the original compound.

More recently, a glyphosate acetylation detoxification system has been developed by molecular evolution (Box 5.5).

Pleiotropic effects of transgenes

It should be noted at this stage that the insertion of a transgene into a plant may result in unforeseen and perhaps undesirable effects. Roundup Ready crops have not been without problems of this type. One phenomenon encountered by Roundup Ready soybeans during hot weather has been splitting of the stems. It has been suggested that this occurs due to the 20% higher lignin content of these plants. A look back to the mode of action of glyphosate should indicate why the introduction of an additional EPSPS gene could result in increased lignin biosynthesis (see Box 5.2). It should be remembered that the plant EPSPS enzyme will still be functional under normal growth conditions in the absence of the herbicide, and that the expression of additional enzyme from the transgene may affect the balance of the relevant metabolic pathways. This lesson will be returned to in subsequent chapters, particularly Chapters 10 and 11 dealing with the manipulation of plant metabolic pathways.

● Transgenes may have unforeseen effects.

CASE STUDY 5.2 **Phosphinothricin**

Glyphosate tolerance is one of the most widespread commercial GM traits. The closest rival to glyphosate in terms of the number and acreage of tolerant crops is the herbicide phosphinothricin (PPT) or glufosinate (Table 5.3). Although both are broad-spectrum herbicides, glyphosate is particularly effective against grasses, and PPT is more effective against broadleafed weeds and least effective against perennials and volunteer cereals. (In this chapter, PPT will be used in preference to glufosinate to avoid confusion between glufosinate and glyphosate. Note also that glufosinate is usually applied as the ammonium salt, and is commonly called glufosinate ammonium.) PPT is unusual among herbicides in being derived from a natural product. Bialaphos is a tripeptide of the form PPT–Ala–Ala, produced by certain *Streptomyces* species. It can be applied directly as a herbicide and has been marketed as such under various trade names, for example Herbiace (Meiji Seika). Bialaphos is converted to the active form L-PPT by proteolytic removal of the alanine residues. PPT was marketed by the German company Hoechst, under the trade name Basta. One point to be aware of in tracking the progress of a particular transgenic crop is the dynamic nature of the agrochemical and biotechnology sector: there have been a number of take-overs and mergers resulting in company name changes. Thus, Hoechst has since undergone a series of mergers such that the Basta brand name has been owned successively by AgrEvo, Aventis, and now Bayer CropScience. A different formulation of PPT is also marketed by Bayer under the brand name Liberty, and complements the LibertyLink lines of transgenic PPT-tolerant crops, which are described below.

● Phosphinothricin is a naturally occurring herbicide.

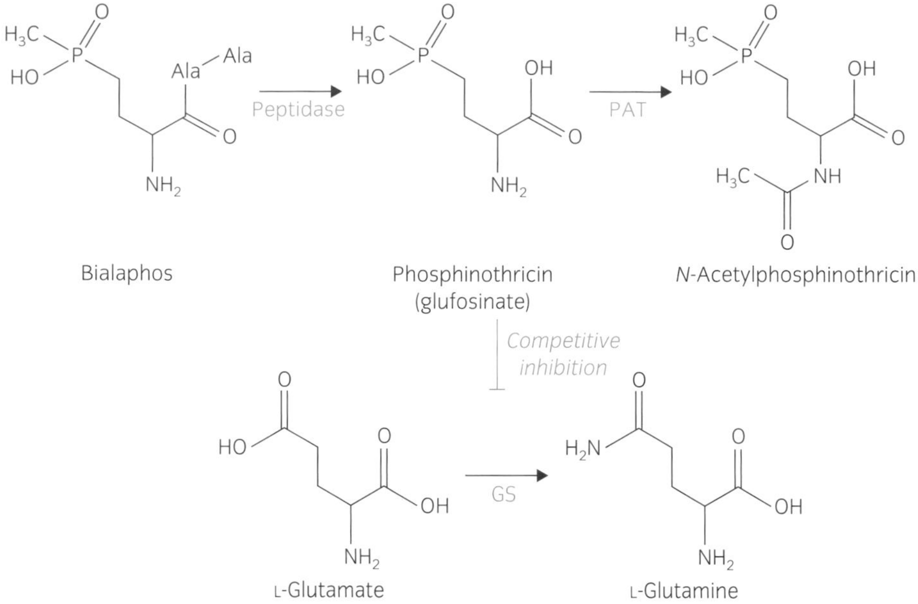

Figure 5.4 The formation, mode of action, and detoxification of PPT. The conversion of bialaphos to PPT involves removal of the two alanine residues by a peptidase. The compound acts as a competitive inhibitor of glutamine synthase (GS), and the figure highlights the structural similarity between PPT and the substrate L-glutamate. The detoxification reaction catalysed by phosphinothricin acetyltransferase (PAT) is also shown.

The herbicidal action of PPT is a result of its competitive inhibition of glutamine synthetase. Figure 5.4 shows the similarity in size and charge between PPT and glutamate, which is one of the substrates of glutamine synthetase. The immediate effect of inhibiting glutamine synthetase is the accumulation of ammonia to toxic levels, which rapidly kills the plant cells. The disruption of glutamine synthesis also inhibits photosynthesis, and it is the combined effects of ammonium toxicity and inhibition of photosynthesis that account for the herbicidal activity of PPT. Uptake of PPT is through the leaf, the speed of which is dependent on many factors, including plant species, stage of growth, air humidity, temperature, and rate of application. Translocation within the plant is limited (unlike glyphosate), and varies according to species. Some limited systemic activity may occur as a result of movement around the leaf, from leaf to leaf, and from leaf to roots. This may be sufficient to suppress the regrowth of perennial weeds that are not killed outright by contact activity. However, it will often not provide the *roots-and-all* kill seen by glyphosate for many perennial grass weeds.

Strategy for PPT tolerance

● Basta tolerance is engineered using a detoxification strategy.

The natural occurrence of bialaphos provides a lead to follow in devising a strategy for engineering tolerance. Toxic compounds such as PPT, with a simple structural homology to a common substrate such as glutamate, are likely to be toxic to the host organism. The fact that the compound is synthesized as an inactive precursor is indicative of this fact. It is therefore not surprising that *Streptomyces* species also contain a detoxification gene that protects the organism from the toxic effects of PPT. The *bar* gene of *Streptomyces hygroscopicus* and the closely related *pat* gene of *Streptomyces viridochromogenes* code for the enzyme phosphinothricin acetyltransferase. The addition of an acetyl group to the amino group of PPT inactivates the compound (Figure 5.4). Thus, transferring this gene to a plant should, in theory, provide resistance against PPT. This approach to the engineering of PPT tolerance was developed by Plant Genetic Systems under contract from Hoechst (Plant Genetic Systems was subsequently acquired by AgrEvo). The *bar* gene has now been integrated into many different plants, usually under the control of the 35S promoter. The current major LibertyLink crop lines are oilseed rape and maize.

The *bar* gene has also proved to be useful as a selectable marker for the transformation and regeneration of transgenic plants. It provides an alternative to selection with antibiotics such as kanamycin, to which different species have a highly varied response (see Chapters 3 and 4). Transgenic plants can be selected directly on PPT medium, but caution is required. Inhibition of glutamine synthetase by PPT causes ammonia accumulation in non-transgenic material and hence death of the plant tissue, but accumulation of ammonia in the non-transformed tissue can also cause problems of toxicity to neighbouring transformed cells.

Prospects for plant detoxification systems

- Plant detoxification mechanisms involve hydroxylation, conjugation, and transport processes.

One other strategy that has yet to be fully exploited is the possibility of enhancing endogenous plant detoxification mechanisms. Many xenobiotic (foreign) compounds are detoxified in plants but the pathways may involve more than one step, such as hydroxylation, conjugation, and transport stages, so it may prove difficult to identify single-gene mechanisms to engineer tolerance. The hydroxylation of compounds involves enzymes such as the cytochrome P450 monooxygenases, which form a large gene family. For example, the analysis of weeds resistant to the herbicide bromoxynil (Table 5.4) revealed that the bromoxynil was being detoxified by an endogenous cytochrome P450 monooxygenase. This offers the opportunity to use endogenous plant genes to enhance resistance against a range of herbicides. The concept has been validated by the demonstration that certain plant cytochrome P450 enzymes confer resistance to phenylurea herbicides. However, more research is required to identify which members of the cytochrome P450 gene family are specific for particular classes of xenobiotics, and which have roles in normal metabolic pathways.

Plant detoxification pathways often involve conjugation to glutathione by glutathione S-transferase (GST) activity, and specific transport of the conjugate into the vacuole. GSTs also comprise a large gene family, some members of which are known to be involved in endogenous metabolic reactions. In some cases, the hydroxylation and conjugation pathways operate in concert. Hence, the resistance of maize to atrazine is ascribed to a two-step pathway involving both 2-hydroxylation and conjugation to glutathione (Figure 5.5). Thus, there is the potential here to enhance endogenous systems, or transfer systems between plant species, once more

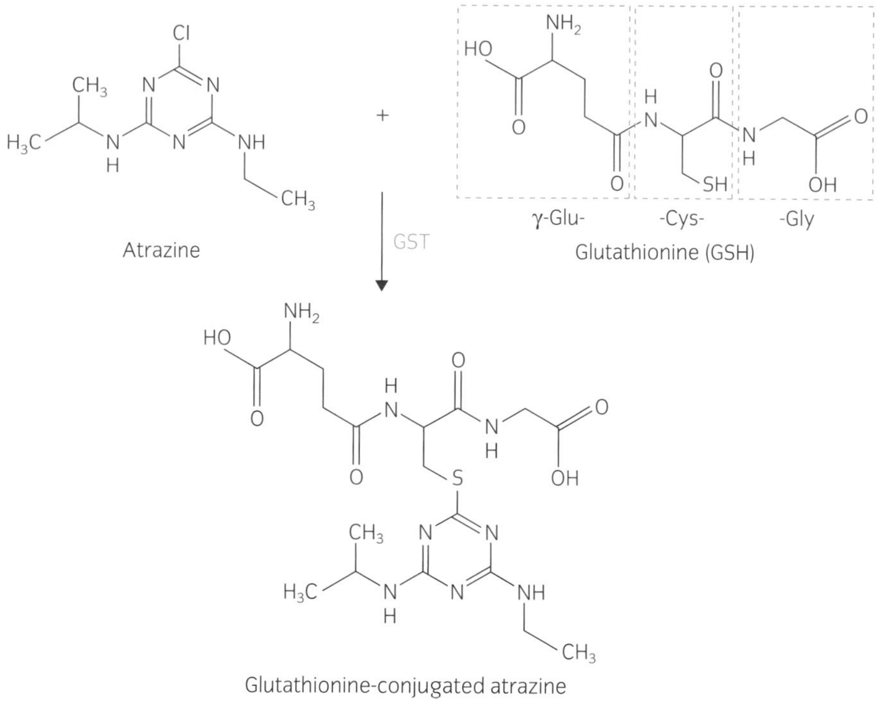

Figure 5.5 The conjugation of atrazine to glutathione. The detoxification of atrazine is a two-stage process involving a 2-hydroxylation step (removing the chlorine residue) prior to the addition of glutathione. Glutathione is a tripeptide in which the key residue is the middle cysteine. The -SH group is involved in a number of redox and conjugation reactions (see also Chapter 9).

information about the functions of these large gene families is available. The exploitation of functional genomics techniques (Chapter 1) will accelerate the acquisition of this knowledge.

Commercialization of herbicide-tolerant plants to date

This chapter has described the two most widespread examples of genetically manipulated herbicide tolerance. Table 5.5 indicates those crops resistant to a number of other herbicides that have been developed to the field-trial stage. It is important to note that some of these have not been widely planted commercially. The atrazine-tolerant crops, for example, have not been developed further, given the environmental concerns about the use of this class of persistent herbicide.

The table gives the strategy employed in each case. All these examples use either the mutant-target-enzyme approach, or the prokaryotic detoxification-gene approach, which have been described in detail when considering the glyphosate and PPT case studies. However, one more example is worth exploring further, because it demonstrates the use of homologous recombination to modify an endogenous plant gene to engineer herbicide tolerance.

Table 5.5 Herbicide-resistant crops grown in the field

Class of herbicide	Compound/herbicide	Transgene/mechanism	Companies	Crops
Glycine	Glyphosate (Roundup)	*Agrobacterium* CP4-resistant gene	Monsanto	Alfalfa, cotton, oilseed rape, soybean, wheat
			Monsanto/Syngenta	Sugar beet
	Glyphosate (Roundup)	Maize resistant gene	Monsanto	Maize
	Glyphosate (Roundup)	Oxidoreductase detoxification	Monsanto	Maize, oilseed rape, soybean
	Glyphosate (Roundup)	Acetylation detoxification	DuPont-Pioneer Hi-Bred	Soybean, maize
Phosphinic acid	Phosphinothricin (Basta), (Liberty)	*bar* gene; PAT detoxification	Bayer CropScience	Maize, rice, wheat, cotton, oilseed rape, potato
			Syngenta	Tomato, sugar beet
Sulphonylurea	Chlorsulphuron (Glean)	Mutant plant acetolactate synthase	DuPont-Pioneer Hi-Bred	Rape, rice, flax, tomato, sugar beet, maize
Imidazolinone	(Lightening, Beyond)	Mutant plant acetolactate synthase	BASF	Canola, maize, rice, sunflower, wheat (Clearfield non-GM technology)
S-Triazine	Atrazine (Lasso)	Mutant plant chloroplast *psbA* gene	DuPont	Soybean
Nitriles	Bromoxynil (Buctril)	Nitrilase detoxification	Calgene/Bayer CropScience	Cotton, oilseed rape, potato, tomato
Phenoxy-carboxylic acids	2,4-D	Monooxygenase detoxification	Bayer CropScience	Maize, cotton

2,4-D, 2,4-dichlorophenoxyacetic acid; PAT, phosphinothricin acetyltransferase.

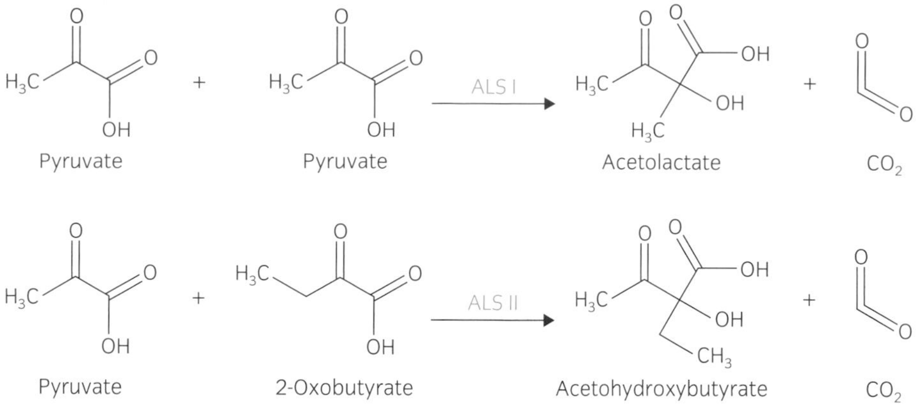

Figure 5.6 The reactions catalysed by ALS forms I and II. The formation of acetolactate by ALS I is the first step in the biosynthesis of leucine and valine, whereas the similar reaction catalysed by ALS II is a key step in isoleucine synthesis.

CASE STUDY 5.3 Engineering imidazolinone tolerance by targeted modification of endogenous plant genes

As stated near the beginning of this chapter, two important classes of herbicide (sulphonyl-ureas and imidazolinones) have the same mode of action; that is, they inhibit the enzyme ALS (see Table 5.3 and Figure 5.6). This blocks the synthesis of the aliphatic amino acids isoleucine and/or leucine and valine. Herbicide-tolerant forms of this enzyme have been isolated from a range of species, from bacteria and yeast to plants, and some of these have been used to engineer herbicide tolerance by standard Strategy 2-type approaches. The sequence analysis of various resistant ALS genes also provided information about the precise changes in amino acid sequence responsible for the resistant character. Researchers at Pioneer Hi-Bred International used this type of information to predict that a single base change in the endogenous maize ALS gene family would be sufficient to change Ser-621 (encoded by AGT) to Asn-621 (AAT) and that this should confer tolerance to the herbicide Lightning (a mixture of imazathapyr (Table 5.3) and imazapyr). This was actually achieved by bombarding maize cells with an oligonucleotide made up of a combination of DNA and RNA bases, with a 32-base section having exact homology to the target sequence of the endogenous maize genes, apart from the single-base mismatch to give the desired mutation. Cells that were able to grow and develop callus on medium containing imazathapyr were selected, and plants were regenerated at an estimated transformation efficiency of 10^{-4}.

● The very low rate of homologous recombination in plants is a barrier to the manipulation of endogenous plant genes.

Some nine independently transformed plants were tested for Lightning tolerance: three were resistant to fourfold the normal field dose, whereas four others were only slightly injured by the fourfold dose and were able to tolerate the normal field dose. The remaining two plants were as susceptible to the herbicide as control plants. There was a direct correlation between Lightning tolerance and the sequence of ALS genes in the transformed plants. The highly resistant plants contained the predicted AGT→AAT mutation, whereas the moderately resistant plants showed mutations within a few bases of the target site. In contrast, the sequence of the ALS genes was not altered in the susceptible plants.

The conclusion drawn from this work is that although homologous recombination is a rare event in plants, it can be used to change endogenous gene sequences under certain circumstances. Does this therefore mean that many more traits will be engineered by this type of procedure in the future? There are certainly many advantages to the manipulation of an endogenous gene *in situ*, since it avoids many of the problems associated with transgene cloning, modification, expression, and targeting, as well as the requirement for a selectable marker gene, unpredictable effects following random insertion into the genome, etc. However, this procedure will only be suitable for certain types of genetic modification. First, the precise base change needed to produce the desired mutation must be known. Second, the targeted gene mutation must confer a trait that is selectable at this very low frequency of transformation. This is ideal for herbicide tolerance, but generally not for the types of trait such as pest and disease resistance that are discussed in the next three chapters. This point is demonstrated by the fact that the majority of examples of gene targeting have focused on the ALS genes. Thus, chlorsulphuron-resistant ALS genes have been produced in tobacco, changing Pro-196 to Ala, Thr, Glu, or Ser and by a Try-573→Leu substitution. Similarly, in rice, three different chimaeric RNA/DNA oligonucleotides designed to modify Pro-171→Ala, Trp-548→Leu, and Ser-627→Ile conferred chlorsulphuron resistance.

The environmental impact of herbicide-tolerant crops

● The adoption of herbicide-tolerant crops could have positive or negative environmental consequences.

The predominance of herbicide-tolerant crops in terms of their development and commercial growing was highlighted at the beginning of this chapter. Some of the reasons for this trait maintaining its position in league tables of GM crops have been discussed. Scientifically, herbicide tolerance has a number of advantages that facilitated its rapid development. A number of different single-gene strategies using accessible genes are possible, and the trait itself can be used as a selectable marker. These factors, combined with the research and development impetus provided by the agrochemical industry, inevitably led to their prime position. However, it should be noted that herbicide-tolerant crops do not top the league table just for these reasons. They have maintained their leading position because of the rapid adoption of these crops in countries where the technology has been accepted.

In the USA, where the greatest area of GM crops are grown, the proportion of herbicide-tolerant soybean rose from 17% of the total soybean planting in 1997 to 68% in 2001. Similarly, herbicide-tolerant cotton expanded from 10% of cotton in 1997 to 56% in 2001. This rapid adoption of herbicide-tolerant crops begs a number of questions. The first one is whether farmers are adopting these crops so rapidly because there is a clear commercial advantage, and if so, what is it? The second one relates to the environmental impact of this rapid switch in growing patterns, and hence herbicide-usage patterns. Box 5.6 presents one analysis of the economic benefits of growing glyphosate-tolerant soybean, and its impact on herbicide use.

BOX 5.6

Roundup Ready soybean: glyphosate usage

It has been calculated that from 1997 to 1998 there was an 81% increase in glyphosate use (up by 5540 t) in line with an increase in the proportion of GM soybean from 13 to 36% of the crop. However, this was more than compensated for by a reduction in the use of other herbicides on soybean of 8990 t, resulting in a net reduction in total herbicide use of 3360 t (−9.7%). Since glyphosate has better properties than some of the other herbicides in terms of toxicity, environmental persistence, and biodegradability, this is claimed to be of net benefit to the environment. This also has an impact on the production costs of the crop. For example, typical 1999 production costs (in US dollars) of Roundup Ready and non-GM soybean have been calculated to be approximately $225 and $250 per hectare, respectively, showing a net saving of $25 per hectare. The large saving in herbicide costs more than compensates for the technology fee charged for the Roundup Ready crop.

It is clear that there is a significant reduction in production costs associated with the use of Roundup Ready soybeans, derived principally from the reduced use of herbicides. In addition, it has been estimated that most of the economic benefits accrue to the farmer, as demonstrated in the table.

Beneficiaries	Estimated benefits (million $US)	Distribution of benefits (%)
Seed companies	32	3
US consumer	42	4
Technology inventor	74	7
US farmer	769	76
Total benefits	1061	100

(Data from Falck-Zepeda *et al.* (2000).)

This indicates that although the use of glyphosate increased compared with that for non-transgenic soybean, the total use of herbicides decreased. Indeed, the major beneficiary of this technology was the farmer rather than the producer, mainly by way of a marked overall reduction in herbicide costs. Thus, the assumption that herbicide-tolerant crops will inevitably lead to a greater use of herbicides needs to be qualified. The use of the target herbicide may well increase, but this may displace or reduce the requirement for other herbicides, such that there is a net reduction in their use. Since the environmental impact of herbicides such as glyphosate and PPT is thought to be lower than many of the compounds they are displacing, the producers of herbicide-tolerant crops can claim that the adoption of their crops will have a positive effect on the environment. However, there are other environmental concerns about the rapid adoption of herbicide-tolerant crops. One area of concern is the reduction of biodiversity as a result of the more efficient removal of weed species from arable land. These concerns relate more widely to current intensive agricultural practices, and lie outside the direct scope of this book. Another area of concern relates to the possibility of encouraging the development of herbicide-resistant weeds, and is discussed below.

The development of super-weeds

One of the major concerns about the use of herbicide-tolerant crops is that their introduction will lead to the appearance of so-called super-weeds, which could evade control by the commonly used herbicides. It is important to put these fears into context. Previously we have discussed the fact that some crops are already tolerant to certain herbicides, and that herbicide-resistant weeds may appear naturally wherever there is a selective pressure created by the repeated use of the same herbicide. Thus, the problem of herbicide-resistant weeds is not a new phenomenon unique to GM technology and, bearing in mind the large armoury of herbicides listed in Table 5.2, is not one that need pose a major threat. Nevertheless, it is sensible to ask whether the widespread use of herbicide-tolerant crops will make a difference to the number, rate of appearance, and type of resistant weeds because: (1) they encourage the repeated use of the same herbicide, and (2) the herbicide-resistance gene is transferred to a weed population by one of the processes of gene flow described below.

Herbicide-resistant weeds could theoretically arise by three types of mechanism.

● Super-weeds could arise by three different mechanisms.

1 The herbicide-tolerant crop itself appears as a 'volunteer' weed in fields where rotational crops are grown. This volunteer population could reproduce outside of cultivation and form a self-sustaining weed population.

2 Pollen from the herbicide-tolerant crop fertilizes weedy relatives of the crop plant, producing herbicide-tolerant hybrids. Subsequent backcrossing of these hybrids with the weed species could lead to introgression of the herbicide-tolerant trait into the weed population.

3 There is horizontal gene transfer (as opposed to the vertical transmission in points 1 and 2) by other mechanisms (e.g. viruses) that spread the herbicide-tolerant trait into a much wider range of plant species.

Just as the environmental impact of each herbicide should be considered separately, it is also important to treat different crops on a case-by-case basis. Thus, certain crops are normally more prone to form volunteer weeds than others, and the presence of a herbicide-tolerant strain could be a problem if that herbicide is normally used to clear the volunteers.

However, the large arsenal of herbicides shown in Table 5.2 should ensure that volunteer weeds can be cleared from cultivated land for rotational crops. It must be borne in mind that the concept of a weed is the product of a complex mix of characteristics that do not normally include herbicide tolerance. When considering gene flow from a GM crop to other crops and weed species, the nature of each individual crop must also be considered. For example, the spread of pollen from out-crossing crops is more of a problem than with crops that are self-crossing. Another factor to consider is the proximity of closely related weed species. Thus, maize, soybean, and cotton have no compatible wild-type relatives in the countries where GM varieties are widely grown (USA and Canada). Wheat, oilseed rape, and sugar beet are grown in close proximity to related species, but only the latter two are out-crossing. For this reason, oilseed rape and sugar beet have been used as crops in worst-case analyses for the study of gene flow into weeds. The studies that have been carried out to date tend to indicate that there is detectable gene flow from crops to weed species, but that transgene introgression under field conditions is probably rare. In order to minimize this potential gene flow, it is necessary to predict the distance that pollen can travel from each GM crop, and to establish appropriate buffer distances to prevent the fertilization of weedy relatives. However, many of the studies to determine buffer distances have been on a relatively small scale. A recent large-scale study using herbicide-tolerant oilseed rape (produced by non-transgenic means) does indicate that oilseed rape pollen can travel much further than had been suspected, but that nevertheless, the amount of gene flow to non-resistant plants was minimal.

There is as yet little evidence that horizontal gene flow between plants poses a particular problem for the transfer of transgenes into unrelated species. What is clear about the entire issue of environmental impact of herbicide-tolerant crops is that much more research is required to establish the full extent of all types of gene flow between plants in the environment. Thus, a recent review of the environmental risks and benefits of GM crops concluded that:

1 the risks and benefits of GM crops are not entirely certain or universal;
2 the ability to predict ecological impacts of any introduced species (GM or not) is difficult, and available data have limitations;
3 some benefits and risks may exist that have not yet been identified or addressed in published literature;
4 the quantity and quality of different GM crops that may eventually be developed merit special consideration for risk assessment;
5 better evaluation of potential benefits will help risk managers know how these balance with potential risks;
6 measures developed to prevent gene transfer to wild plants can reduce the potential environmental impacts and prolong potential benefits.

SUMMARY

This chapter has given an overview of the genetic manipulation of herbicide-tolerant plants. The wide range of different chemical families of herbicides and their modes of action have been described. Four basic strategies for the genetic engineering of herbicide tolerance have been discussed with reference to a number of case studies. It has been concluded that Strategy 2 (resistant target protein) and Strategy 3 (specific detoxification) are currently the most effective approaches. However, future developments may well include the more precise enhancement of endogenous detoxification mechanisms and the targeted mutation of endogenous resistance genes. Concerns about the environmental impact of herbicide-tolerant crops have been touched upon, with a view to returning to these wider issues in Chapter 12.

The next four chapters deal with the genetic manipulation of resistance to other stresses: pests, diseases, and abiotic stress. Some of the concepts introduced in this chapter will be recurring themes in the subsequent chapters. However, we will also introduce new ideas and techniques in a logical sequence as we progress through the next chapters.

FURTHER READING

Visit www.oxfordtextbooks.co.uk/orc/slaterplants2e/ for hyperlinked references that take you directly to online abstracts for the journal articles listed below:

Plant biochemistry

Buchanan, B. B., Gruissem, W., and Jones, R. L. (2002) *Biochemistry and Molecular Biology of Plants*. American Society of Plant Physiologists, Rockville, MD.

Heldt, H.-W. (1997) *Plant Biochemistry and Molecular Biology*. Oxford University Press, Oxford.

Herbicide-tolerant crops

Cole, D. J. (1994) Molecular mechanisms to confer herbicide resistance. In *Molecular Biology in Crop Protection* (G. Marshall and D. Walters, eds), pp. 146–176. Chapman & Hall, London.

Falck-Zepeda, J. B., Traxler, G., and Nelson, R. G. (2000) Rent creation and distribution from biotechnology innovations: the case of *Bt* cotton and herbicide-tolerant soybeans in 1997. *Agribusiness* **16**, 21–32.

Mullineaux, P. M. (1992) Genetically engineered plants for herbicide resistance. In *Plant Genetic Manipulation for Crop Protection* (A. M. R. Gatehouse, V. A. Hilder, and D. Boulter, eds), pp. 75–107. CAB International, Wallingford.

Directed evolution

Castle, L. A., Siehl, D. L., Gorton, R., Patten, P. A., Chen, Y. H., Bertain, S., Cho, H.-J., Duck, N., Wong, J., Liu, D., and Lassner, M. W. (2004) Discovery and directed evolution of a glyphosate tolerance gene. *Science* **304**, 1151–1154.

Siehl, D. L., Castle, L. A., Gorton, R., Chen, Y. H., Bertain, S., Cho, H.-J., Keenan, R., Liu, D., and Lassner, M. W. (2005) Evolution of a microbial acetyltransferase for modification of glyphosate: a novel tolerance strategy. *Pest Management Science* **61**, 235–240.

Homologous recombination

Iida, S. and Terada, R. (2005) Modification of endogenous natural genes by gene targeting in rice and other plants. *Plant Molecular Biology* **59**, 205–219.

Kovchenko, A. and Willmitzer, L. (2003) Chimeric RNA/DNA oligonucleotide-based site-specific modification of the tobacco acetolactate synthase gene. *Plant Physiology* **132**, 174–184.

Zhu, T., Peterson, D. J., Tagliani, L., St Clair, G., Baszczynski, C. L., and Bowen, B. (1999) Targeted manipulation of maize genes *in vivo* using chimeric RNA/DNA oligonucleotides. *Proceedings of the National Academy of Sciences USA* **96**, 8768–8773.

Zhu, T., Mettenburg, K., Peterson, D. J., Tagliani, L., and Baszczynski, C. L. (2000) Engineering herbicide-resistant maize using chimeric RNA/DNA oligonucleotides. *Nature Biotechnology* **18**, 555–558.

Environmental impact

Rieger, M. A., Lamond, M., Preston, C., Powles, S. B., and Roush, R. T. (2002) Pollen-mediated movement of herbicide resistance between commercial canola fields. *Science* **296**, 2386–2388.

Senior, I. J. and Dale, P. J. (2002) Herbicide-tolerant crops in agriculture: oilseed rape as a case study. *Plant Breeding* **121**, 97–107.

Wolfenbarger, L. L. and Phifer, P. R. (2000) The ecological risks and benefits of genetically engineered plants. *Science* **290**, 2088–2093.

Web-links

AgBiotechInfonet, web-link 5.1: www.biotech-info.net/herbicide-tolerance.html

AgBioTechNet, web-link 5.2: www.agbiotechnet.com/

Agricultural Biotechnology Council, web-link 5.3: www.abcinformation.org/benefits.php

Bayer CropScience BioScience, web-link 5.4: www.bcsbioscience.co.uk/

Biotechnology Industry Organisation, web-link 5.5: www.bio.org/foodag/

Council for Biotechnology Information, web-link 5.6: www.whybiotech.com

CropGen, web-link 5.7: www.cropgen.org/

DuPont Biotechnology, web-link 5.8: www2.dupont.com/Biotechnology/en_US/

Felsot, A. *Agrichemical and Environmental News* 2000–2001 [articles in issues 173, 175, 176, and 178], web-link 5.9: www.tricity.wsu.edu/aenews/Feb01AENews/Feb01AENews.htm#anchor5338542

GEO-PIE Genetically Engineered Organisms, Public Issues Education Project, Cornell University, web-link 5.10: www.geo-pie.cornell.edu/gmo.html

Herbicide Resistance Action Committee and the Weed Science Society of America, web-link 5.11: www.plantprotection.org/HRAC/MOA.html

James C. (2001, 2005) *International Service for the Acquisition of Agri-biotech Applications*, web-link 5.12: www.isaaa.org/

Monsanto Biotechnology Knowledge Centre, web-link 5.13: www.biotechknowledge.monsanto.com

Montsanto company, web-link 5.14: www.monsanto.com/monsanto/layout/default.asp

Pesticide Action Network Pesticide Database, web-link 5.15: www.pesticideinfo.org

Pioneer Hi-Bred, web-link 5.16: www.pioneer.com/biotech/default.htm

USDA Economic Research Service, web-link 5.17: www.ers.usda.gov/Briefing/Biotechnology/

6

The genetic manipulation of pest resistance

INTRODUCTION

In Chapter 5, the scale of the problem of world crops lost to weeds, pests, and diseases was discussed. It has been estimated that 13% of the potential world crop yield is lost to pests (see Figure 5.1). Plant pests range from nematodes to birds and mammals, but insect pests cause a major proportion of the total pest damage to crops. This chapter will therefore focus upon biotechnological solutions to the problem of insect pest damage to crops.

Far from reducing pest damage, modern agricultural practices have, if anything, exacerbated the problem. One of the contributory factors to this exacerbation has been the breeding out of endogenous pesticidal traits. Some defensive characters may have been lost accidentally during the selection process for other properties, but pesticidal traits may also have been removed deliberately because they affect other characteristics like yield and quality. For whatever reason, the fact remains that many current elite cultivars have less natural resistance to pests than did their predecessors. This problem is compounded by the practice of growing monocultures, since the growing of a single crop over a wide area in repeated years encourages the build-up of pests. These negative effects of modern agriculture on pest damage have encouraged an increasing reliance on chemical pesticides, but this has tended to drive a cycle in which each new pesticide just keeps pace with the appearance of resistance to the previous pesticide in the insect population. Against this background, the possibility of developing insect-resistant crops that could help to reduce insect damage, while reducing the reliance on chemical insecticides, has a number of attractions. Here we will consider the scientific strategies for achieving pest resistance before discussing the potential benefits and drawbacks of approaches using genetically modified (GM) crops.

● Modern agriculture tends to exacerbate pest damage.

Table 6.1 Common insect pests of major crops

Insect species	Common name of pest	Order	Crops affected
Helicoverpa armigera	Cotton bollworm	Lepidoptera	Cotton, maize
Helicoverpa zea	Corn earworm, tomato fruitworm	Lepidoptera	Cotton, maize, tomato
Heliothis armigera	Old World bollworm, tomato fruitworm	Lepidoptera	Cotton, tomato
Heliothis virescens	Tobacco budworm	Lepidoptera	Tobacco, cotton
Manduca sexta	Tobacco hornworm	Lepidoptera	Tobacco, tomato, potato
Ostrinia nubilalis	European corn borer	Lepidoptera	Maize
Pectinophora gossypiella	Pink cotton bollworm	Lepidoptera	Cotton
Spodoptera spp.	Armyworm, leafworm	Lepidoptera	Maize, rice, cotton, tobacco
Leptinotarsa decemlineata	Colorado beetle	Coleoptera	Potato
Callosobruchus maculatus	Cowpea seed beetle	Coleoptera	Cowpea, soybean
Tribolium confusum	Confused flour beetle	Coleoptera	Cereal flours
Locusta migratoria	Locust	Orthoptera	Grasses
Nilaparvata lugens	Brown plant hopper	Homoptera	Rice

The nature and scale of insect pest damage to crops

Before examining GM strategies for developing insect resistance, it is useful to consider some of the characteristics of the insects causing the damage. The first point to make is that, whereas some adult insects feed off plants and can damage crops (think of a plague of locusts), most of the problems are caused by insect larvae. The major classes of insect that cause crop damage are the orders Lepidoptera (butterflies and moths), Diptera (flies and mosquitoes), Orthoptera (grasshoppers and crickets), Homoptera (aphids), and Coleoptera (beetles). Table 6.1 lists a few of the common pests that cause extensive damage to some of the major crops of the world.

GM strategies for insect resistance: the *Bacillus thuringiensis* approach

- There are two main GM strategies for insect resistance.

In this chapter, two approaches will be compared: first, the use of bacterial insecticidal genes to provide protection from pest damage; and, second, the potential for using endogenous plant-protection mechanisms (the Copy Nature approach). By far the most widespread example of the first approach is the use of the *cry* endotoxin genes from *Bacillus thuringiensis*.

B. thuringiensis was discovered by Ishiwaki in 1901 in diseased silkworms, and was subsequently classified and named after its isolation from the gut of diseased flour moth larvae in Thuringberg, by Ernst Berliner. The adverse effect of the bacteria on the insect larvae was subsequently identified as arising from a number of toxins produced by the bacteria. The bacterium produces an insecticidal crystal protein (ICP; also called Cry proteins, encoded by *cry* genes), which forms inclusion bodies of regular bipyramidal or cuboidal crystals during sporulation. ICPs are one of several classes of endotoxins produced by the sporulating bacteria; hence they were originally classified as δ-endotoxins, to distinguish them from the other classes of α-, β-, and γ-endotoxins.

The structure of the *cry* genes and their δ-endotoxin products have been well characterized. The *cry* genes are carried on plasmids and belong to a superfamily of related genes. Table 6.2 shows the classification of the *cry* gene superfamily according to size

Table 6.2 Classification of the ICP genes of *B. thuringiensis*

Cry protein family holotypes	Protein size (kDa)	Susceptible insect class
Cry1A(a–i)	133	Lepidoptera
Cry1B(a–g)	140	Lepidoptera
Cry1C(a, b)	133–134	Lepidoptera
Cry1D(a, b)	131–132	Lepidoptera
Cry1E(a, b)	133–134	Lepidoptera
Cry1F(a, b)	132–134	Lepidoptera
Cry1G(a–c)	132–133	Lepidoptera
Cry1H(a, b)	133	
Cry1I(a–f)	81	Lepidoptera
Cry1J(a–d)	133	Lepidoptera
Cry1Ka	137	Lepidoptera
Cry1La	133	
Cry2A(a–e)	71	Lepidoptera and Diptera
Cry3Aa, Cry3B(a, b), Cry3Ca	73–75	Coleoptera
Cry4Aa, Cry4Ba	135, 128	Diptera
Cry5A(a, b)	142–152	Nematodes
Cry5Ac	135	Hymenoptera
Cry5Ba	140	Hymenoptera
Cry6Aa, Cry6Ba		Nematodes
Cry7A(a, b)	129–130	Coleoptera
Cry8Aa, Cry8B(a–c), Cry8(Ca–Ha)	131	Coleoptera
Cry9Aa, Cry9B(a, b), Cry9Ca, Cry9D(a, b), Cry9E(a–d)	130	Lepidoptera
Cry10Aa	78	Diptera
Cry11Aa	72	Diptera
Cry11B(a, b)	81	Diptera
Cry12Aa–Cry17Aa, Cry18(Aa–Ca), Cry19(Aa, Ba), Cry20Aa, Cry21(Aa, Ba), Cry22A(a, b), Cry22Ba, Cry23Aa, Cry24(Aa, Ba), Cry25Aa–Cry29Aa, Cry30(Aa, Ba), Cry31(Aa, Ba), Cry32(Aa–Da), Cry33Aa, Cry34A(a–c), Cry34Ba, Cry35A(a–c), Cry35Ba, Cry36Aa–Cry39Aa, Cry40(Aa, Ba) Cry41A(a, b), Cry42Aa, Cry43(Aa, Ba), Cry44Aa, Cry45Aa, Cry46A(a, b), Cry47Aa, Cry48A(a, b), Cry49A(a, b), Cry50Aa, Cry51Aa	Various	Various

Table 6.3 The range of ICPs in individual *B. thuringiensis* strains

B. thuringiensis subspecies and strains	Crystal protein
aizawai	Cry1Aa, Cry1Ab, Cry1Ad, Cry1Ca, Cry1Da, Cry1Eb, Cry1Fa, Cry9Ea, Cry39Aa, Cry40Aa
entomocidus	Cry1Aa, Cry1Ba, Cry1Ca, Cry1Ib
galleriae	Cry1Ab, Cry1Ac, Cry1Da, Cry1Cb, Cry7Aa, Cry8Da, Cry9Aa, Cry9Ba
israelensis	Cry10Aa, Cry11Aa
japonensis	Cry8Ca, Cry9Da
jegathesan	Cry11Ba, Cry19Aa, Cry24Aa, Cry25Aa
kenyae	Cry2Aa, Cry1Ea, Cry1Ac
kumamotoensis	Cry7Ab, Cry8Aa, Cry8Ba
kurstaki HD-1	Cry1Aa, Cry1Ab, Cry1Ac, Cry1Ia, Cry2Aa, Cry2Ab
kurstaki HD-73	Cry1Ac
kurstaki NRD-12	Cry1Aa, Cry1Ab, Cry1Ac
morrisoni	Cry1Bc, Cry1Fb, Cry1Hb, Cry1Ka, Cry3Aa
tenebrionis	Cry3Aa
tolworthi	Cry3Ba, Cry9Ca
wuhanensis	Cry1Bd, Cry1Ga, Cry1Gb

and sequence similarities. This classification was introduced in 1998 and supersedes previous nomenclature that may still appear in the literature. The sequence comparison indicates a large number of distinct families (*cry1–cry51* at the time of publication) and within each family there may be three further levels or ranks of subfamily. Hence, *cry1Aa1* is very closely related to the other *cry1Aa* genes, which are all closely related to the other *cry1A* genes, which all form part of the *cry1* family. Table 6.3 shows that the strains of *B. thuringiensis* produce a wide range of different crystal proteins. Thus, it is not just a case of each strain containing one specific *cry* gene encoding one particular crystal protein.

● The Cry proteins share a common core of three domains.

Apart from sequence similarity, it is apparent that there are large differences in size among different Cry proteins, though they tend to cluster as either large (≈130 kDa) or small (≈70 kDa) proteins. Although there is a considerable difference in size among the different subfamilies, they share a common active core comprising three domains. Figure 6.1 shows a simple alignment of different classes of Cry protein, showing the common core domains. It can be seen that the N-terminal end of each gene has a similar organization, despite the great difference in overall length. In fact, the larger proteins such as the Cry1 group are inactive, and are activated by proteolytic removal of the C-terminal sequence.

Figure 6.2 shows the structure of the active core of the Cry1A protein, exemplifying the three discrete domains, each with a quite different folding structure. Domain I at the N-terminal end comprises a series of α-helices arranged in a cylindrical formation. The open cylindrical structure of this domain is thought to be responsible for creating a pore through the membrane of the insect midgut. Domain II comprises a

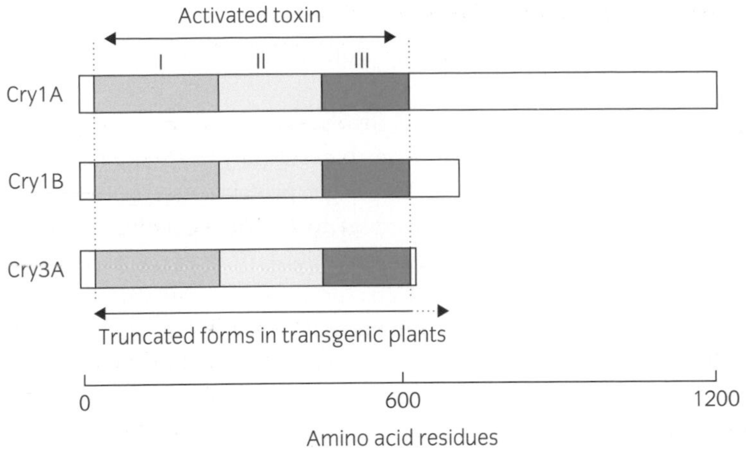

Figure 6.1 Comparison of the structures of different classes of Cry protein. Three different classes of Cry protein are compared to demonstrate the differences in overall size, but the alignment of the common core region that comprises the activated toxin. The organization of the core into domains I, II, and III (see Figure 6.2) is also indicated. The N- and C-terminal extensions are trimmed by insect gut proteases to release the active toxin. In some transgenic plants, a truncated, active form of the protein is produced directly. (Redrawn with permission from de Maagd *et al.* (1999).)

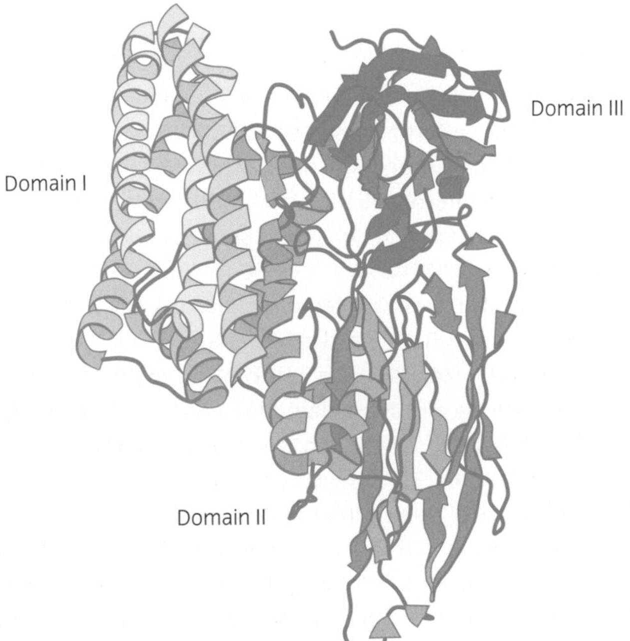

Figure 6.2 Ribbon model of the Cry1Aa toxin molecule. The structure of the activated toxin is shown, demonstrating the three distinct domains. The α-helical cylinder that forms domain I is involved in membrane insertion and pore formation. Domains II and III are involved in recognition and binding to receptors in the insect midgut wall. (Redrawn with permission from de Maagd *et al.* (1999).)

triple β-sheet and is involved in receptor recognition. Domain III is a β-sandwich, with a number of putative roles including protection from degradation, toxin–bilayer interactions, and receptor binding.

- Cry proteins bind to specific receptors in the insect larva midgut, causing selective pores in the midgut membrane to open.

The mode of action of δ-endotoxins involves a specific interaction between the protein and the insect larva midgut. After ingestion by an insect larva, the protein crystals are solubilized in its midgut. The larger proteins such as the 130-kDa Cry1 group are proteolytically cleaved at this stage to release the active 55–70-kDa active fragment of the protein. This interacts with high-affinity receptors in the midgut brush-border membrane. The result of this binding is to open cation-selective pores in the membrane. The flow of cations into the cells results in osmotic lysis of the midgut epithelium cells, causing their destruction. Thus, the δ-endotoxins are extremely toxic and can be lethal to susceptible insect larvae at relatively low concentrations. On the other hand, their toxicity to other animals (e.g. mammals) is extremely low.

The conditions in the insect larva midgut vary according to insect class. The midgut of Lepidoptera and Diptera is mildly alkaline, whereas the coleopteran gut is generally either more alkaline or acidic. These different conditions favour the solubilization and activation of different Cry subfamilies. Furthermore, the specificity of the interaction between the endotoxin and the midgut receptor means that individual Cry proteins are active against particular insect larvae. Table 6.2 indicates the specificity of some of the best characterized groups of Cry proteins.

The use of *B. thuringiensis* as a biopesticide

Preparations of *B. thuringiensis* spores or isolated crystals have now been used as an 'organic' pesticide for half a century. The isolated crystals have a limited persistence on foliage of a few days, whereas the spore preparations are effective for about 40 days on foliage and up to 2 years in soil. Neither method of application has a particularly effective penetration, with regard to concealed surfaces and organs of the plant, or against sap-sucking insects; that is, they are not systemic. However, the agronomic experience of using *B. thuringiensis* spores or isolated crystals as a biopesticide, coupled with the associated safety data and regulatory approvals, has been useful for the rapid development of the *Bt* strategy to genetic modification.

- *Bt*, Cry, ICP, and δ-endotoxin are different terms used to describe the *Bt* proteins.

Note that *Bt* is generally used as a shorthand for a crop transformed with a *cry* gene (hence *Bt* cotton, etc.), and also for the Cry proteins (hence *Bt* protein). This can be confusing, bearing in mind that we have now met four different terms for effectively the same group of proteins: ICP, δ-endotoxin, Cry, and now *Bt*.

Bt-based genetic modification of plants

- The sequences of bacterial *cry* genes must be modified to optimize expression in plants.

Although the GM approach to using the *cry* genes to obtain pest resistance in plants is conceptually simple, it does provide an object lesson in the detailed molecular biology that may be required to achieve high levels of expression of a bacterial gene in a transgenic plant. This goes beyond the obvious requirements for plant promoter and terminator sequences to regulate transcription as described in previous chapters.

The first attempts to express Cry1A and Cry3A proteins under the control of the cauliflower mosaic virus (CaMV) 35S or *Agrobacterium* T-DNA promoters resulted in very low levels of expression in tobacco, tomato, and potato plants. It was realized that the prokaryotic gene sequence itself would need to be modified extensively to obtain high levels of stable expression. These modifications are described in detail in Box 4.5. The eventual result was that expression was enhanced by 100-fold to give much better levels of expression, of the order of 100 ng of *Bt* protein per mg of total protein. Subsequent laboratory tests showed that the effect of producing a specific Cry protein at this level was to provide a considerable degree of protection against damage by susceptible insect larvae. Further confirmation of the effectiveness of these plants in providing protection against insect damage in small-scale field trials resulted in the first *Bt* crops gaining approval for commercial planting in the USA in the mid-1990s.

The success of the *Bt* approach led to the development of *Bt* crops by several of the major biotechnology companies involved in crop protection (Table 6.4). This table demonstrates the point that the specificity of Cry proteins permits the targeting of specific pests by particular transgenes, and that different crops may have different *cry* genes inserted. We will see later that the nature of the *cry* gene construct can be important for the success or failure of a particular transgenic crop, and that some of the lines shown in Table 6.4 have already been discontinued. Indeed, maize and cotton are the only *Bt* crops that are currently grown commercially in the USA, since the NewLeaf *Bt* potato has now been discontinued.

Table 6.4 Examples of the commercialization of *Bt* technology

Company	Trade name	*Bt* protein	Crop	Insect pests
Monsanto	NewLeaf	Cry3A	Potato	Colorado beetle
Monsanto	Bollgard	Cry1Ac	Cotton	Tobacco budworm, cotton bollworm, pink bollworm
Monsanto	Bollgard II	Cry1Ac + Cry2Ab	Cotton	Tobacco budworm, cotton bollworm, pink bollworm
Dow	WideStrike	Cry1Ac + Cry1Fa	Cotton	Tobacco budworm, cotton bollworm, pink bollworm
Monsanto	YieldGard	Cry1Ab	Maize	European corn borer
Syngenta	Agrisure CB	Cry1Ab	Maize	European corn borer
Aventis	StarLink (discontinued)	Cry9C	Maize	European corn borer
Mycogen (Dow) Pioneer (DuPont)	Herculex 1	Cry1Fa	Maize	European corn borer
Dow/Pioneer	Herculex RW	Cry34Ab + Cry35Ab	Maize	Corn rootworm
Dow/Pioneer	Herculex Xtra	Cry1Fa/Cry34Ab/Cry35Ab	Maize	European corn borer + corn rootworm
Monsanto	YieldGard Rootworm	Cry3Bb	Maize	Corn rootworm
Monsanto	YieldGard Plus	Cry1Ab/Cry3Bb	Maize	European corn borer + corn rootworm
Syngenta	Agrisure RW	mCry3Aa	Maize	Corn rootworm

CASE STUDY 6.1 **Resistance of *Bt* maize to the European corn borer and other pests**

The European corn borer (*Ostrinia nubilalis*) is a major pest of maize. As the name suggests, the larvae damage maize crops by tunnelling into the central pith of the stalks and ears. However, despite its name, it is not confined to Europe, and causes considerable damage to the maize crop worldwide. Developing a European corn borer-resistant maize has therefore become one of the main targets of the agricultural biotechnology industry. Table 6.4 shows that a number of different companies have each produced a *Bt* maize line that is resistant against European corn borer. The rate of adoption of *Bt* corn has been rapid, such that in 2004 it accounted for 14% of global maize production.

The development of commercial lines involves testing a large number of transformed plants to find the optimal line in terms of the quality of the desired trait, the copy number, and stability of integration of the transgene, as well as the stability, level, and pattern of transgene expression. Regulatory approval is normally given only for the varieties ultimately descended from a single, fully characterized transformation event.

Three different transformation events with the *cry1Ab* gene (Bt176, Bt11, and Mon810) were developed by different companies and subsequently licensed to, and marketed by, a number of others. In all three cases, a truncated *cry1Ab* gene with modified codon usage optimized for maize expression was used. The transgene in the Bt11 (Agrisure CB, Syngenta) and Mon810 (YieldGard, Monsanto) events is regulated by a constitutive promoter. On the other hand, the Bt176 event was produced by bombardment with two separate *cry1Ab* constructs: one controlled by a maize phosphoenolpyruvate carboxylase (PEPC) promoter (specific for green tissue), and the other under the control of a promoter from a pollen-specific protein kinase gene. Varieties derived from the Bt176 event were marketed under trade names including Knockout (Novartis) and NaturGard (Mycogen), but proved to be less effective against the second brood of European corn borer than the other two events, and have now been phased out. Not surprisingly, the Bt176 event was also shown to produce the Cry1Ab toxin in pollen at much higher levels than the other two, with consequences for the potential toxicity of the pollen to non-pest lepidopterans (see Box 6.3, below). Other *cry* genes have also been targeted against the European corn borer, including *cry1Ac* (Bt-Xtra, DeKalb), *cry1F* (Herculex, Mycogen), and *cry9C* (StarLink, Aventis; see Table 6.4). The particular problems experienced by the StarLink lines are described in Box 6.1.

The current *Bt* maize hybrids have generally provided better control against the European corn borer than could be achieved with a single, well-timed insecticide treatment. Bt11 and Mon810 also provide some control against corn earworm. On the other hand, Cry1F provides additional protection against the fall armyworm and western bean cutworm. However, these Cry1 proteins are not effective against other maize pests such as corn rootworm beetles and their larvae or spider mites. A number of different transgenic maize lines have been developed for the control of corn rootworm larvae, by integration of genes active against coleopteran pests such as Cry3Bb (Monsanto), a modified Cry3Aa (mCry3Aa, Syngenta) and Cry34Ab plus Cry35Ab (Dow AgroSciences) (see Table 6.4).

BOX 6.1 *The rise and fall of StarLink corn*

The Aventis *Bt cry9C* maize event, marketed as StarLink, has been the topic of considerable controversy in the USA and Europe. The problem arose when AgrEvo (subsequently Aventis) sought approval for registration of StarLink in the USA. There were doubts raised by the US Environmental Protection Agency (EPA) about the suitability of the Cry9C protein for human consumption, because it is more stable in acid than other approved Cry proteins (hence more slowly digested in the stomach), and because of a lack of data about allergenicity. AgrEvo accepted a limited and conditional approval for StarLink in order to get the product launched into what was a very competitive marketplace for GM maize. Under the terms of this limited registration, StarLink products could only be used for animal feed, and had to be prevented from contaminating food for human consumption. AgrEvo proceeded with the commercial planting, given that the bulk of US maize is used for animal feed and industrial purposes, while carrying out further allergenicity tests on the protein. Unfortunately, in 2000, before StarLink products had been approved for human consumption, traces of StarLink corn were identified in taco shells manufactured by Kraft Foods. There was an immediate recall of these products, followed by costly recalls of other maize products from other suppliers.

The case raises a number of issues around the decision to give conditional approval, and the mechanisms required for ensuring that the growers and processors of GM products complied with the conditions placed upon the biotechnology company that registered the transgenic crop. The case has also demonstrated the requirement to investigate the allergenicity of GM products, and highlighted the need for test procedures to detect the presence of GM materials in non-GM products.

● StarLink products registered only for use in animal feed were found in the human food chain.

The problem of insect resistance to *Bt*

One of the problems encountered by the accelerating use of *Bt* technology to provide insect resistance has been the equally rapid appearance of resistant pests. This problem initially attracted widespread attention during the first commercial growing season of the *Bt* cotton crop (see Box 6.2, below).

The mechanism of insect resistance can be related directly to the specific binding involved in the mechanism of action of the Cry proteins. It may only require a small number of significant mutations in the insect gene coding for the receptor protein to greatly reduce the binding of a particular Cry protein. This is one of the consequences of single gene–gene interactions that will be discussed again in the subsequent chapters on disease resistance. Thus, insect pests resistant to the *Bt* crop could, in theory, appear within a few generations, and repeated growing of *Bt* crops in the same area would provide the selective advantage to accelerate the appearance of a resistant pest population. There are a number of strategies for countering the build-up of insect resistance.

● The specificity of *Bt* protein-receptor binding provides a mechanism for the development of insect resistance to *Bt*.

One way is to think of additional GM approaches to tackle the problem caused by the first GM approach. For example, it is possible to use more than one transgene, a process called pyramiding, in which transgenes are successively stacked by conventional

● Pyramiding involves stacking more than one transgene in a crop.

crosses between different transgenic lines. For example, Monsanto's Bollgard II is a new line of transgenic cotton developed by stacking of a second *Bt* gene (*cry2Ab*) with *cry1Ac* to prevent the development of resistant cotton bollworm and to provide added protection against beet armyworm and fall armyworm. The WideStrike cotton under development by Dow has a similar approach, using the *cry1Fa* gene stacked with *cry1Ac*. Stacked *Bt* traits have also been developed for maize, such that resistance against the major lepidopteran and coleopteran pests is now available in Monsanto YieldGard Plus (Cry1Ab plus Cry3Bb) and Dow Herculex Xtra (Cry1Fa plus Cry34Ab plus Cry35Ab) lines. The stacked Syngenta Agrisure CB (cornborer; Cry1Ab) and RW (rootworm; mCry3Aa) traits are also in the pipeline (Table 6.4). Many of the commercial *Bt* traits are also available stacked with herbicide-tolerant traits (see Chapter 5).

Another GM approach is to further enhance the effectiveness and range of activities of *cry* genes by domain engineering to produce chimaeric proteins. For example, a hybrid gene *SN19* was created by fusing domains I and III of *cry1Ba* with domain II of *cry1Ia*. Transformation of potato plants with this gene conferred resistance against three significant pests: Colorado potato beetle (a coleopteran), European corn borer, and potato tuber moth larvae (both Lepidoptera).

One other approach is to look beyond the large family of *Bt* genes to find other bacterial insecticidal genes that could be used in place of, or to augment, the current suite of *Bt* toxins. Candidates already under development include members of the vegetative insecticidal protein (Vip) family. These proteins are secreted into the medium during vegetative growth of *B. thuringiensis*, as opposed to the Cry proteins that accumulate during sporulation, and have a different mode of action. Syngenta has developed the COT102 line of transgenic cotton containing the *vip3A* gene to resist attack by a broad spectrum of Lepidoperan pests. Another bacterial gene that has attracted interest is the toxin A (TcdA) gene of *Photorhabdus luminescens*. This bacterium lives symbiotically within a parasitic nematode that uses the bacteria to kill its insect host. The broad host range of the nematode is reflected in the broad range of toxicity of the 283-kDa toxin A protein. Transformation of *Arabidopsis* with a construct comprising the large *tcdA* gene bounded by the 5′ and 3′ untranslated regions of the plant osmotin gene resulted in expression of the protein to levels greater than 7000 ppm. The plants were highly resistant to tobacco hornworm (*Manduca sexta*) and extracted protein also inhibited the growth of southern corn rootworm (*Diabrotica undecimpunctata howardi*).

However, it is still important to realize that *Bt* and other GM approaches to insect resistance cannot be viewed as a magic bullet to permanently eliminate the threat of insect damage. Rather, the GM crops just alter the balance of all the interactions between plant and environment that occur in the field, and the build-up of resistance in the insect population must be managed by good agricultural practices.

● Integrated pest management considers the wider context of pest–crop interactions.

The current practice of integrated pest management takes into consideration the wider context of a pest–crop interaction (natural predators, adjacent plant species, crop rotation, and chemical sprays). For example, selective usage of *Bt* crops (rotating *Bt* crops with non-*Bt* crops, avoiding growing different *Bt* crops within the migration distance of pests common to both, etc.) may prevent the build-up of resistance in the insect population, which would otherwise occur where there is a continuous selective advantage created by repeated use of *Bt* crops. One example of pest-resistance management favoured for managing the build-up of *Bt* resistance is the high-dose/refuge approach (Figure 6.3). In this scheme, transgenic crops expressing a high dose of *Bt* protein are grown alongside a smaller refuge of non-transgenic crop, or any other plant that the insect pest feeds upon. The high dose is important to ensure that only homozygous resistant insects are able to tolerate feeding on the GM crop. Homozygous resistant insects are likely to be rare in the early stages of the development of

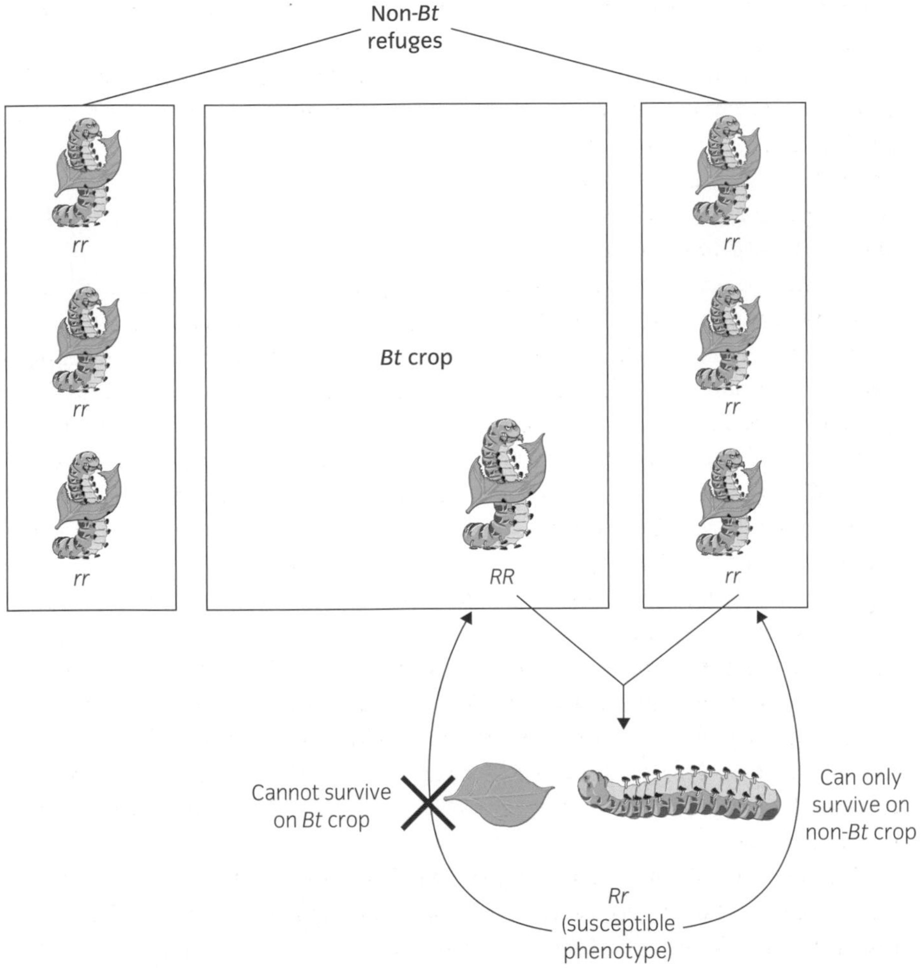

Figure 6.3 High dose/refuge-resistance management scheme. The rationale behind the high-dose/refuge strategy is demonstrated. The high dose is required to maintain resistance against heterozygote insects carrying one copy of a resistance gene *R*. Thus, only homozygote *RR* insects will be able to survive on the crop. The refuge provides a mechanism for ensuring that the proportion of *RR* insects is kept very low. The majority of *RR* insects will mate with the much larger population of *rr* insects in the refuge, producing susceptible progeny of genotype *Rr*.

resistance. The refuge ensures the presence of a much larger population of susceptible insects in the vicinity of the *Bt* crop than the homozygous resistant insects. Therefore, most mating events between insects carrying one or two resistance genes will probably be with a homozygous susceptible insect, producing heterozygous resistant offspring, which cannot survive on the crop.

● The high-dose/refuge strategy is one approach to inhibiting the build-up of insect resistance to *Bt*.

Box 6.2 gives an example of refuge options approved by the US Environmental Protection Agency (EPA) for the growing of Bollgard cotton in 2001. It is obviously critical that the approved refuge strategy is correct and monitored, and that the growers are fully informed and comply with the requirements of the resistance-management plan. Any deviation from the planting scheme could create the conditions for insect resistance to become established. Another problem resulting from the refuge strategy is that it may provide additional opportunities for the mixing of transgenic and non-transgenic materials. The problems caused by contamination of non-GM foods with GM materials are discussed in Box 6.1, and in Chapter 12.

BOX 6.2 *The development of resistance to* Bt *cotton*

The cotton crop in the USA is prone to damage by lepidopteran larvae, particularly cotton bollworm (*Helicoverpa zea*), pink bollworm (*Pectinophora gossypiella*), and tobacco budworm (*Heliothis virescens*). The development of insect-resistant cotton has been one of the major targets for US plant biotechnology.

Following a considerable amount of laboratory and precommercial field testing, a single transformation event, Monsanto 531, was chosen as the event from which cotton varieties marketed by Monsanto as Bollgard are descended. Bollgard lines contain a synthetic *cry1Ac*-like gene driven by an enhanced CaMV 35S promoter (see Chapter 4). The proportion of Bollgard cotton grown in the USA has risen from about 12% of the total area in 1996 to approximately 35% in 2000, and considerable information about its performance in diverse conditions and varieties has accumulated. It has been found that Bollgard cotton consistently provides a high level of protection against tobacco budworm and pink bollworm damage. On the other hand, protection against cotton bollworm is sufficient for moderate infestations, but applications of insecticide spray may be required for heavy infestations. In the first year of commercial-scale planting in 1996, there was an unexpected mid-season outbreak of cotton bollworm, requiring supplemental control with insecticidal sprays. This raised concerns that, in the haste to introduce the crop as widely as possible, there had been insufficient consideration given to the management of *Bt* resistance in the cotton bollworm population. In particular, if the dose of Cry1Ac was insufficient to kill virtually all the cotton bollworm insects, then this could accelerate the development of a resistant population.

As part of the US commercial licensing agreement with Monsanto, farmers are required to follow the terms of a resistance-management plan. Resistance-management plans are required by the EPA when a GM variety is approved for commercial growing. The conditional resistance-management plans for Bollgard cotton approved by the EPA in 1995 were updated in 2000 and amended for the 2001 season to allow three options for growers, as follows.

BOX 6.2 *Continued*

1 *80:20 external sprayed refuge*. For every 100 acres (40.5 ha) of Bollgard cotton, plant 25 acres (10.1 ha) of non-*Bt* cotton that can be treated with insecticides that control tobacco budworm, cotton bollworm, and pink bollworm. This refuge should be planted within 1 mile (1.6 km) of the edge of the Bollgard field (see Figure 6.3).

2 *95:5 external structured unsprayed refuge*. For every 95 acres (38.4 ha) of Bollgard cotton, plant 5 acres (2.0 ha) of non-*Bt* cotton. This refuge cannot be treated with any insecticides specified for tobacco budworm, cotton bollworm, and pink bollworm. The refuge must be managed similarly to the Bollgard cotton and be planted within 0.5 mile (0.8 km) of the edge of the Bollgard field.

3 *95:5 embedded refuge*. Plant at least 5 acres (2.0 ha) of non-Bollgard cotton for every 95 acres (38.4 ha) of Bollgard cotton, embedded as a contiguous block within the Bollgard field. This refuge may be treated with insecticides for the control of tobacco budworm, cotton bollworm, and pink bollworm whenever the entire field is treated.

These three options are based on a target refuge size of approximately 5%. (If it is assumed that insecticide spraying in Option 1 will kill 75% of the insects in the non-*Bt* cotton, the area of this sprayed refuge is 20% of the total acreage, so the proportion of susceptible insects in the refuge will be ≈5% of the total.)

Overall, it is estimated that the amount of Bollgard cotton planted in the USA in 1998 accounted for a reduction in weight of 0.9 million kg of insecticide sprayed. The economic benefits to growers have been estimated at approximately $US20 per hectare, with a yield advantage of approximately 10% over the non-transgenic crop. Thus, there is an incentive to all parties to implement the resistance-management plans to ensure that the pest resistance does not collapse due to the appearance of resistant insects.

The environmental impact of *Bt* crops

Although the first environmental concerns regarding *Bt* crops related to the build-up of resistance in the pest population, a separate issue was brought to the fore by a report that appeared in 1999 indicating that pollen from *Bt* maize might be toxic to the larvae of the Monarch butterfly. This particular report and its consequences are described in Box 6.3, and, to some extent, the particular fears about the Monarch butterfly have now been allayed. The affair does still beg the question of how event Bt176 was approved for commercial growing, given the deliberately high levels of expression of the *Bt* protein in the pollen of these plants. This highlights the need to assess the risks and monitor the environmental impact of GM crops more thoroughly than perhaps has been the case. However, this point also applies to the positive effects of these crops. For example, to what extent have non-pest insect populations benefited from the reduction in chemical pesticides that have certainly resulted from the adoption of *Bt* cotton?

BOX 6.3 *The Monarch butterfly affair*

One of the major controversies to hit *Bt* crops was the publication in 1999 of a paper in *Nature* by scientists at Cornell University, showing that the larvae of Monarch butterflies fed on milkweed leaves covered in pollen from *Bt* maize did not grow as well as those on control leaves. The Monarch larvae feed only on milkweed leaves, which are commonly found in cultivated areas and other habitats, so even though they are lepidopterans, and therefore susceptible to the Cry proteins produced in *Bt* maize, there should be no direct risk from them eating the maize plants. However, there is the potential for them to come into contact with pollen from GM maize, and this paper reported an attempt to test the toxicity of GM maize pollen to the larvae. The method was to dust pollen from *Bt* maize on to milkweed in a laboratory and feed these to Monarch caterpillars. The larvae fed on *Bt* pollen-covered leaves ate less and grew more slowly than controls on leaves dusted with non-GM pollen, or with no pollen.

The paper sparked off considerable public concern, given that the Monarch butterfly is one of North America's most colourful and familiar natives. Opponents of GM crops made the most of the report and claimed that it demonstrated there were major environmental problems with *Bt* crops. On the other hand, the proponents of GM technology criticized the methodology and dismissed the paper as just a preliminary finding. A major research collaboration between six groups in the USA and Canada was funded to research the question of *Bt* pollen toxicity in more detail, and to investigate the likelihood of exposure of Monarch caterpillars to *Bt* maize pollen under natural conditions. The findings of the six groups were published together in the *Proceedings of the National Academy of Sciences USA* (Sears *et al.*, 2001).

One of the conclusions of the group was that there was a significant difference between the levels of expression of *Bt* in pollen from event Bt176 and other events containing the same *cry1Ab* gene (see Table 6.4) (1.1–7.1 $\mu g\ g^{-1}$ in Bt176, compared with 0.09 $\mu g\ g^{-1}$ in the other two events). Given the pollen-specific promoter used to regulate the expression of one of the *cry1Ab* constructs in Bt176, such a significant difference in protein levels in the pollen is not surprising. There was also a correspondingly marked difference in the toxicity of pollen from different *Bt* maize plants. Pollen from Bt176 had an LD_{50} (the median lethal dose) of the order of 100–400 grains cm^{-2} when tested on Monarch butterfly larvae, and 600 grains cm^{-2} on black swallowtail butterfly larvae, whereas the other two showed no effect at concentrations of over 1600 grains cm^{-2} (the highest concentration of pollen actually measured in a maize field).

Since Bt176 is no longer available in US maize varieties, the overall conclusion is that the risk to Monarch butterfly populations from current *Bt* maize varieties is low.

The Copy Nature strategy

Copy Nature is an alternative philosophy to the *Bt* magic bullet approach.

Some of the problems encountered with *Bt* have led some scientists to contrast the single, highly toxic, heterologous gene approach epitomized by *Bt* with what has been termed a Copy Nature approach (Boulter, 1993). The strategy involves a rational approach to the development of pest-resistant crops and is characterized by the following steps.

1 *Identification of leads.* The point here is to look in nature for plants that show resistance to insect pests. These leads can be found in the literature, world seed collections, or observation in the field. The identification of resistant plants provides the starting point for the discovery of plant genes that could confer resistance to insect damage.

2 *Protein purification.* The next stage of the process is the purification of proteins with insecticidal properties. Proteins are the primary target for this type of screen because they provide the most direct way to isolate corresponding genes. (It should be remembered that plants produce a large number of insecticidal secondary products, but it would be a much longer process to engineer the synthesis of a secondary metabolite in a transgenic plant than of an insecticidal protein. Starting from the discovery of a novel compound, the biosynthetic pathway would need to be determined, the key enzymes identified, and the genes isolated in order to produce a transgenic plant.) Partial sequencing of the purified protein is one method for isolating the corresponding gene. Characterization of the protein may also provide a means of identifying the gene in a model organism such as *Arabidopsis*, as well as using the known DNA sequence as a probe to screen for the orthologous gene in the plant of interest.

3 *Artificial-diet bioassay.* It is important to determine the activity of the isolated protein against the target insect pests by performing feeding assays in the laboratory. In other words, is the protein going to be an effective deterrent to insect feeding at achievable levels of expression in a transgenic plant?

4 *Mammalian toxicity testing.* Prior to any insertion of the gene into a crop plant, the toxicity of the protein against mammals (and hence potentially humans) should be tested. The point is that it would be a waste of time inserting a gene into plants that subsequently was found to cause concerns about food safety.

5 *Genetic engineering.* It is only at this stage that the isolated gene would be considered valuable for transfer to crop plants. The typical techniques for transformation technique, vector design, and construct building are described in Chapters 3 and 4. One particular consideration for pest-resistant constructs is the choice of promoter. Although it is possible to rely on a strong, constitutive promoter such as CaMV 35S, there are certain advantages to using either a tissue-specific promoter or a wound-inducible promoter. This is also a Copy Nature strategy, since many insect defence genes are produced site-specifically or are induced by insect damage. Expression of the insecticidal protein at the site of insect damage limits the exposure only to the insects causing the damage. It also reduces the biosynthetic burden on the plant, and reduces the amount of transgenic protein in those parts of the plant destined for human consumption. Some of the promoters used to drive the expression of insect resistance genes are shown in Table 6.5.

● Site-specific and inducible promoters have several advantages.

6 *Selection and testing.* After transformation, the selection of transgenic plants, confirmation of transformation, inheritance of transgenes in T_1, T_2, etc. generations, and testing of expression levels need to be carried out. The effectiveness of the construct then has to be evaluated by insect feeding assays.

Table 6.5 **Promoters used with insect resistance genes**

Promoter	Origin	Expression site	Insecticidal protein	Plant
Mannopine synthase TR	*Agrobacterium* Ti plasmid	Most plant tissues	Cry1Ab	Tobacco, potato
Phytohaemagglutinin (PHA-L)	Bean	Seed	a-AI-Pv	Pea, adzuki bean, tobacco
CaMV 35S	Cauliflower mosaic virus	Most plant tissues	Most proteins	Most plants
Sucrose synthase (RSs1)	Rice	Phloem	GNA	Tobacco
Metallothionein-like (MT-L)	Maize	Root preferred	Cry1Ab	Maize
PEPC	Maize	Green tissue	Cry1Ab	Maize, rice
Pollen-specific	Maize	Pollen	Cry1Ab	Maize
Tryptophan synthase α-subunit (trpA)	Maize	Pith preferred	Cry1Ab	Maize
Ubiquitin-1 (Ubi-1)	Maize	All plant organs	Cry1Ac	Rice
Proteinase inhibitor II (Pot PI-II)	Potato	Wound-inducible	Pot PI-II, ipt	Rice, tobacco, tomato
rRNA operon (Prrn)		Chloroplasts	Cry1Ac	Tobacco
Actin-1 (Act-1)	Rice	All plant organs	CpTI	Rice
Pathogenesis-related protein-1a (PR-1a)	Tobacco	Chemically induced	Cry1Ab	Tobacco

PEPC, phosphoenolpyruvate carboxylase.

7 *Biosafety*. The effect of the transgene on crop yield, insect damage, and the wider ecosystem should be properly evaluated in field trials, rather than after several years of commercial planting, as occurred with *Bt* crops (Box 6.3).

The aim of the Copy Nature strategy has been stated as: 'insect pest control which is relatively sustainable and environmentally friendly'. The strategy recognizes a complex interplay in biological communities between plants, animals, microbes, the soil, and the physical environment. It is not just a case of crop plant versus insect pest. The strategy also takes account of the fact that host-plant resistance to pests is universal and falls into two major categories.

1 *Horizontal resistance*. This type of trait is usually polygenic; that is, it results from the cumulative effect of several minor gene characters. This form of resistance does not involve gene–gene matching (as is the case with Cry proteins and their receptors in the insect midgut, for example) and is usually durable.

2 *Vertical resistance*. In contrast, vertical resistance typically arises from one major gene with a high level of expression. The mechanism of resistance may well involve gene–gene matching. The important point to note is that this form of resistance normally occurs in plants to provide a buffer during short-term changes in the level and type of insect damage, and is unlikely to be durable.

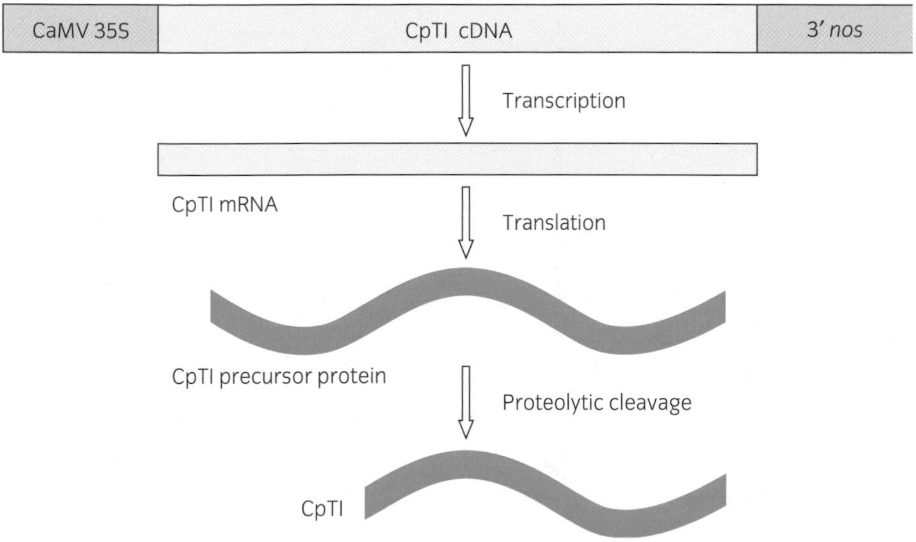

Figure 6.4 **Construction of the CpTI expression vector for plant transformation. The CpTI cDNA from cowpea with CaMV 35S promoter and 3′ *nos* terminator was inserted into a pBIN19 derivative. Expression of the gene in plants produces a CpTI precursor protein that is subsequently cleaved by specific proteases to yield the active CpTI protein.**

CASE STUDY 6.2 **Cowpea trypsin inhibitor**

The first stage in the development of cowpea trypsin inhibitor (CpTI) as an anti-insect, pest-control mechanism came from the identification of strains of cowpea growing in Africa that were resistant to attack from a range of insect pests. The isolated insecticidal protein was found to be a trypsin inhibitor. In an artificial-diet bioassay, the protein was shown to be effective against Lepidoptera, Orthoptera, and Coleoptera. Further tests showed that this protease inhibitor did not affect mammalian trypsins and was not toxic to mammals. The gene was isolated and inserted into the transformation vector pROK2 (a pBIN19 derivative) (Figure 6.4).

For the initial trials, the gene was transferred into tobacco (see Chapter 3), and the transgenic plants produced were tested for their ability to withstand attack from a number of different lepidopteran pests of tobacco. It was found that the transgenic plants were visibly much more resistant to damage under laboratory conditions than were the control plants. There are different ways of quantifying the extent of resistance of a plant to insect damage, either in terms of measuring the amount of damage (loss of weight or leaf area), or assaying the effect of eating the transgenic plant on the insect. Both types of bioassay were used to show that lepidopteran pests (tobacco budworm, cotton bollworm, and cotton leafworm) fed on the CpTI transgenic tobacco plants did less damage and grew less well than those on control plants (Figure 6.5).

Having tested the concept in the laboratory, it is important to confirm the findings in the field. The results of field tests of CpTI tobacco showed a significant level of protection against damage by lepidopteran pests, and a reduction in the number of feeding larvae (Figure 6.6). CpTI has subsequently been transformed into a range of crop plants including rice, potato, wheat, cotton, strawberry, and pigeonpea (Table 6.6). However, none of these crops has yet been produced on a commercial scale.

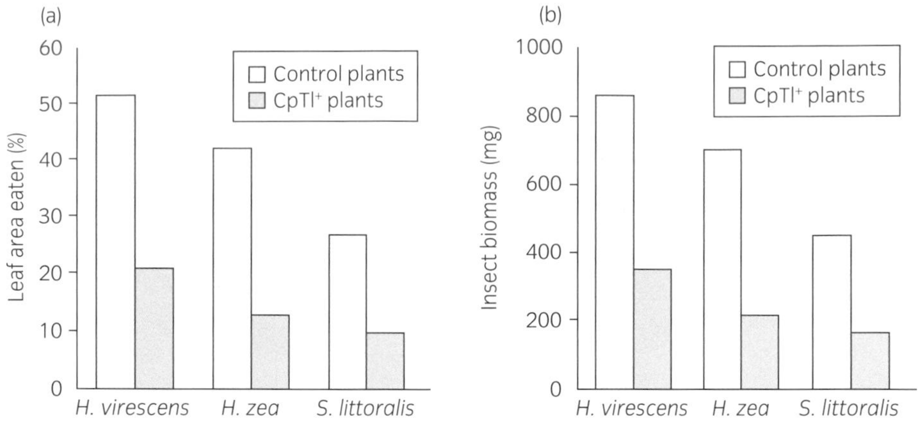

Figure 6.5 Bioassay data for feeding trials of lepidopteran pests on CpTI transgenic tobacco plants. The activity of three different lepidopteran pest larvae (*Heliothis virescens*, *Helicoverpa zea*, and *Spodoptera littoralis*) on transgenic tobacco plants carrying the CpTI gene was assayed in the laboratory in two different ways: (a) the extent of damage to the tobacco leaves was determined by measuring the leaf area eaten; (b) the growth of the larvae on the tobacco was determined by measuring insect biomass. In both assays, there was a significant protective effect from the CpTI transgene compared with control plants. (Redrawn with permission from Gatehouse *et al.* (1992).)

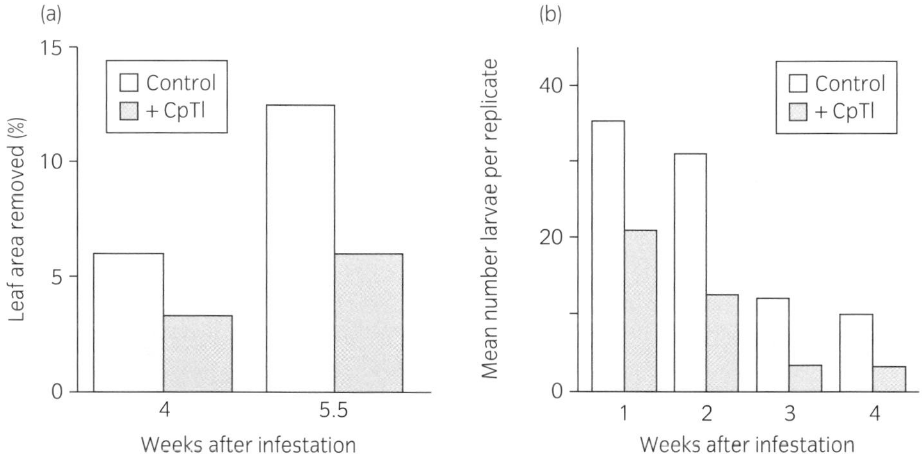

Figure 6.6 The insect resistance of transgenic tobacco carrying the CpTI gene in the field. The ability of transgenic tobacco to resist damage from *Helicoverpa zea* larvae in the field was assessed by two different assays: (a) the leaf area removed was determined; and (b) the number of larvae per plant was counted. There was a significant control of insect damage provided by the CpTI gene. (Redrawn from Gatehouse *et al.* (1992).)

CpTI is not the only gene that has been isolated from plants and used to engineer insect pest resistance. Plant resistance genes that have been tested include other protease inhibitors, α-amylase inhibitors, and lectins. Table 6.6 shows the different genes that have been tried and the range of different transgenic crops that have been

Table 6.6 Plant insecticidal genes used to engineer pest resistance

Plant gene	Encoded protein	Plant of origin	Target insects	Transformed plants
Protease inhibitors	*Inhibited protease*			
C-II	Serine protease	Soybean	Coleoptera, Lepidoptera	Oilseed rape, poplar, potato, tobacco
CMe	Trypsin	Barley	Lepidoptera	Tobacco
CMTI	Trypsin	Squash	Lepidoptera	Tobacco
CpTI	Trypsin	Cowpea	Coleoptera, Lepidoptera	Apple, lettuce, oilseed rape, potato, rice, strawberry, sunflower, sweet potato, tobacco, tomato, wheat
14K-CI	Bifunctional serine protease and α-amylase	Cereals		Tobacco
MTI-2	Serine protease	Mustard	Lepidoptera	*Arabidopsis*, tobacco
OC-1	Cysteine protease	Rice	Coleoptera, Homoptera	Oilseed rape, poplar, tobacco
PI-IV	Serine protease	Soybean	Lepidoptera	Potato, tobacco
Pot PI-I	Proteinase	Potato	Lepidoptera, Orthoptera	Petunia, tobacco
Pot PI-II	Proteinase	Potato	Lepidoptera, Orthoptera	Birch, lettuce, rice, tobacco
KTi3, SKTI	Kunitz trypsin	Soybean	Lepidoptera	Potato, tobacco, rice
PI-I	Proteinase	Tomato	Lepidoptera	Alfalfa, tobacco, tomato
PI-II	Proteinase	Tomato	Lepidoptera	Tobacco, tomato
α-Amylase inhibitors				
α-AI-Pv	α-Amylase	Common bean	Coleoptera	Azuki bean, pea, tobacco
WMAI-I	α-Amylase	Cereals	Lepidoptera	Tobacco
14K-CI	Bifunctional serine protease and α-amylase	Cereals		Tobacco
Lectins				
GNA	Lectin	Snowdrop	Homoptera, Lepidoptera	Grapevine, oilseed rape, potato, rice, sweet potato, sugarcane, sunflower, tobacco, tomato
p-lec	Lectin	Pea	Homoptera, Lepidoptera	Potato, tobacco
WGA	Agglutinin	Wheat germ	Lepidoptera, Coleoptera	Maize
Jacalin	Lectin	Jack fruit	Lepidoptera, Coleoptera	Maize
Rice lectin	Lectin	Rice	Lepidoptera, Coleoptera	Maize
Others				
BCH	Chitinase	Bean	Homoptera, Lepidoptera	Potato
Peroxidase	Anionic peroxidase	Tobacco	Lepidoptera, Coleoptera, Homoptera	Sweet gum, tobacco, tomato
Chitinase	Chitinase			Oilseed rape
TDC	Tryptophan decarboxylase	*Catharanthus roseus*	Homoptera	Tobacco

produced. However, very few of these examples have progressed from the laboratory to the field trial and none is currently in commercial production. The Copy Nature approach has considerable promise as a complementary method to current pest-management strategies. However, one of the factors that has slowed down the implementation of this strategy is that of food safety.

BOX 6.4 *The Pusztai affair*

The Pusztai affair precipitated a public backlash against GM food.

In the mid–late 1990s, tins of GM tomatoes and processed foods containing GM soya were clearly labelled and freely available on the shelves of UK supermarkets. Within the space of a year, most supermarkets had followed an apparently huge swing in public opinion against GM foods, and had removed all genetically manipulated products from their shelves. What had happened to cause this shift in the UK public's perception of GM food? One of the most influential factors was the announcement on television by Dr Pusztai of the Rowett Institute, Aberdeen, Scotland, in 1998, that GM potatoes caused intestinal changes when fed to rats.

There was considerable confusion about the nature of his initial claims, which were not immediately published in the peer-reviewed literature. In the storm of publicity that followed, Dr Pusztai was removed from his job. The Royal Society reviewed the unpublished work at the Rowett Institute, and published a report stating that the work was flawed. However, this did little to allay public concerns about food safety and the reliability of scientific advice in the aftermath of the bovine spongiform encephalopathy (BSE) crisis in the UK. By the time the work was subsequently published as a research letter in *The Lancet* in 1999, UK public concerns about GM food (and the environmental impact of GM crops) had reached a peak, and the supermarket shelves were being cleared of GM products.

The report itself concluded that feeding transgenic potatoes carrying the snowdrop lectin GNA (*Galanthus nivalis* agglutinin) gene to rats resulted in cell proliferation of the gastric mucosa. This effect could be attributed to the presence of GNA, since GNA added to normal potatoes had the same effect. However, increases in crypt length in the jejunum and decreases in caecal mucosal thickness were suggested to result from the genetic transformation of the potato, rather than the presence of GNA per se. Subsequently, it was suggested that the viral origin of the CaMV 35S promoter could, in some way, be responsible for this transformation-related toxicity. In addition to the Royal Society report, a number of critiques of the study have been published, pointing to problems of experimental design (lack of comparability between test and control diets, overall protein deficiency in all the diets), technique (measurement of crypt length, no actual measurement of cell proliferation), and amount of data (no time courses, no dose–response curves). However, finding a study to be inadequate is not the same as proving its findings to be wrong, and clearly further investigation of this issue was required. However, to date there have been no convincing reports of a change in toxicity of plant material as a direct result of the act of genetic transformation, or of the integration of the CaMV 35S promoter. Nevertheless, the Pusztai affair catalysed a reappraisal of the adequacy of food safety testing and regulations regarding GM foods. This topic will be covered in more detail in Chapter 12.

Insect-resistant crops and food safety

The wide range and type of genes shown in Table 6.6 raise understandable questions about their safety for human consumption, given that certain protease inhibitors and lectins are known to have toxic effects in mammals. One of the genes in the table, the snowdrop lectin GNA (*Galanthus nivalis* agglutinin), was the first such gene to attract considerable attention following the suggestion that potatoes carrying this transgene might be responsible for changes to the gut lining of rats (Box 6.4). A resolution of this issue is required, since the value of GNA has recently been demonstrated as a means of developing rice resistant to sap-sucking insects, which are not susceptible to *Bt* toxins or plant proteases. The brown leafhopper is responsible for annual rice crop damage equivalent to US$250 million and the green leafhopper is responsible for transmission of rice tungro virus, and hence rice crop damage equivalent to $340 million.

SUMMARY

In this chapter, we have looked in detail at the *cry* genes of *B. thuringiensis* and their exploitation for the production of insecticidal proteins in transgenic crops. The current situation in North America is that two major *Bt* crops, maize and cotton, have been approved for commercial planting since 1996. The widespread growing of these crops following approval has highlighted a number of issues, some of which have generated considerable controversy. Two major environmental concerns have been considered: the management of insect resistance to *Bt*, and the effects of *Bt* crops on non-pest insects. Two food safety issues have also emerged. The StarLink corn affair raises a series of questions about regulatory approval, allergenicity testing, and the contamination of non-GM foods. General issues of food safety testing and assurance are also raised by the use of insecticidal proteins from plants in transgenic crops. The different approach to pest resistance offered by such proteins has been contrasted with *Bt* crops in terms of a Copy Nature strategy.

FURTHER READING

Visit www.oxfordtextbooks.co.uk/orc/slaterplants2e/ for hyperlinked references that take you directly to online abstracts for the journal articles listed below:

Insect-resistant crops

Babu, R. M., Sajeena, A., Seetharaman, K., and Reddy, M. S. (2003) Advances in genetically engineered (transgenic) plants in pest management—an overview. *Crop Protection* **22**, 1071–1086.

Estruch, J. J., Carozzi, N. B., Desai, N., Duck, N. B., Warren, G. W., and Koziel, M. G. (1997) Transgenic plants: an emerging approach to pest control. *Nature Biotechnology* **15**, 137–141.

Moar, W. J. (2003) Breathing new life into insect-resistant plants. *Nature Biotechnology* **21**, 1152–1155.

Schuler, T. H., Poppy, G. M., Kerry, B. R., and Denholm, I. (1998) Insect resistant transgenic plants. *Trends in Biotechnology* **16**, 168–175.

cry genes and other bacterial insecticidal proteins

Crickmore, N., Zeigler, D. R., Feitelson, J., Schnepf, E., Van Rie, J., Lereclus, D., Baum, J., and Dean, D. H. (1998) Revision of the nomenclature for the *Bacillus thuringiensis* pesticidal crystal proteins. *Microbiology and Molecular Biology Reviews* **62**, 807–813.

Lee, M. K., Walters, F. S., Hart, H., Palekar, N., and Chen, J.-S. (2003) The mode of action of the *Bacillus thuringiensis* vegetative insecticidal protein Vip3A differs from that of Cry1Ab δ-endotoxin. *Applied and Environmental Microbiology* **69**, 4648–4657.

Liu, D., Burton, S., Glancy, T., Li, Z.-S., Hampton, R., Meade, T., and Merlo, D. J. (2003) Insect resistance conferred by 283-kDa *Photorhabdus luminescens* protein TcdA in *Arabidopsis thaliana*. *Nature Biotechnology* **21**, 1222–1228.

Naimov, S., Dukiandjiev, S., and de Maagd, R. A. (2003) *Plant Biotechnology Journal* **1**, 51–57.

Bt crops

de Maagd, R., Bosch, D., and Stiekma, W. (1999) *Bacillus thuringiensis* toxin-mediated insect resistance in plants. *Trends in Plant Science* **4**, 9–13.

Peferoen, M. (1997) Progress and prospects for field use of *Bt* genes in crops. *Trends in Biotechnology* **15**, 173–177.

Zhao, J.-Z., Cao, J., Li, Y., Collins, H. L., Roush, R. T., and Earle, E. D. (2003) Transgenic plants expressing two *Bacillus thuringiensis* toxins delay insect resistance evolution. *Nature Biotechnology* **21**, 1493–1497.

Bt cotton

Perlak, F. J., Oppenhuizen, M., Gustafson, K., Voth, R., Sivasupramanian, S., Heering, D., Corey, B., Ihrig, R. A., and Roberts, J. K. (2001) Development and commercial use of Bollgard cotton in the USA—early promises versus today's reality. *Plant Journal* **27**, 489–501.

Gene stacking

Halpin, C. (2005) Gene stacking in transgenic plants—the challenge for 21st century biotechnology. *Plant Biotechnology* **3**, 141–155.

Monarch butterfly

Losey, J. E., Rayor, L. S., and Carter, M. E. (1999) Transgenic pollen harms monarch larvae. *Nature* **399**, 214.

Sears, M. K., Hellmich, R. L., Stanley-Horn, D. E., Oberhauser, K. S., Pleasants, J. M., Mattila, H. R., Siegfried, B. D., and Dively, G. P. (2001) Impact of *Bt* corn pollen on monarch butterfly populations: a risk assessment. *Proceedings of the National Academy of Sciences USA* **98**, 11937–11942.

Shelton, A. M. and Sears, M. K. (2001) The monarch butterfly controversy: scientific interpretations of a phenomenon. *Plant Journal* **27**, 483–488.

Protease inhibitors and other plant genes

Boulter, D. (1993) Insect pest control by copying nature using genetically engineered crops. *Phytochemistry* **34**, 1453–1466.

Gatehouse, A. M. R., Boulter, D., and Hilder, V. A. (1992) Potential of plant derived genes in the genetic manipulation of crops for insect resistance. In *Plant Genetic Manipulation for Crop Protection* (A. M. R. Gatehouse, V. A. Hilder, and D. Boulter, eds), pp. 155–181. CAB International, Wallingford.

Gatehouse, A. M. R., Shi, Y., Powell, K. S., Brough, C., Hilder, V. A., Hamilton, W. D. O., Newell, C. A., Merryweather, A., Boulter, D., and Gatehouse, J. A. (1993) Approaches to insect resistance using transgenic plants. *Philosophical Transactions of the Royal Society of London Series B* **342**, 279–286.

Lawrence, P. K. and Koundal, K. R. (2002) Plant protease inhibitors in control of phytophagous insects. *Electronic Journal of Biotechnology* **5**, 1–9.

Nagadhara, D., Ramesh, S., Pasalu, I. C., Rao, Y. K., Krishnaiah, N. V., Sarma, N. P., Bown, D. P., Gatehouse, J. A., Reddy, V. D., and Rao, K. V. (2003) Transgenic indica rice resistant to sap-sucking insects. *Plant Biotechnology Journal* **1**, 231–240.

The Pusztai affair

Ewen, S. W. B. and Pusztai, A. (1999) Effect of diet containing genetically modified potatoes expressing *Galanthus nivalis* lectin on rat small intestine. *Lancet* **354**, 1353–1354.

Horton, R. (1999) Genetically modified foods: 'absurd' concern or welcome dialogue? *Lancet* **354**, 1314–1316.

Web-links

Bacillus thuringiensis Toxin Nomenclature, web-link 6.1: www.lifesci.sussex.ac.uk/home/Neil_Crickmore/Bt/

Bt Corn & European Corn Borer, University of Minnesota Extension Service, web-link 6.2: www.extension.umn.edu/distribution/cropsystems/DC7055.html

Corn pests BASF, web-link 6.3: http://basfagproducts2.com/information/insect-id-guide.asp/

Genetically Engineered Corn and the Monarch Butterfly Controversy, web-link 6.4: http://pewagbiotech.org/resources/issuebriefs/monarch.pdf

NAPIS PestTracker, web-link 6.5: www.ceris.purdue.edu/napis/

Peairs, F. B. (2001) Managing corn pests with *Bt* corn: some questions and answers, web-link 6.6: http://cls.casa.colostate.edu/TransgenicCrops/BtQnA.html

Royal Society (1999) Review of data on possible toxicity of GM potatoes, web-link 6.7: www.royalsoc.ac.uk/displaypagedoc.asp?id=6170

Royal Society (2002) Genetically modified plants for food use and human health—an update, web-link 6.8: www.royalsoc.ac.uk/displaypagedoc.asp?id=11319

The *Bacillus thuringiensis* Toxin Specificity Database, web-link 6.9: www.glfc.cfs.nrcan.gc.ca/science/research/netintro99_e.html

The StarLink case: issues for the future, web-link 6.10: http://pewagbiotech.org/resources/issuebriefs/starlink/starlink.pdf

Transgenic crops: an introduction and resource guide, web-link 6.11: http://cls.casa.colostate.edu/TransgenicCrops/index.html

Union of Concerned Scientists, web-link 6.12: www.ucsusa.org/food_and_environment/genetic_engineering/

US Environmental Protection Agency, web-link 6.13: www.epa.gov/pesticides/biopesticides/pips/index.htm

7

Plant disease resistance

INTRODUCTION

With the world's population continuing to rise, the major target of modern agriculture has to be a sustainable increase in yield that keeps pace with the increasing number of mouths to feed. This may be achieved by several approaches, not least by a significant reduction in pre- and post-harvest losses due to disease.

- Using just monocultures has lead to serious problems with disease sensitivity.

It is difficult to quantify the damage done by all the different plant diseases. It has been estimated that crop losses in the USA alone cost some US$33 billion per annum. In human terms, the loss of a crop by a subsistence farmer is as damaging, if not more so, than the loss of several thousand hectares in the mid-west of the USA. These huge losses are partly due to the use of monocultures, which encourage epidemics of pests and diseases. This is exemplified by the case of southern corn leaf blight caused by the fungus *Bipolaris maydis*. Maize hybrids, the main form of the crop, are produced by breeding female, cytoplasmic male sterility (CMS) lines with male pollinators, under controlled conditions. By the late 1960s, maize varieties developed from the same CMS line (CMS-T) were being used in large areas of the USA. Unfortunately, this germplasm was sensitive to the blight. Large areas of maize crop were decimated by what had previously been a minor form of leaf blight, but was now a very virulent form of the disease. Although this is an extreme case, it highlighted the problems of monocultures. Recent work has begun to explain why dispersed plant regimes are better than monocultures. One particular point, which is discussed later in this chapter in the section on inducible resistance, is that wild or mixed populations have a better chance of surviving attack from a virulent pathogen because they are likely to display polymorphisms in their disease-resistance systems. Monocultures are less likely to do so. There are two other important points against monocultures: (1) that different plant types add to the diversity of environmental conditions, making them less beneficial for particular pathogens; and (2) that the increased distance between plant genotypes dilutes the inoculum of a given pathogenic race as it is dispersed among compatible host varieties.

This chapter will describe some of the progress that has been made by plant biotechnology in combating plant diseases. It will briefly review the types of diseases, the organisms involved, and the costs due to these diseases, and then move on to consider the mechanisms of disease resistance found in normal plants and how these are being enhanced using genetic manipulation.

Plant–pathogen interactions

Interactions between plants and microorganisms can be viewed simply as being of four different types:

1 microorganisms can form symbiotic relationships with the host; these include bacteria like *Rhizobium* spp. and the rhizo-fungi;

2 the microorganism may cause disease in the host plant;

3 the host plant may be resistant to the pathogen, and no infection develops;

4 the host plant may show some tolerance to infection; in this case the pathogen is able to grow and replicate but symptoms of infection are minimal.

Within this broad sweep of interactions is a whole range of others that are now beginning to be understood.

- There are two major types of resistance: host and non-host resistance.

The pathogens themselves are of two basic types: (1) necrotrophs, which kill the host and feed on the contents, and which are often associated with toxin production; and (2) biotrophs, which require a living host to complete their life cycle. Bacteria and fungi may be of either type, whereas viruses are obligate biotrophs. In between these two extremes are hemibiotrophs; these are initially biotrophic but then switch to being necrotrophic. It is also clear that some organisms are pathogenic to some plants but not others. Why this is the case is beginning to be understood at the molecular level. There are two extremes of disease resistance exhibited by plants: non-host resistance and host resistance. Non-host resistance is when an entire plant species is resistant to a specific parasite or pathogen. The resistance mechanism that brings this about is conserved across the species and is complex—involving many inheritable markers—but is consequently robust in nature. This durable resistance is perhaps the ultimate goal of the biotechnologist, contrasting favourably with host resistance. Host resistance is exhibited by a single plant cultivar to a specific pathogen. Other members of the species may be susceptible to the infection. This type of resistance is due to interaction between a single plant protein and a single protein from the pathogen (gene–gene interaction). Normally it leads to a hypersensitive response (HR; see inducible resistance) that blocks the spread of the pathogen. It is not as robust as non-host resistance but at present it is achievable although possibly only for a short-term benefit. As with the pathogen, between the two extremes of resistance there is race-non-specific resistance or a general response. This is also polygenic response but individually the genes would probably not provide a successful defence. As we will see below, there is a great deal of overlap exhibited by these defence systems and perhaps they should be viewed as a continuum of measures that plants can put up against attack.

In the following sections, a brief summary of the main types of organisms and their effects will be given. Then the molecular biology of resistance will be discussed in order to understand the transgenic approaches so far attempted.

Prokaryotes

Only a limited number of bacterial genera are plant pathogens.

Among the prokaryotes, the most economically important pathogenic organisms are members of the phytoplasma group (previously known as mycoplasma-like organisms) and organisms from a small number of bacterial genera. The phytoplasma belong to a group of prokaryotes that lack cell walls and survive as either saprophytes (organisms that live on dead organic matter, which they help to break down) or as intracellular pathogens. Diseases include aster yellows, in which the phytoplasma is spread by the aster leafhopper. The pathogen infects over 300 host plant species, including flax, oilseed rape (canola), wheat, barley, and potato. Killing the leafhopper vector with chemicals controls the disease.

The main groups of bacterial plant pathogens include members of the genus *Agrobacterium*, which cause crown galls and hairy roots by transferring T-DNA into plant cells (see Chapter 3). Some pathogens can enter the host through natural openings such as stomata (see Box 7.4, below) but, like many other pathogens, *Agrobacterium* enters the plant through wound sites. Other bacterial pathogens are found among the *Corynebacterium*, *Erwinia*, *Pseudomonas*, *Xanthomonas* (see Case study 7.2, below), and *Streptomyces* genera. Many of the symptoms caused by the different infestations of bacteria are similar: spots, galls, cankers (localized death of an organ), blights (rapid discoloration, wilting, death), soft rots (caused by secreted enzymes that break down plant cell walls and reduce storage time), and wilts (caused by bacteria blocking the vascular tissues). These infections can be localized and have a minimum commercial effect, disfiguring the fruit or leaves, or they can have major effects on yield or commercial value. For instance, the bacterial disease fireblight, caused by *Erwinia* spp., can kill whole fruit trees in a single season.

Fungi and water moulds

Fungi are the most important plant pathogens.

The vast majority of cellular pathogens of plants are found among the fungi. The losses of food caused by fungal diseases have had profound effects on humanity. This is in a number of ways. They can damage plants in the field and after harvest (rotting etc.) and they can contaminate foodstuffs with extreme toxins such as aflotoxins. There are over 100 000 known species of fungi. Most are examples of saprophytes, but the fungi also include symbionts, such as the mycorrhiza, and perhaps more than 8000 species that cause plant diseases. Modes of entry include wound sites and natural openings such as stomata, and the enzymes that break down dead organic matter can be used by pathogens to gain entry into living plants by degrading cell-wall macromolecules. There are too many fungal pathogens to go into any detail of the diseases they cause or their modes of action, but some specific examples are given in Table 7.1. Included is *Phytophthora infestans* (which is a classic example of an

Table 7.1 Examples of pathogens causing serious economic loss

Pathogen	Major crops infected (disease caused)	Notes
Fungi and water moulds		
Bipolaris maydis	Maize (southern corn leaf blight)	A minor pathogen until a major outbreak occurred in 1969/1970 on maize lines with certain cms backgrounds
Fusarium spp.	Wheat, barley (scab)	Multibillion dollar losses worldwide
	Cotton (wilt)	Multimillion dollar losses in USA annually
Phytophthora spp.	Potato (blight)	*Phytophthora* caused the Irish potato famine of 1846
	Cocoa (e.g. black pod)	Loss >450 000 tons annually in Africa/Brazil/Asia corresponds to $430 million
	Soybean (e.g. root rot)	15% of Ohio land area is covered with a crop worth $1 billion; losses reach $120 million annually
		Total crop losses due to *Phytophthora* spp. of >$10 billion
Puccinia graminis	Wheat (rust)	A major problem controlled with chemicals. Losses could be in the order of billions of dollars annually
Rhizoctonia solani	Many major crops (rots, scabs)	Serious fungal pathogen
Viruses		
Barley yellow dwarf virus	Major small-grain cereals	Annual losses in USA estimated at >$300 million
Cacao swollen shoot virus	Cocoa (swollen shoot)	Losses in Africa of >$50 million
Cassava African mosaic virus	Cassava	Losses attributed to this and other cassava viruses of >$2 billion annually
Rice tungro virus complex	Rice	Losses in south-east Asia of >$1.5 billion
Tomato spotted wilt virus	Peanuts (spotted wilt disease)	Losses of $20–40 million annually in Georgia, USA
Bacteria		
Agrobacterium spp.	Range of dicotyledonous plants	See Chapter 3
Pseudomonas syringae	Beans (blight, brown spot)	Florida's climate makes the $2 billion-worth of susceptible crops particularly vulnerable to the disease and losses can run into many millions
Xanthomonas campestris	Tomato, pepper, beans (bacterial spot)	
Xanthomonas fragariae	Strawberries (angular leaf spot)	

Costs are in US dollars.

Oomycetes-group pathogen), the organism responsible for the potato blight that caused the great Irish famine of 1846. Until recently, Oomycetes were classified as fungi; however, biochemical and molecular studies indicate that these water moulds should now be classified separately. The cell walls of these organisms are composed of cellulosic compounds and glycan, not chitin, and the nuclei within the hypha (filaments) are diploid, not haploid, as in fungi. Another member of the Oomycete

BOX 7.1 *Emerging diseases in plants*

As with human pathogens, there are examples in plant pathology of what could be called emerging pathogens. A significant example here is the oomycete *Phytophthora ramorum*, the causative agent of what is known as sudden oak death. In the 1990s, this organism caused the death of millions of oak trees in the USA and it was only in 2000 that this previously unknown forest pathogen was identified. *P. ramorum* has now been detected in many countries, including those of Europe, and in many other plant species. In 2004, it had been found to cause disease in 34 different genera of trees, including ash and beech, as well as ornamental plants such as rhododendrons, viburnum, pieris, and camellia.

An interesting emerging bacterium is *Pantoea ananatis*. This organism has been identified as a human pathogen and has recently been linked to various plant diseases such as bacterial blight and dieback in *Eucalyptus* species and centre rot of onion. This bacterial disease is economically important, causing serious losses to onion crops in the US state of Georgia, and it also infects cantaloupe melons, honeydew melons, muskmelons, pineapple, and tomatoes.

group, *Phytophthora ramorum*, is responsible for a recently recognized disease that is discussed in Box 7.1.

Viruses

Biotechnological approaches to plant virus resistance will be dealt with in Chapter 8. Here, it is important to note that viruses constitute a major group of pathogens, and that many of the natural host defences discussed below are induced by the cellular and viral pathogens.

Some effort has been made in recent years to classify plant viruses (families, genus, species, etc.; for the most up-to-date taxonomy, see web-link 8.5). Full details of 950 plant viruses, their hosts, and the diseases they cause can be found at Plant Viruses Online (web-link 7.1). The vast majority of plant viruses are single-stranded, positive-sense RNA viruses, but other types such as circular, double-stranded DNA viruses exist. Interestingly, one of the examples given in Table 7.1, tungro virus disease, often involves two viruses: rice tungro bacilliform virus (a DNA badnavirus) and rice tungro spherical virus (an RNA waikavirus). The badnavirus causes the tungro symptoms (yellow–orange leaves and stunting), but it depends on the waikavirus for transmission by leafhoppers (the waikavirus also increases the severity of the tungro symptoms), whereas the waikavirus alone causes mild stunting.

Existing approaches to combating disease

At present, many diseases are controlled by the large-scale use of expensive chemical treatments that kill either the pathogen or the vectors that carry them. At present, US$700 million is spent on fungicides in the USA annually. These chemicals may, by their very nature, be detrimental to human health and the environment. An example of this is the fumigant methyl bromide, which has been used to control pathogens, insects, and nematodes. In the USA, some 27 200 t (30 000 tons) of the chemical have been used each year to fumigate the soil prior to planting crops, and to fumigate harvested crops during storage and prior to export and import (to kill any potential pathogens). Methyl bromide has been identified as a chemical that contributes to depletion of the ozone layer, and as such its use is now being phased out. This has left the agricultural industry with the task of developing methyl bromide-free farming practices. Another problem is that the pathogens can become resistant to the chemicals being used. In this chapter, we deal with some examples of strategies being developed to reduce the use of these chemicals (see also Chapter 6).

An alternative cleaner strategy that has been investigated is using biological control strategies. Certain natural strains of *Agrobacterium*, including *Agrobacterium rhizogenes* 84, have been used commercially to limit crown gall disease (see Chapter 3). There is some limitation on the range of plants that this approach is successful with, so other strains are being developed that have an extended host range, such as *Agrobacterium vitis*. This bacterium has been modified to express a ribosomally synthesized peptide antibiotic trifolitoxin (TFX), so it is not a genetic-modification (GM)-free approach.

● Biological control may possibly replace chemical pesticides.

Agricultural seed companies have been successful in developing disease-resistant lines. However, this is a continual battle as the pathogens usually mutate to overcome the resistance. In the past, this may have taken many years of intensive breeding programmes to produce the plants, which may not necessarily have all the desired properties or the required level of resistance. Box 7.2 gives an example of a product produced by conventional breeding.

Transgenic approaches have significant value in this area. To understand the approaches being used and the opportunities available, we need to consider the natural mechanisms of disease resistance that can be optimized by gene manipulation.

BOX 7.2 *Traditional breeding—a long haul*

It can take many years to breed crops with desirable traits. In 2000, Cornell University in New York announced the development of a potato, New York 121. This potato is resistant to the oomycete *Phytophthora infestans* (the cause of late blight), golden nematodes, scab, and potato virus Y. The history of this potato dates back 30 years when the breeding programme first started with seeds obtained from the Andes mountain range in South America. The Cornell group developed varieties that were adapted to the New York region and had resistance to certain viruses. One of these lines was used as the mother of New York 121. This was crossed with lines that had been developed from seeds obtained from Peru with resistance to golden nematode. It took 9 years to develop and test the progeny New York 121. The potato has one major drawback: it is good for boiling but not for making French fries.

Natural disease-resistance pathways: overlap between pests and diseases

● Plants defence systems are complex but effective.

It would be wrong to give the impression that plants have no resistance against pathogen attack, for it is clear that they do. They do not have an immune system that produces specific cells to attack invading microbes; rather, they have adopted more general defence systems and these show many similarities with the innate defence systems of mammals. There is also a lot of overlap between the plant's response to pathogens and plant pests. The general cellular damage caused by both can act as a signal to trigger general defence systems. There are advantages to the plant having such general systems because pests often act as vectors for pathogens. As we shall see below, plants also have systems that respond to specific signals. However, much of the recent molecular work has been done with model plants such as tobacco and *Arabidopsis* and it is possible that the observations are not applicable to all plants. In this section, four different levels of defence will be considered:

1 anatomical defences;

2 pre-existing protein and chemical protection;

3 inducible systems;

4 systemic responses.

All four levels can contribute to resistance; the amount will depend on how successful the pathogen is at overcoming the preceding barrier. Levels 1 and 2 form the basis of the general defences of the plant, and the inducible systems also contribute to these. However, the inducible systems also show more-specific defence responses.

Anatomical defences

As indicated above, many pathogens have to invade plants through wounds. This is because plants have developed morphological and structural systems that preclude pathogen access to living cells (the first line of defence). These can be thick layers of protective material, such as cuticle, bark, or waxes. Once this defence is breached then cascades of defence systems come into play.

Pre-existing protein and chemical protection

The second line of defence is made up of a range of antimicrobial proteins produced by the plant during growth and development, including defensins and defensin-like proteins. The defensin proteins are similar to those found in insects and mammals where they play an important role in defence against infectious agents. Their structure has a conserved three-dimensional folding pattern, which suggests that they represent a superfamily of peptides that pre-dates the divergence of plants and animals. Some defensins cause increased branching in fungi, while others simply slow growth. They are frequently associated with seeds at the time of germination, when they may be released into the environment and create a microenvironment around

BOX 7.3 *Secondary products as antimicrobial compounds*

Plant cells make huge numbers of small molecules as secondary products. Many of them, such as terpenes, phenolics, and alkaloids, have antipathogen properties. Presynthesized compounds have been given the name of phytoanticipins. These may be directly toxic or become so following release from, for example, a conjugate state. For instance, plant glycosides are often hydrolysed following insect damage or pathogen invasion that releases vacuolar glycosidases. This enzyme reaction releases aglycones that are toxic to the invader and adjacent cells, producing a local 'fire break'. Antipathogenic chemicals synthesized *de novo*, on infection, are known as phytoalexins. Some of these chemicals can be classified as phytoanticipins in one plant and phytoalexins in others. Such is the case for the methylated flavanone sakuranetin, which is constitutively accumulated in blackcurrant, but is a major inducible antimicrobial in rice.

the seed that suppresses fungal growth. Many of the large number of small chemicals made as secondary products may also have antimicrobial properties. These proteins and chemicals may simply deter pest or pathogen growth, or they may actually be toxic to them (see Box 7.3).

Inducible systems

Elicitor response

The third level of defence is a switch to counterattack that relies on protein synthesis *de novo*. It would be costly for the plant to have its defence system permanently switched on, so there are mechanisms in place to detect the infection and then turn on the defence system. These mechanisms will be discussed below (see Box 7.4).

When a pathogen arrives and gains entry to a living cell, it may induce resistance to infection. The inducers, or elicitors, are of a number of different types, and the responses are complex and overlapping in nature. First, there may be a local response that involves interactions with molecules released by the pathogen (exogenous elicitors) or the host itself, in response to pathogen attack (endogenous elicitor), with receptor molecules on the host. These interactions may stimulate general defence systems in both host and non-host plants, or they may be race-specific interactions, inducing defence responses only in specific cultivars. Both these interactions may lead to a cascade of reactions that may include the hypersensitive response (HR). These interactions are summarized in Figure 7.1. Table 7.2 shows a range of elicitors and their effects. The next part of the defence response is the induction of a systemic resistance (and even the passage of signals to other plants).

General response

In the first stages of an infection, there is normally damage to the plant cell wall. Fungi or bacteria release proteolytic enzymes that cause the release of wall fragments such as pectic-oligomers (e.g. oligogalacturonides). These often act as general signals —endogenous elicitors—which bind to specific receptors, setting off a cascade of

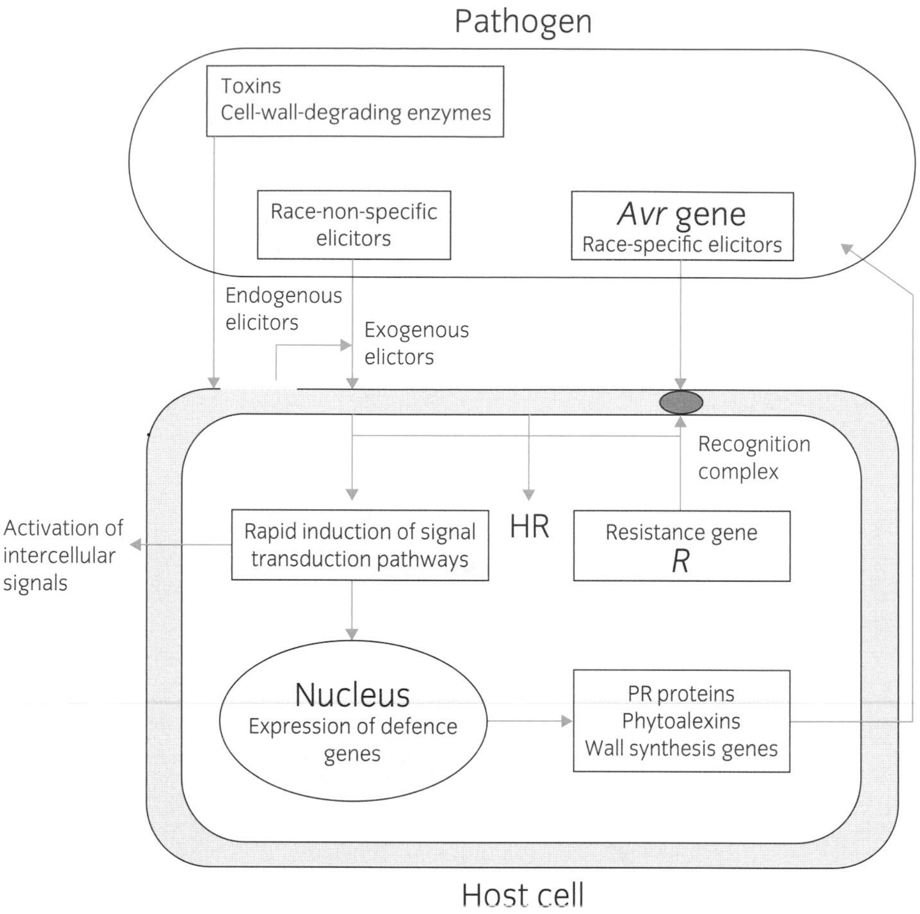

Figure 7.1 Cellular response to pathogen attack. Two of the major plant pathogen interactions are highlighted in this diagram. Host defence systems are induced by endogenous and exogenous elicitors and by the interaction between plant *R* (resistance) genes and the pathogen avirulence genes (*Avr*).

reactions that lead to the induction of specific defence genes. These defence genes code for enzymes that synthesize structural components for cell-wall thickening (to repair the damage), enzymes of secondary metabolism, lectins (multimeric sugar-binding proteins that agglutinate cells), and many so-called pathogenesis-related (PR) proteins (see Table 7.3). PR proteins include chitinases and β-1,3-glucanases, protease inhibitors, non-specific lipid-transfer proteins, ribosomal inhibitor proteins, and various antimicrobial proteins. These antimicrobial proteins include defensins such as SN1, which is active against bacterial and fungal pathogens in potato. One important point of relevance to the use of the genes in transgenic experiments is that a number of the PR gene products have been identified as latex allergens. This is a family of proteins found in many plant species and responsible for serious medical conditions in many people who show allergenic response to latex products and a

Table 7.2 Examples of elicitors

Elicitor	Pathogen	Type	Biological response
β-Glucans	Fungi	PAMP/general	Induction of defence response in rice, tobacco, and legumes Phytoalexins in soybean
Chitin	Fungi	PAMP/general	Induction of defence response in tomato, *Arabidopsis*, rice, wheat, and barley Phytoalexin in rice Lignification in wheat leaves
Pectolytic enzymes releasing endogenous elicitors, e.g. oligogalacturonides	Fungi/bacteria	General	Protein inhibitors and defence genes in *Arabidopsis* and tomato
Flagellin	Gram-negative bacteria	PAMP/general	Defence gene and ROS induction in tomato and *Arabidopsis*
Endoxylanase	Various fungi	Race-specific	HR and defence genes in tobacco
Avr proteins	Fungi/bacteria	Race-specific	HR in a range of plants
Peptide toxins	Fungi	Race-specific	Programmed cell death in oats

Avr, avirulence protein; PAMP, pathogen-associated molecular pattern; ROS, reactive oxygen species.

Table 7.3 Types of pathogenesis-related (PR) protein

Family	Type member	Properties
PR-1	Tobacco PR-1a	Antifungal, 14–17 kDa
PR-2	Tobacco PR-2	Class I, II, III endo-β-1,3-glucanases, 25–35 kDa
PR-3	Tobacco P, Q	Class I, II, and IV endochitinases, 30 kDa
PR-4	Tobacco R	Antifungal, *win*-like proteins, endochitinase activity, similar to prohevein C-terminal domain, 13–19 kDa
PR-5	Tobacco S	Antifungal, thaumatin-like proteins, osmotins, zeamatins, permeartins, similar to α-amylase/trypsin inhibitors
PR-6	Tomato inhibitor I	Protease inhibitors, 6–13 kDa
PR-7	Tomato P	Endoprotease
PR-8	Cucumber chitinase	Class III chitinases, chitinases/lysozyme
PR-9	Lignin-forming peroxidase	Peroxidases, peroxidase-like proteins
PR-10	Parsley PR-1	Ribonucleases, Bet v 1-related proteins
PR-11	Tobacco class V chitinase	Endochitinase activity
PR-12	Radish Ps-AFP3	Plant defensins
PR-13	*Arabidopsis* THI2.1	Thionins
PR-14	Barley LTP4	Non-specific lipid-transfer proteins

(Based on table in web-link 7.2: http://dmd.nihs.go.jp/latex/defense-e.html with permission of site owner/maintainer.)

number of fruit species. More information can be found in the web-links listed in the Further reading section at the end of this chapter.

The effect of synthesizing the defence proteins depends, in part, on the pathogen. For instance, if fungal hyphae breach the cells defences, then the chitinases and glucanases may cause some degradation of the pathogen's cell wall. This will lead to the production of chitin and β-1,3-glucan oligomers. These compounds may act as signal molecules (exogenous elicitors), which bind to membrane receptors and re-enforce the induction of the defence systems. The cells may also produce phytoalexins (phenolic compounds or terpenes) that kill any pathogens as well as the cells in the vicinity of the infection, therefore limiting the spread.

The identification of cell-wall components (lipopolysaccharide, flagellin, chitins) as general elicitors of resistance is not unexpected, as they are at the interface of the interaction. What is perhaps more surprising is that plants and animals respond to the same elicitors and that the receptors of the signals have many structural similarities. Bacterial flagellin, a well-studied example of a pathogen-associated molecular pattern (PAMP), is discussed in Box 7.4.

BOX 7.4 *Elicitor–R-protein interactions*

Flagellin, the subunit of bacterial flagella, is a well-studied general elicitor. Its PAMP activity has been located to a 15–22-amino acid conserved sequence in the subunit from the Gram-negative bacterium *Pseudomonas syringae*. The receptor for this molecule is FLS2. It has a leucine-rich repeat (LRR) module linked via a transmembrane section to a protein kinase. As shown in Figure 7.3, this R protein traverses the plasma membrane. The binding of the elicitor to the external receptor is signalled to the nucleus by the internal protein kinase and then via a mitogen-activated protein kinase (MAPK) cascade. This has recently been identified for FLS2 in *Arabidopsis*. The cascade includes a classic series of kinases (MAPKKK, a MAPKK and a MAPK) and a transcription factor, WRKY. The transcription factor is involved in regulating the expression of various defence genes. Analysis of the *Arabidopsis* genome indicates that these are part of a large family of MAPK genes to be found in the organisms that function in stress and disease resistance.

Analysis of the FLS2 system in guard cells has recently shown that the stomata are an active participant in the defence systems of the plant. Experiments with the *P. syringae* strain pv. tomato (DC3000) have shown that it can swim towards the stomatal opening to gain entry into the leaf. The guard cells perceive the presence of the bacterium through the PAMPs of the flagellin molecules (possibly lipopolysaccharide and others) leading to a cascade of reactions involving abscisic acid (ABA), nitric oxide, and the OST1 kinase. This leads to stomatal closure, preventing the pathogen entering the leaf. However, the pathogen can fight back and it has been found that it can produce an inhibitor of this process, a phytotoxin called coronatine.

The Pto protein (an R protein) in tomato is a serine/threonine kinase that confers resistance to strains of the *Pseudomonas* genus carrying the *avrPto* gene. This is an interesting example to compare with FLS2 for several reasons. First, the R protein is cytoplasmic, which suggests that the bacterial AvrPto protein must be introduced into the plant cell. The trafficking of pathogen proteins into the host cell via the type III protein-secretion system (TTSS) is a common feature of

BOX 7.4 *Continued*

pathogen attack. Second, Pto is a kinase, a common and direct mechanism for the initiation of a signal transduction cascade. The third reason is that Pto lacks the LRR recognition domain. This domain is found in most of the R proteins identified to date. So how does Pto function? It has been shown that this reaction depends on a second plant protein, Prf, which does contain an NB–LRR domain. The Pto protein forms a complex with AvrPto and Prf, which initiates the HR reaction and a signal transduction cascade (via the kinase activity) that leads to the expression of various defence proteins. It has also recently been discovered that the AvrPto protein can be phosphorylated. It mimics the substrate for a kinase that is found in the tomato leaf. This phosphorylation enhances the virulence activity of the protein (see figure).

Several models have been proposed to account for these observations. One is that the Avr protein interferes with the R protein's function in an innate non-specific defence system. In this scenario, the role of the NB–LRR domain (or in the Pto system, the Prf protein) is to recognize the malfunction of the innate defence system and guard the R protein by activating the cell's inducible defences. This is known as the **Guard hypothesis**.

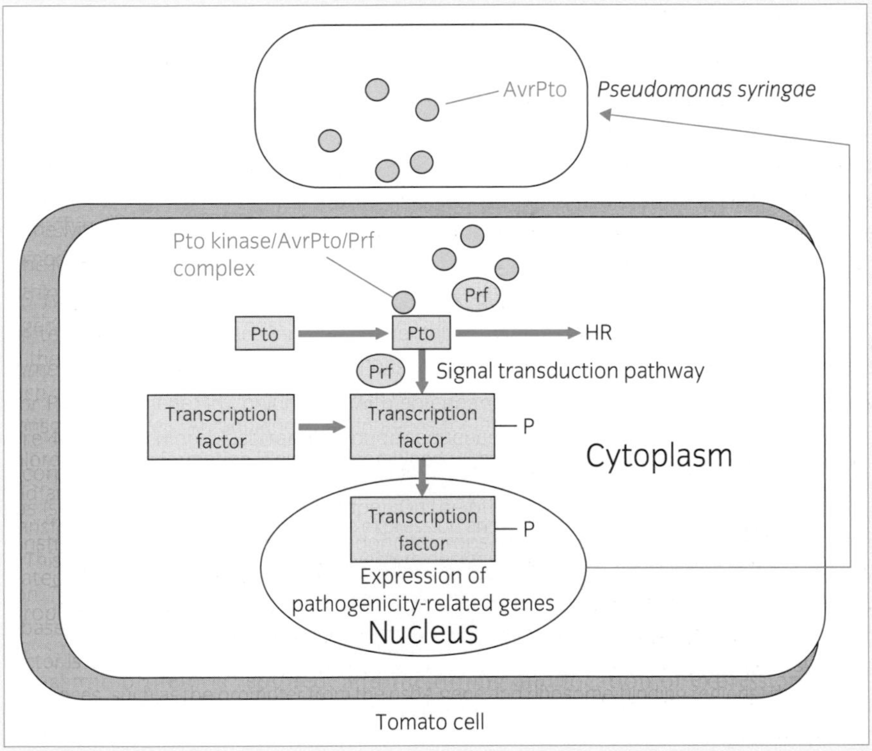

The Pto kinase pathway. When *Pseudomonas syringae* producing the AvrPto protein infects a tomato plant, the protein can form a complex with the Pto kinase and the Prf protein. The complex activates the HR and a signal transduction pathway that induces the expression of a range of pathogenicity-related genes.

Race-specific response

• The hypersensitive response (HR) is a suicide system to prevent the spread of pathogens in the plant.

Using a genetic analysis of the interactions between flax and flax rust, Harold Flor developed the gene-for-gene hypothesis more than 50 years ago. The idea stems from the observation that disease resistance requires two complementary genes. The pathogen carries the avirulence gene (*Avr*), which codes for a protein that is recognized by a specific receptor protein in the plant cell, encoded by the resistance gene (*R*). This interaction induces the HR, which is manifested as a local necrosis that develops through a NADPH (reduced nicotinamide adenine dinucleotide phosphate)-dependent oxidative burst and/or the release of phenolics and nitric oxide. The activation of signalling pathways also leads to the induction of many of the pathogenesis-related proteins. The most important feature of this recognition system is that if either of the proteins is absent, then the pathogen will cause disease in the plant; this is depicted in Figure 7.2.

The gene–gene system is one of the most important ways plants have of switching on resistance systems, and it has been shown to function for interactions between plants and aphids, nematodes, fungi, viruses, and bacteria. It is becoming clear that the many of the bacterial avirulence proteins fall into a class of proteins transported into the plant cell by the type III protein-secretion system (TTSS). It is postulated that they might function as virulence factors, subverting cellular functions through interactions with plant-encoded pathogenicity targets. For example a type III effector from *Pseudomonas syringae* pathovar (pv.) *phaseolicoli*, VirPphA, blocks other masked Avr proteins from inducing the HR response. Other types of effectors are able to suppress PAMP-induced defences. AvrPto has been shown to suppress the expression of a set of genes that encode cell wall and defence proteins.

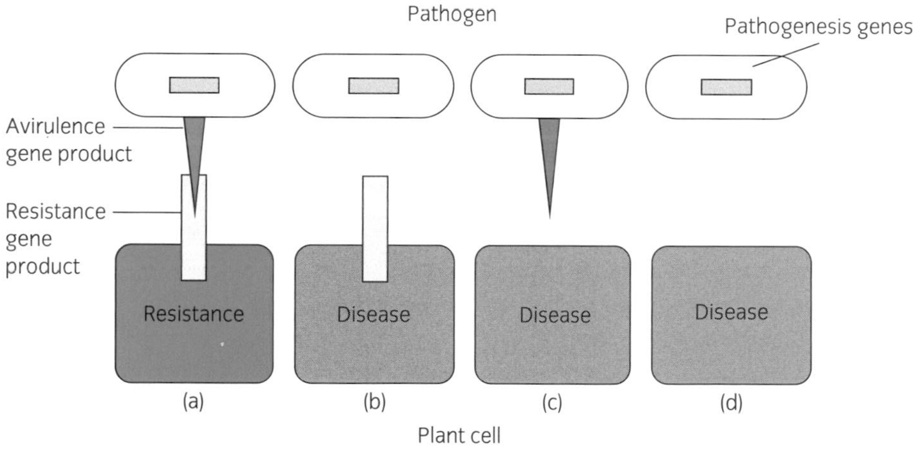

Figure 7.2 HR interactions. In (a) the plant cell contains the resistance gene (*R*) and the pathogen produces the avirulence gene product (Avr), so the defence systems are activated and the plant shows resistant to infection. In (b) the pathogen is not producing the Avr protein, a local HR reaction is not induced and so the interaction results in disease. In (c) the host cell is not producing the R protein so disease results. In (d) neither the R protein nor the Avr protein is being produced so disease results.

When fungal hyphae break through the cell wall into a cell, they produce a complex cellular structure, the haustoria, that drains the required nutrients from the cell. Interestingly, the haustoria does not actually penetrate the host cell's membrane, yet the fungus is able to highjack the cell's contents. Recent work with fungi and oomycete pathogens indicates that these organisms inject small proteins into the host plant cells that are designed to alter the host's metabolism, to the benefit of the pathogen, and to disrupt any defence pathways. It is these proteins that can stimulate the HR response. They may be recognized as avirulence proteins by corresponding R proteins (see below), inducing host defence systems.

Elicitor receptors (R proteins)

Many examples of the R proteins have now been identified: the annotation of the *Arabidopsis* genome sequence has indicated there are about 100 *R* loci distributed throughout the genome. PAMP receptors and TTSS proteins show many similarities in structure to mammalian and insect receptor proteins required for the onset of the innate immune response and involved in the sensing of pathogen-derived factors. They fit into five basic structural groups, which are shown in Figure 7.3. In the biggest group, the proteins are characterized by having a nucleotide-binding site and a leucine-rich repeat region (these are termed NB–LRR proteins). The other four groups are structurally diverse, but examples have been found in which the R proteins also contain an LRR region or, as discussed in Box 7.4, the R protein interacts with another protein that contains the LRR domain.

- PAMPs and the TTSS are different types of elicitors with similar receptors.

Xa21 is an R protein from rice that is an LRR receptor kinase. It confers cultivar-specific resistance to *Xanthomonas oryzae* pv. *oryzae* strains that express AvrXa21 avirulence protein. It resembles PAMP receptors in structure: both human Toll-like

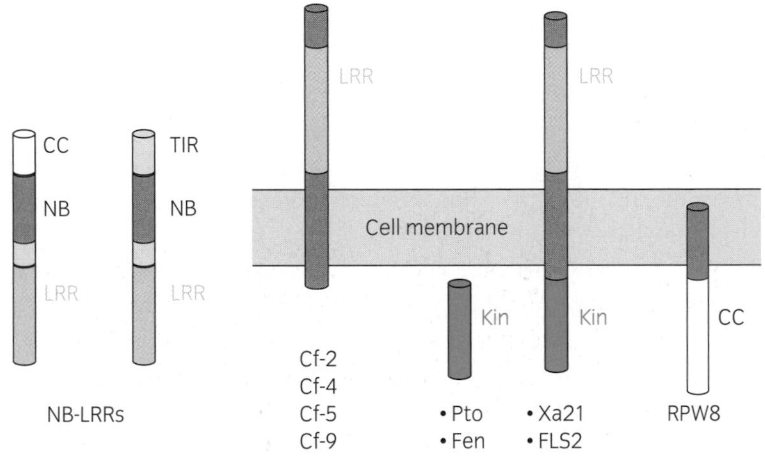

Figure 7.3 A representation of the location and structure of the five main classes of plant disease protein. NB–LRRs and Pto are believed to be cytoplasmic, although they may be associated with membranes. The other types (represented by Cf-2, Xa21, and RPW8, etc., R proteins) are all membrane-associated. Abbreviations not in the text: TIR, Toll-interleukin-resistance; Kin, kinase; CC, coil–coil domain. (Redrawn with permission from Dangl and Jones (2001).)

receptors and the flagellin receptor FLS2 from *Arabidopsis*. Comparisons like these indicate that plants use receptors of similar structure for the sensing of different types of elicitor and that these have evolved from a limited number of ancient receptors. It is intriguing to consider that such a small number of R proteins can give the diversity required to deal with the large number of plant pathogens, and how they evolve to deal with evolutionary changes in the elicitor proteins. One factor seeming to favour plant defence mechanisms is that many elicitor proteins are key factors in the biology of the pathogen, and as such they do not evolve at high rates. There is also evidence indicating that to meet with the changes that do occur, some variability in the plant R proteins comes from the fact that the LRR domain does mutate, within a highly variable region, and that mutations encoding new amino acids are selected for. The *R*-gene loci are very polymorphic in wild populations, so with each *R*-gene allele being present at a low frequency there is limited selection for virulence in the pathogen population. In conventional agriculture, with the use of monocultures, the balance between polymorphisms within the resistance genes is disrupted, therefore making the crops potentially more susceptible to virulent infections. The mechanisms by which interactions between R and Avr bring about the HR and induce the expression of the defence proteins are now being uncovered, and several of these points are discussed in Box 7.4.

Systemic responses

● Messages are passed from sites of infection systemically and even to other plants.

The induction of local defence pathways may lead to the induction of intercellular signals that produce a systemic response, termed systemic acquired resistance (SAR). Both avirulent and virulent pathogens may result in the induction of SAR, although it is usually a slower process in the case of virulent pathogens. SAR has two phases: the initiation phase and the maintenance phase. In the initiation phase, cells at the foci of the infection release signal molecules, typically salicylic acid, into the phloem. These are transported to target cells in other parts of the plant where SAR genes (such as certain PR proteins etc.) are expressed, thus giving the plant some level of resistance against infection. In the longer term, a maintenance phase is reached that may last for weeks or even the full life of the plant, in which there is quasi-steady-state resistance against virulent pathogens (see Figure 7.4).

Other states of disease resistance, such as induced systemic resistance, which are independent of salicylic acid, are also induced by avirulent pathogens through pathways that seem to include ethylene or jasmonic acid as messengers (see Figure 7.5). These may result in the induction of different classes of defence proteins. Evidence now suggests that signals carried by volatile chemicals, such as methyl jasmonate, can travel between plants to warn of an attack. Many of the proteins involved in these pathways have been identified and are providing further insights into disease resistance.

One interesting development from transcriptomic studies is that there is significant levels of crosstalk between biotic (pathogen wounding, mechanical wounding by insects, etc.) and abiotic (drought, high salinity, cold, etc.) stress signalling networks. It is clear that many of the same genes are expressed in response to these different

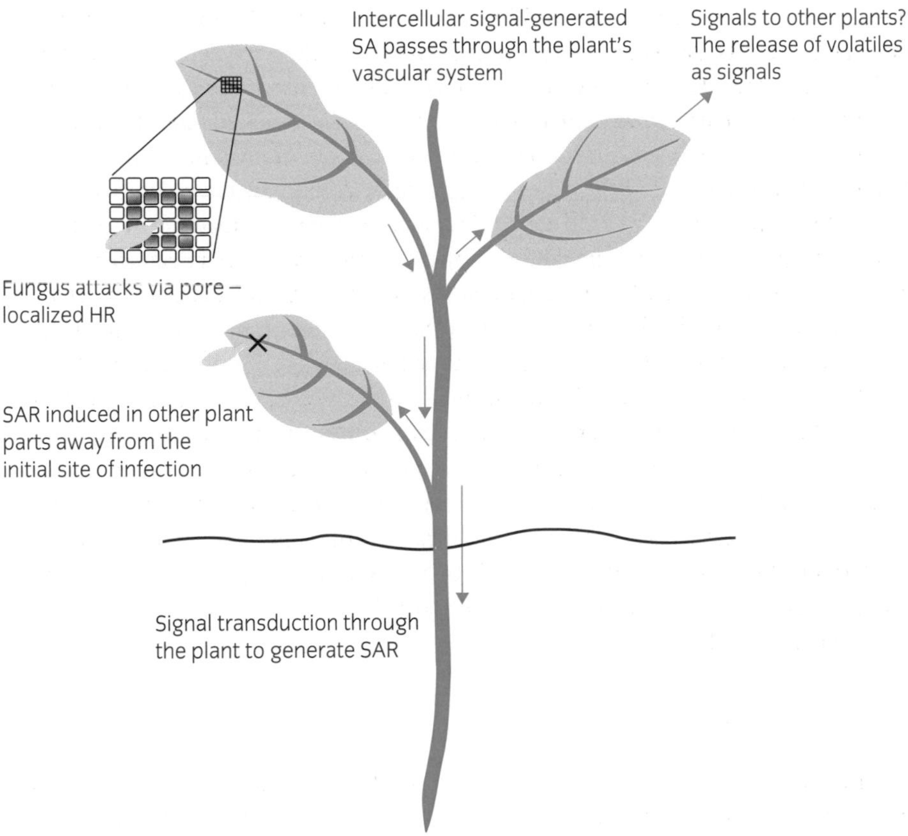

Figure 7.4 Induction of systemic acquired resistance (SAR). SA, salicylic acid.

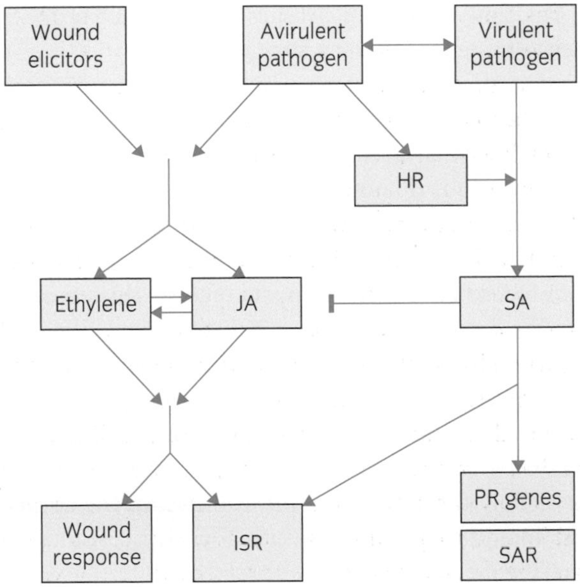

Figure 7.5 Network of disease-response pathways. ISR, induced systemic resistance; JA, jasmonic acid; SA, salicylic acid; SAR, systemic acquired resistance. (Based with permission on Dong (1998).)

stresses and many of the same hormones are involved in regulating these processes. Recently there has been a lot of interest in the role of reactive oxygen species (ROS) in regulating these phenomena. There has been much evidence for ROS interacting with nitric oxide in the establishment of the HR but now there is evidence that ROS are a common signal that trigger the downstream processes in many stress responses.

It is becoming clear that the different plant defence systems have significant levels of overlap such that they could be termed the plant innate immune system. Host resistance is dependent upon the interaction of an *R* gene with a TTSS protein, normally leading to an HR reaction. Non-host resistance is more complex and seems to be composed of various defence genes including R proteins. It has been proposed that there are two types of non-host resistance: type one does not induce an HR reaction and involves general defence mechanisms such as the physical barriers of the cell wall, antimicrobial compounds, and secondary metabolites followed by the response to PAMPs. The pathogen cannot get further in its attack. The second type of non-host resistance does produce a non-host HR reaction. If the pathogen overcomes the preliminary defences and enters the cell, then there can either be a response to cell-surface elicitors, as in the case of fungi or oomycetes, or there can be a form of TTSS–*R*-gene interaction in the case of bacteria.

Biotechnological approaches to disease resistance

A great deal of progress has been made in converting the fundamental information obtained on resistance into strategies for enhancing plant defence systems. As one would imagine, these range from strategies in which individual pathogens are targeted to approaches designed to enhance general resistance systems to inhibit a range of pathogens. Many have been tested on model systems, some have been introduced into crop plants, but none (except for virus resistance, which is dealt with in Chapter 8) have so far been commercialized on a large scale.

In the following section, the strategies and targets for engineering will be considered. Initially genes for proteins with specific actions will be discussed. These include PR proteins, which have shown some activity at inhibiting disease but have limited activity and generally do not switch on disease-resistance pathways. The use of antimicrobial proteins is also discussed; these have been used singularly or in combination with other genes, and some examples of engineered forms are highlighted. A more advanced approach has been to attempt to regulate clusters of disease-resistance genes by switching on regulatory pathways. This has been achieved by using R proteins, elicitors, hormone expression, or regulatory proteins that coordinate a number of pathways. The *R* genes are particularly favoured as they have been used for traditional breeding approaches so they may be more accepted in the industry for transgenic approaches.

The following section will highlight some examples of the approaches used. It is not a complete list; many other examples can be found by consulting the papers listed at the end of the chapter.

Protection against pathogens

PR proteins

Initial studies for the enhancement of resistance have been based upon the plant's own defence systems, in particular the transformation of plants with genes coding for PR proteins. To combat fungal pathogens, many plants have been transformed with genes that code for chitinase and glucanase enzymes. These degrade polymers in the cell walls of many, but not all, fungi, without affecting the host plant or any animals. The genes for these enzymes have been isolated from a number of sources, including plants (rice, barley), bacteria (*Serratia marcescens*), and even fungi (*Trichoderma harzianum*). The choice of enzyme used has partly depended on its availability and on its effectiveness against the pathogen being addressed, since successful pathogens are already likely to have resistance against plant systems. The 35S promoter (Chapter 4) has often been used to express these genes. Constitutive expression of the defence protein is designed to provide a barrier against the initial attack and not allow the pathogen to establish, as may be the case during the induction period of the host defence system. Wound-inducible promoters (Box 4.1), such as the potato *prp1-1* promoter, have also been used, as they mediate rapid and localized expression in response to pathogen attack. This type of promoter restricts gene expression to where the protein is required and may reduce any potential hazards associated with the proteins. As gene expression is only turned on during the infection, therefore limiting gene expression, it means there is no serious drain on the plant's biosynthetic capability nor a serious amount of yield reduction (yield drag).

● Transgenic approaches using single defence genes are simple but not necessarily durable.

The efficiency of these systems has been tested in many plants, and in many cases some level of resistance/tolerance to specific fungal pathogens has been demonstrated. One of the earliest demonstrations of tolerance was exhibited by field-grown transgenic tobacco lines, containing a chitinase gene from the bacterium *S. marcescens*. Both intracellular and extracellular forms of the enzyme (glycosylated or unglycosylated forms) gave the plants increased tolerance to the pathogen *Rhizoctonia solani*. In another example, transgenic cucumber plants showed a range of responses to the grey mould *Botrytis cinerea* when transformed with a rice chitinase sequence (RCC2) driven by a 35S promoter. Interestingly, high-level resistance was observed in a number of lines where spread of the disease was blocked completely. However, the block in fungal attack occurred at different stages in the infection process. Some lines prevented hyphal penetration, while others allowed penetration but restricted the spread.

Perhaps the most ambitious work with chitinase genes has been carried out with fruit trees. Apple tree plants expressing an exochitinase and/or an endochitinase isolated from the fungus *T. harzianum* have been produced using *Agrobacterium*-mediated transformation. A positive correlation was found to exist between resistance to scab produced by the fungus *Venturia inaequalis* and the level of expression of the proteins. If both genes were introduced into plants, the proteins acted synergistically. One drawback was that expression of the endochitinase reduced plant growth. One important question that remains to be answered is how long the resistance will be maintained. A further disadvantage of the chitinase approach is that major oomycete pathogens such as *Phytophthora* spp. do not have chitin in their cell walls. To overcome these difficulties, combinations of PR proteins have been used (described below) and other antimicrobial proteins have been tested. This is a sensible strategy anyway, as the use of more than one gene will help to prevent the development of resistance and provide longer-term protection.

Glucanases, which are also one of the groups of PR proteins, have been used as a tool to enhance resistance to fungal infection. There are different forms of the enzyme with different substrate specificities, but one of the most widely used is a class II β-1,3-glucanase from barley. When expressed in transgenic tobacco plants under control of the 35S promoter, the gene provided increased protection against the soil-borne fungal pathogen *R. solani*. In plants also expressing chitinase transgenes, there is a synergistic effect and resistance is further enhanced.

A third type of protein, ribosome-inactivating proteins (RIPs), has also been used as a defence against fungal infections. RIPs are enzymes that remove an adenine residue from a specific site in the large rRNA of eukaryote and prokaryote ribosomes, thereby inhibiting protein synthesis. The specificities of these proteins vary, but the critical point is that certain examples do not inhibit plant ribosomes. Type 1 barley RIP, expressed constitutively, has been used to provide resistance to fungal attack and has been tested in tandem with the chitinase gene. Again, a synergistic effect was found against *R. solani* in transgenic tobacco plants.

The major hurdle with these approaches is their transfer from greenhouse to field. This was tested in a major study with transgenic wheat that stably expressed genes for thaumatin-like protein, chitinase, and glucanase, which gave resistance to the scab-causing organism *Fusarium graminearum*. In the greenhouse, a line co-expressing the chitinase and the glucanase showed a delay in the spread of the infection. This had also been found for a line expressing the thaumatin gene. However, when grown in the field, neither line showed resistance to scab when met with heavy inoculum loads of the fungus. Clearly further work is required to improve this type of approach. Different combinations of genes may be required under field conditions to bring about broad-range durable resistance. Several other approaches are discussed below.

Antimicrobial proteins

Many other genes that code for proteins with both antifungal and antibacterial activity have been introduced into plants with varying success. The proteins used have included lysozyme (this enzyme degrades chitin as well as peptidoglycan), which delayed fungal infection for a short time. Potato plants expressing lysozyme have been studied with respect to their resistance to *Erwinia carotovora*. Various thionin proteins (see Table 7.3) have been introduced into plants with mixed success; barley α-thionin expressed in transgenic tobacco was shown to increase resistance to *P. syringae*, but other attempts did not enhance resistance.

An exciting area of development is in the area of defensins. These proteins are found in all living cells. They have various properties, but are small antimicrobial peptides (26–50 amino acid residues) that are ancient mediators of innate defences and some 500 different ones have been reported. Defensins display lytic activity by binding within microbial plasma membranes. This sort of attack may prove difficult for the pathogen to develop resistance to. As the integrity of the membrane is critical, it is unlikely that changes in the structure of the membrane could occur to prevent defensin infiltration. An alfalfa defensin gene, *alfAFP*, has been introduced into potatoes and its performance monitored in the field. The defensin was processed and secreted into extracellular spaces of leaves and roots in transgenic plants. In the greenhouse and in several field trials, the transgenic plants were shown to have a significant resistance toward the fungus *Verticillium dahlia*. The level of resistance was comparable to that found with a naturally resistant potato. However, they were also expected to be resistant to *Alternaria solani*, but this proved not to be the case.

An artificial defensin gene has been made that incorporates the cecropin and melittin genes (*CEMA*), obtained from a giant silk moth and bee, respectively. The active portions of these two genes have been fused to remove haemolytic activity from the mellitin (which is the major lytic component of bee sting venom) and because experiments with native cecropin have shown that it is readily degraded by proteases in the host. The *CEMA* gene has been introduced into potatoes, and transgenic plants challenged in tissue-culture conditions with the bacterium *E. carotovora*. Tubers were then tested for *E. carotovora* soft rot; during the course of the experiment the transgenic tubers remained firm, whereas control plants rotted. This type of study also shows the power of transgenic approaches where genes with the required characters can be moved across kingdoms. The fused *CEMA* gene, linked to a wound-indicible promoter, has also been introduced into tobacco where it has been shown to confer resistance against the ubiquitous fungus *Fusarium solani*.

Some of the most impressive studies in this area have been carried out to combat the bacterial disease fireblight. Many of the most important commercial apple cultivars and rootstocks grown in Europe are sensitive to fireblight. This disease is caused by *Erwinia amylovora*, which, once established, can kill a tree within a growing season. It is also a very serious disease of pears. A number of independent groups are addressing this problem by producing transgenic trees containing the genes for antimicrobial proteins. One strategy used has been the lytic approach, to produce transgenic fruit trees (via *Agrobacterium*-mediated transformation) containing the gene for T4 lysozyme and genes for the insect antimicrobial proteins attacin E and cecropin. A second approach being used involves the inclusion of a bovine lactoferrin gene to compete with microbial siderophores and a depolymerase gene whose product is capable of degrading specific exopolysaccharides.

These experiments are at various stages, but field trials have been carried out with apple plants. It is claimed that plants expressing attacin E under the control of a constitutive promoter show increased field resistance to fireblight. Experiments were carried out in 1998 in which 2- and 3-year-old plants were inoculated with *E.*

amylovora and the percentage of the current season's shoot-length-blighted tissue determined. Many of the transgenic lines showed less blight than the control plants. One line in particular had only 5% of shoot length blighted compared with 56% in non-transgenic controls.

Another group of peptides that are being tested are the temporins. These are small cationic antimicrobial peptides isolated from skin secretions of European red frog (*Rana temporaria*). They have the advantage of not being haemolytic and, hence, are unlikely to be toxic. These non-cysteine-containing peptides are 10–13 amino acids long and the most potent is temporin A, which has both antibacterial and antifungal activity. An engineered form of the gene for this peptide has been used to transform potatoes where its expression makes the plants resistant to several pathogens including *E. carotovora* ssp. *carotovora* (bacterial soft rot), *P. infestans* (late blight), which is costing US$4 billion annually in losses, and *Phytophthora erythroseptica* (pink rot). The latter organism causes serious post-harvest losses and has recently developed resistance to the only effective fungicide, mefanoxam.

There are many more examples of genes for antimicrobial proteins being introduced into plants, such as *Ace-AMP1* in rose and wheat, gallerimycin, a moth protein, expressed in tobacco, and Esculentin-1 from *Rana esculenta*, also expressed in tobacco, all of which bring about a change in the resistance patterns of the host plants.

The examples given above highlight the use of antimicrobial proteins to interact directly with pathogens and stop infection. One can ask, though, will they ever be commercialized? It is likely that some will, but so far these approaches have been disappointing, as even when using non-specific enzymes the spectrum of resistance achieved has been limited. This may be a problem, as one view is that a broad spectrum of resistance in the plant lines is more likely to be acceptable to farmers than if only single pathogens are being dealt with. The alternative view, as exemplified by the approach to fireblight, is that some diseases are so important that strategies that deal with these alone are acceptable. Other attempts have been made to introduce broad-spectrum resistance; for instance, strategies to switch on resistance pathways have been developed. These are discussed below.

Transgenic crops for food safety

The antifungal approaches listed above not only protect the plant from disease but also protect us from food mycotoxins. Recently, transgenic approaches have been used to combat the effects of fungal infection rather than the infection itself. One interesting example is *Fusarium verticillioides*, which causes Fusarium ear mould in maize and produces fumonisin mycotoxins, which are acutely toxic to certain livestock and are also carcinogenic. It has not yet been possible to incorporate ear-mould resistance into the maize crop, but it has been possible to incorporate either a fungal esterase gene or a fungal aminc oxidase gene into maize and detoxify the maize to some extent.

The use of *Bacillus thuringiensis (Bt)* genes to minimize insect damage also has the effect of reducing the instance of ear rot as it reduces wound sites that can be infected.

It is not a long-term solution but raises the possibility of interesting avenues to be explored. Another approach has used the synthetic, 17-amino-acid-long linear peptide D4E1, which has a very broad antimicrobial activity, inhibiting the growth of fungal and bacterial pathogens. The peptide has been introduced into cotton to inhibit the growth of *Aspergillus flavus* with the aim of reducing the production of aflotoxins produced by the fungus.

Induction of HR and SAR in transgenic plants

• Switching on defence pathways may be a more durable approach than using single defence genes.

From the information given above, it is clear that the most obvious approach for switching on a general resistance pathway would be to use components of the elicitor-receptor system. One possible implication of the Guard hypothesis (see Box 7.4) is that the R protein has a function in the cell that relates to the general resistance mechanism. This would suggest that an increase in the production of the R protein will bring about an increase in the expression of the resistance systems. This hypothesis has been tested in a series of experiments in which the *pto* gene, under the control of the 35S promoter, has been introduced into tomato plants.

Close examination of three transgenic lines indicated that overexpression of the Pto protein activated a defence response in the absence of the Pto–AvrPto interaction. Some 600 genes were differentially expressed in plants expressing the Pto protein as compared with sibling plants lacking the protein and, of these, cDNA microarray analysis has shown that 223 were Pto-overexpression-responsive genes. Also 40% of the induced genes previously have been shown to be differentially expressed during a human and/or fruit fly immune response. Again, this is evidence for the innate immune system being in existence in some ancestral form. At the cellular level, it was seen that cell death due to an HR reaction was highly localized to palisade mesophyll cells. The mesophyll cells also showed accumulation of autofluorescent compounds, callose deposition, and lignification. Leaves also exhibited salicylic acid accumulation and increased expression of pathogenesis-related genes. The important outcome of this work is that this constitutive expression of the resistance systems is not pathogen-specific. Plants that normally would be sensitive to *P. syringae* pv. *tomato*, without *avrPto*, were shown to be resistant. They were also tested with *Xanthomonas campestris* pv. *vesicatoria*, a leaf pathogen that causes bacterial spot disease, and *Cladosporium fulvum*, which causes leaf-mould disease. The transgenic plants were also more resistant to these pathogens.

Another approach, adopted to switch on the HR-related defence system, has been devised, in which a *Phytophthora cryptogea* gene that codes for a highly active elicitor, termed crytogein, has been fused to a pathogen-inducible promoter from tobacco and introduced into transgenic tobacco. Under non-induced conditions (i.e. in the absence of the pathogen), no elicitor is made. On infection by the virulent fungus *Phytophthora parasitica* var. *nicotianae*, the crytogein gene was expressed, and the elicitor-induced defence gene expression was detected around the infection site. Localized necrosis, similar to what would have been produced by non-virulent pathogens, occurred around the infection site. The plants also exhibited resistance to other fungi unrelated to the *Phytophthora* genus. Broad-spectrum resistance, at least to fungi, has been produced without the constitutive expression of the transgene.

An approach that attempts to enhance disease resistance by triggering the expression of a range of genes that are regulated by different stimuli has been the introduction of the *Arabidopsis* gene *NPR1* into crop plants such as tomato, wheat, and rice. The product of this gene lies at a node in the regulatory pathways that link SAR, induced systemic resistance, *R*-gene-mediated resistance, ethylene, salicylic acid, and jasmonic acid. It activates gene expression by interacting with members of the TGA family of basic leucine zipper (bZIP) transcription factors. The introduction of the gene enhanced resistance to a range of plant pathogens in tomato, but in rice the gene had a detrimental effect. The mode of action for the transgene seems to be a priming function; the plant responds more rapidly to attack rather than having its defence system switched on constitutively, which would have an effect on yield.

Another approach to switching on a cascade of defence genes has been used in sunflower. The involvement of oxygen species such as hydrogen peroxide as a messenger in the HR reaction has been tested by transforming the plant with the gene for oxalate oxidase from wheat. This enzyme converts oxalic acid and O_2 to CO_2 and hydrogen peroxide. The transgenic plant could degrade oxalic acid and generate hydrogen peroxide, and an HR-like response along with salicylic acid synthesis and defence-gene expression were recorded. The genes were identified as being PR and defensin proteins, and it was shown that the transgenic sunflowers exhibited enhanced resistance against the oxalic acid-generating fungus *Sclerotinia sclerotiorum*.

CASE STUDY 7.1 **The BASF potato**

The use of resistance genes for transgenic approaches towards late blight (*P. infestans*) resistance in potatoes has reached an interesting point with the case of the BASF potatoes. The commercial potato *Solanum tuberosum* has a range of dominant genes that code for resistance to the disease. However, the pathogen mutates continually to overcome this resistance. Over many years, resistance genes have been introduced into the edible potato from wild-type varieties by hybridization approaches, but the effect is soon overcome by the pathogen. This approach has recently been used between *S. tuberosum* and a wild Mexican species, *Solanum bulbocastanum*, which is sexually incompatible, but which is resistant to late blight. The problem is that hybridization approaches transfer a vast array of background genes, so to target specifically the resistance genes transgenic techniques have to be used. They have transferred two *R* genes into the potato, coding for two NB–LRR receptors (see Figure 7.3). These receptors are cytoplasmic and will trigger the HR once bound to the stimulating ligand.

BASF has applied to do field trials with this new line, but not surprisingly has met a lot of opposition in the countries of the European Union. There are a number of criticisms, the main one being that the *R* genes used are from a novel background and not from *Solanum demissum*, the usual source of genes used in BASF potatoes. The concern is that the proteins have not been tested for allergenic response and that they need to be before release.

Developments for the future

Other transgenic approaches

Several research groups have tried other approaches to bring about resistance to infection by methods not directly associated with defence mechanisms. One interesting approach has been the use of RNA interference (RNAi) techniques. *Arabidopsis* and tomato plants containing sequences complementary to the *iaaM* and *ipt* oncogenes have rendered the plants resistant to *Agrobacterium* infection.

A commercial enterprise (Integrated Plant Genetics) has designed an antibody approach. They have built single-chain variable fragment antibodies (scFvs) that will bind to pathogenicity-related proteins produced by pathogens. They use scFvs instead of full-size monoclonal antibodies because the latter tend to be exported to the cell membrane or the apoplastic space. The target molecules tend to be found in the cell cytoplasm where the scFvs are found. In the early stage of its development, disease-block technology has been used in a number of plants and has been shown to reduce the symptoms shown but did not necessarily lead to immunity. Further development work is being carried out on the system.

Future prospects for breeding

In the past, breeding strategies relied on basic observation of the phenotype in mature plants, but now they are being enhanced by the use of markers and molecular biology techniques. The use of comparative genomics and transcriptomics can provide huge amounts of information on the genes and signalling pathways involved in disease resistance and allow the location of resistance genes to be mapped. This is not just for model plant systems; in the major crop plants, restriction fragment length polymorphisms (RFLPs) and quantitative trait loci (QTL) maps, augmented with the results of the genome-sequencing projects and polymerase chain reaction (PCR)-based molecular markers such as random-amplified polymorphic DNA (RAPD), amplification fragment length polymorphism (AFLP), microsatellites, and sequence-tagged sites (STSs) provide a battery of very powerful tools that allow genes/traits to be followed during the breeding process. Recently, marker-based breeding has been used to pyramid multiple disease-resistance genes into plants. These combined approaches offer rapid and cleaner routes to broad-spectrum, durable disease resistance, which could lead to a significant improvement in yield with a concomitant reduction in chemical usage. Marker-assisted breeding strategies have a lot of merit. However, they depend upon the existence of resistance genes in the germplasm and the molecular information being available. In the following case study, we consider how traditional breeding strategies enhanced with molecular approaches can be used with transgenic techniques to develop disease-resistant plants.

CASE STUDY 7.2 *Xanthomonas* spp.

Xanthomonas is an interesting genus of plant pathogens in that all are associated with plants, and none are free-living or found associated with any other host types. The genus has been found infecting some 240 genera in 68 plant families. Despite this, the different *Xanthomonas* species have very limited host ranges. One of the most important examples is *X. oryzae* pv. *oryzae,* the cause of a very destructive disease in rice, bacterial rice blight. This bacterial disease has caused major losses in the world rice supply, often exceeding 20% of the total crop. In the past, the Indian states of Punjab, Haryana, and Uttar Pradesh have had total crop failures. Although some copper- and mercury-based chemical controls have been used in the past, no effective environmentally safe chemical controls have been developed for this disease. Rather, success in combating the pathogen will come from a combination of approaches: development of safer chemicals, biological control (in this case with strains of *Pseudomonas fluorescens*), conventional breeding, and transgenesis.

Conventional breeding strategies have benefited from the identification and mapping of genes coding for disease resistance and of associated markers. This information has been obtained with techniques such as RFLP-RAPD, RFLP-PCR, and more recently the rice genome project (see Chapters 1 and 13). In rice, some 30 genes have been linked to resistance to bacterial rice blight (*Xa1–Xa29*). The proteins coded for by these genes have a range of functions and properties. For instance, gene *Xa21* is a dominant marker and codes for a transmembrane LRR receptor kinase. These different genes often give resistance to different races of *Xanthomonas*. To broaden the resistance spectrum of commercial rice lines and to develop a more robust resistance, the approach used recently in conventional breeding has been to introduce different resistance genes into plants by the strategy known as pyramiding. Genes are bred into plant lines either singularly or in groups so that the resulting plant contains a range of genes. This strategy has been followed with three genes, the recessive genes *Xa5* and *Xa13* and the dominant gene *Xa21*. Once in the desired breeding line, molecular-based surveillance can be used to select for progeny containing the desired genes. STS-marker-assisted selection is an example of an approach used here. Selection is not based on phenotype, because the recessive genes would be masked by the dominant gene. Rather, sequence markers are detected to show the presence of the genes.

If these approaches are successful, why do we need transgenics?

Transgenic approaches offer several important benefits. First, a single gene can be moved from one plant line to another. This would allow the resistance properties to be confirmed in the new background. Another benefit is that it is possible to introduce the gene directly into elite breeding lines without diluting the background. Some lines of rice may not carry the resistance gene or are susceptible to certain races, and again transgenics can be used here to move genes about. *Xa21* has been used in transgenic approaches, it has been introduced into indica, japonica, and hybrid rice lines with some degree of success. However, some Korean races of *Xanthomonas* are not inhibited by *Xa21*-derived resistance. So another benefit of the transgenic approach is to use genes from other sources. An *Arabidopsis* gene *AtNPR*, the protein product of which interacts with the TGA transcription factors (bZIP transcription factors), which regulate SAR, has been introduced into rice and has been shown to enhance resistance to *X. oryzae*, although as discussed above this had its costs.

SUMMARY

This chapter has introduced the different types of plant pathogens and their effects. It is clear that they cause major financial losses to the agricultural industry, despite the plants having defence mechanisms against them. The different levels of defence have been identified, but in Western agriculture, resistance systems are inhibited by the nature of monocultures, which limit the natural polymorphisms that occur in wild populations. Over the last 50 years or so, a great deal of information has been gathered on the mechanisms used by plants to combat disease. This has allowed the development of a range of different genetic-manipulation strategies.

Initial attempts at enhancing resistance involved the introduction of antifungal or antibacterial genes that were either PR proteins or proteins identified as having antimicrobial properties. These proteins gave a limited spectrum of resistance. However, when attempting to deal with a very serious pathogen such as *E. amylovora*, which causes fireblight, this is probably acceptable. Care will be necessary to produce durable systems that the pathogens cannot overcome easily. The problems associated with the use of *R* genes share some resemblance with the boom-and-bust cycle now faced with antibiotics. If the transgenic approaches all use the same resistance genes (either singly or in pyramid approaches), then the pathogens will evolve rapidly to overcome these genes. As a first attempt to develop strategies for durable, broad-ranging systems of resistance, transgenic plants have been created that contain genes that switch on signal-transduction cascades. These approaches have been shown to give some benefit to the plant, but it may be at a cost. Switching on defence pathways constitutively can, because of the metabolic burden, lead to a yield loss. As more is learnt about the mechanism of disease resistance and its complexities, then other, more durable systems will be developed that consist of multiple/diverse mechanisms.

So far, no plants enhanced for resistance to fungi or bacteria have been commercialized. This may be due to a number of factors. Some of these are intrinsic to the technology, such as the limited range of resistance and yield drag. Some of the reasons are extrinsic, such as anti-GM campaigns (e.g. the BASF potato), or the absence of drive from industry to commercialize the first generation of resistant plants. The molecular techniques being applied to traditional breeding strategies are now allowing resistant genes to be pyramided into the major crop plants. In this post-genomic era, the sequencing of the *Arabidopsis* and rice genomes (Chapter 1) has, in the first instance, allowed large numbers of disease-resistance genes to be identified, and transcription profiling (transcriptomics was looked at briefly in Chapter 1) will allow their interactions to be studied in detail. This will benefit both molecular-marker-based breeding strategies and transgenic approaches. Breeding can only be successful if the resistance genes are already in the germplasm. Transgenic approaches will allow the transfer of resistance genes between species. One long-term target in the post-genomic era is to artificially re-establish polymorphisms by pyramiding numbers of resistance genes from resistant plants into sensitive crop species.

- There have been lots of successful approaches to engineering plant disease resistance but no commercial successes as yet.

FURTHER READING

Visit www.oxfordtextbooks.co.uk/orc/slaterplants2e/ for hyperlinked references that take you directly to online abstracts for the journal articles listed below:

Non-GM approaches to disease control

Chrispeels, M. J. and Sadava, D. E. (1994) *Plants, Genes, and Agriculture*. Jones and Bartlett, Boston.

Dixon, R. A. (2001) Natural products and disease resistance. *Nature* **411**, 843–847.

Disease resistance

Asai, T., Tena, G., Plotnikova, J., Willmann, M. R., Chiu, W.-L., Gomez-Gomez, L., Boller, T., Ausubel, F. M., and Sheen, J. (2002) MAP kinase signalling cascade in *Arabidopsis* innate immunity. *Nature* **415**, 977–983.

Cvetkovska, M., Rampitsch, C., Bykova, N., and Xing, T. (2005) Genome analysis of MAP kinase cascades in *Arabidopsis* defence response. *Plant Molecular Biology Reporter* **23**, 331–343.

Dangl, J. L. and Jones, J. G. (2001) Plant pathogens and integrated defence responses to infection. *Nature* **411**, 826–833.

Dong, X. (1998) SA, JA, ethylene, and disease resistance in plants. *Current Opinion in Plant Biology* **1**, 316–323.

Espinosa, A. and Alfano, J. R. (2004) Disabling surveillance: bacterial type III secretion system effectors that suppress innate immunity. *Cellular Microbiology* **6**, 1027–1040.

Farmer, E. E. (2001) Surface-to-air signals. *Nature* **411**, 854–856.

Feys, B. J. and Parker, J. E. (2000) Interplay of signalling pathways in plant disease resistance. *Trends in Genetics* **16**, 449–455.

Gomez-Gomez, L. and Boller, T. (2002) Flagellin perception: a paradigm for innate immunity. *Trends in Plant Science* **7**, 251–256.

Grant, M. and Lamb, C. (2006) Systemic immunity. *Current Opinion in Plant Biology* **9**, 414–420 (and other papers in this issue).

Lam, E., Kato, N., and Lawton, M. (2001) Programmed cell death, mitochondria and plant hypersensitive response. *Nature* **411**, 848–853.

Melotto, M., Underwood, W., Koczan, J., Namura, K., and He, S. Y. (2006) Plant stomata function in innate immunity against bacterial invasion. *Cell* **126**, 969–980.

Montesano, M., Brader, G., and Tapio Palva, E. T. (2003) Pathogen derived elicitors: searching for receptors in plants. *Molecular Plant Pathology* **4**, 73–79.

Mysore, K. S. and Ryu, C.-M. (2004) Nonhost resistance: how much do we know? *Trends in Plant Science* **9**, 97–104.

Numberger, T., Brunner, F., Kemmerling, B., and Piater, L. (2004) Innate immunity in plants and animals: striking similarities and obvious differences. *Immunological Reviews* **198**, 249–266.

van der Biezen, E. A. (2001) Quest for antimicrobial genes to engineer disease resistant crops. *Trends in Plant Science* **6**, 89–91.

Zaninotto, F., La Camera, S., Polverari, A., and Delledonne, M. (2006) Crosstalk between reactive nitrogen and oxygen species during the hypersensitive disease resistance response. *Plant Physiology* **141**, 379–383.

Engineering disease-resistant plants

Gurr, S. J. and Rushton, P. J. (2005) Engineering plants with increased disease resistance: what are we going to express? *Trends in Biotechnology* **23**, 275–282.

Salmeron, J. M. and Vernooij, B. (1998) Transgenic approaches to microbial disease resistance in crop plants. *Current Opinion in Biology* **1**, 347–352.

Stuiver, M. H. and Custers, J. H. H. V. (2001) Engineering disease resistant plants. *Nature* **411**, 865–868.

Web links

Plant Viruses Online, web-link 7.1: http://image.fs.uidaho.edu/vide/refs.htm

PR proteins as latex allergens, web-link 7.2: http://dmd.nihs.go.jp/latex/index-e.html

Defence-related proteins of higher plants, web-link 7.3: http://dmd.nihs.go.jp/latex/defense-e.html

8

Reducing the effects of viral disease

INTRODUCTION

Virus infections of plants can cause cell necrosis, hypoplasia (retarded cell division and growth), or hyperplasia (excessive cell division or cell expansion) and these effects can lead to plants showing symptoms such as growth retardation, distortion, mosaic patterning of leaves, yellowing, and wilting. In many cases, the overall result of such infections is of minor consequence. Often viruses are endemic to a region and cause only moderate crop losses. However, as we have seen in the preceding chapter, plant viruses can be a major problem to agriculture, with severe epidemics periodically causing losses adding up to billions of dollars. In this chapter, there will be a brief introduction to the different types of viruses, some of their properties, and the methods that have been used previously in agriculture to prevent the spread of virus infections. This will be followed by an in-depth discussion of several biotechnologically based strategies that have been developed to reduce the effects of viral infections in plants. The main approach used has been pathogen-derived resistance (PDR). This is where pathogen sequences are introduced into the host on the basis that expression at an inappropriate time, or level, or in an inappropriate form, would disrupt the ability of the virus to sustain an infection. As will be shown below, this approach has been successful. This line of research has uncovered many interesting aspects of biology, some of which (like gene silencing) will be touched on in this chapter.

Types of plant virus

There has been an initiative, instigated by the International Committee on Taxonomy of Viruses, to formalize the viruses into families and genera and to standardize the nomenclature used. In this book, the names of viruses are based on those presented on the database of the International Committee on Taxonomy of Viruses, or ICTVdB (see web-link 8.5). If family or genus names are not given in the text, these can be obtained in full from the website.

It is impossible to discuss all the different types of virus in any detail, but Table 8.1 highlights the main families and unassigned genera. The virus particle is composed of

Table 8.1 Genome structure of plant viruses

Genome structure	Examples of families and genera	Notes
RNA, single-stranded, positive-sense (acts as mRNA directly)	Families: *Bromoviridae*, *Comoviridae*, *Potyviridae*, *Sequiviridae* and *Tombusviridae*, *Luteoviridae* Examples of unassigned genera: *Tobamo-* and *Tobravirus*	70% of known plant viruses; both non-segmented and segmented (two or more RNAs in different virus particles)
RNA, single-stranded, negative-sense (RNA needs to be copied before it can act as mRNA)	Family: *Bunyaviridae*, genus: *Tospovirus*; Family: *Rhabdoviridae*, genus: *Rhabdovirus* Unassigned genus: *Tenuivirus*	*Bunyaviridae* possess a lipidic envelope in addition to their nucleocapsid
RNA, double-stranded	Family: *Reoviridae*, genera: *Fijivirus*, *Phytoreovirus*, *Oryzavirus* Family: *Partitiviridae*, genera: *Alpha-* and *Betacryptovirus*	The plant members of the *Reoviridae* family have a genome consisting of 10–12 segments of RNA; each has one ORF that produces a protein
DNA, double-stranded	Family: *Caulimoviridae*, genera: *Caulimovirus* and *Badnavirus*	The only plant viruses of this group are the caulimoviruses; their genome consists of one double-stranded circular DNA molecule with specific single-stranded discontinuities in both strands; it codes for six or eight ORFs located on one strand only; DNA replication occurs by a process of reverse transcription (i.e. via an RNA intermediate) similar to that of animal retroviruses
DNA, single-stranded	Family: *Geminiviridae*	The only plant viruses possessing either one or two molecules of single-stranded genomic DNA; DNA replication occurs via double-stranded DNA intermediates; ORFs located on both the viral strand and its complement

ORF, open reading frame.

a nucleic acid genome (five types are indicated in Table 8.1), which can be a single or multipartite structure (segmented). The genome is surrounded by the capsid, or protein coat, forming the nucleocapsid. Plant viruses come in a range of morphologies, from rod shapes to isometric particles, and some have membrane envelopes around the nucleocapsid. Virus genomes code for similar core proteins: the capsid or coat protein(s), replication-related proteins (such as polymerases and helicases), capping enzymes/proteins, and movement proteins (which facilitate the spread of the viruses). The functions of these and other virus proteins will be discussed in various sections of the chapter. Some viruses also have extra, small RNA segments associated with them, called satellite RNA. These can alter the pathology of the virus infection. Where viruses are mentioned in the text, but no specific background is given, further information can be obtained from the web-links listed at the end of this chapter.

To highlight the organization and different modes of gene expression found with plant viruses, two members of the most common group, positive-sense RNA viruses, which demonstrate widely different modes of gene expression, will be used as examples in the next section.

RNA viruses

Tobacco mosaic virus

- Tobacco mosaic virus is the longest-studied of any virus.
- Tobacco mosaic virus has a complex single-stranded genome with subgenomic promoters.

Figure 8.1 shows the genome structure and expression patterns of the tobamovirus, tobacco mosaic virus (TMV). TMV is perhaps the longest-studied of all viruses: it has been the subject of intensive investigation for over 100 years. It has a genome composed of one single-stranded linear RNA molecule (≈6400 bases) and has a tRNA-like structure at the 3′ end. The genome codes for five translated open reading frames (ORFs); of these, at least three ORFs are non-structural proteins: a 183-kDa protein, the 126-kDa protein, and the 30-kDa protein. The two larger proteins are involved

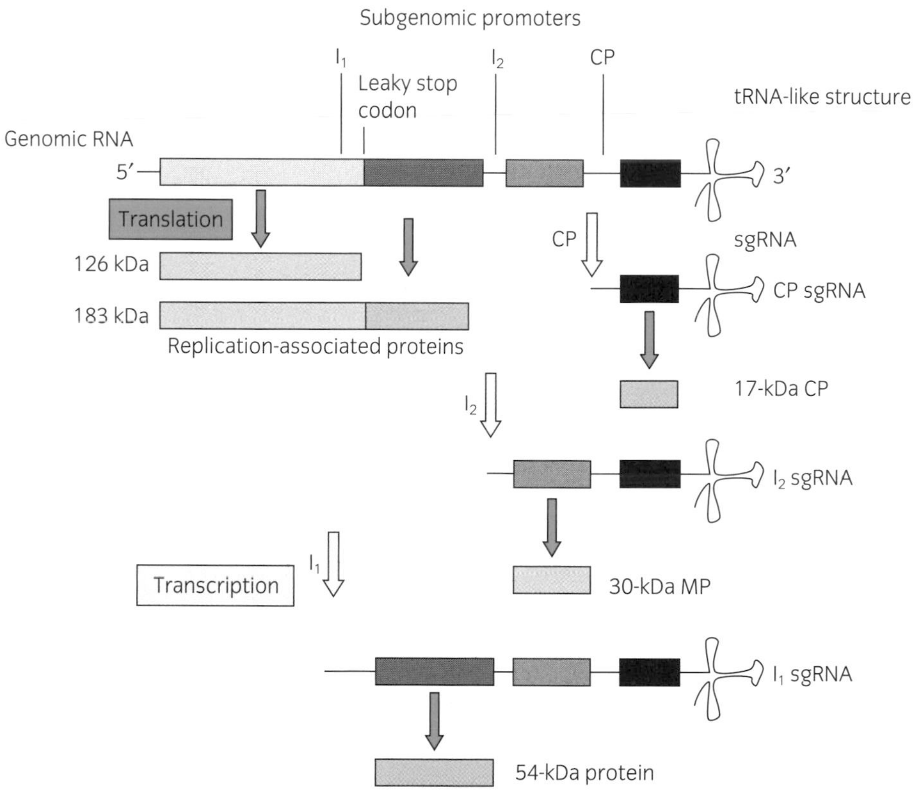

Figure 8.1 Transcription and translation of the tobacco mosaic virus (TMV) genome. Initial translation produces a replicase, which then synthesizes subgenomic mRNAs from the negative (–) strand. Light-green arrows, transcription; dark-green arrows, translation. CP, coat protein, MP, movement protein; sg, subgenomic. (Adapted with permission from web-links 8.1 and 8.4.)

in the replicase complex, the 126-kDa one having methyltransferase activity, which is involved in RNA capping, and a helicase function. The 183-kDa protein is a read-through product of the 126-kDa protein that also has a polymerase function. The 30-kDa protein is a movement protein (MP). The RNA also codes for a putative 54-kDa protein of unknown function, and the capsid or coat protein (CP; 17 kDa). Translation of the TMV ORFs is quite complex (see Figure 8.1). On entry into the cell, the two high-molecular-mass proteins of 126 and 183 kDa are translated directly from the full-length genomic RNA, which acts as mRNA. The 183-kDa protein is produced by read-through of the amber termination codon that terminates translation, to produce the 126-kDa protein. In addition, there is a start codon in this read-through region and an ORF that could potentially encode the putative 54-kDa protein. Alternatively, a subgenomic RNA (I_1) may code for this protein. The replication complex copies the positive-sense RNA to form negative-sense RNA, which subsequently acts as a template for the production of the positive-sense RNA. Subgenomic messages are transcribed from two promoters within the negative-sense strand, and these act as templates for the synthesis of the MP and the CP from the I_2 RNA and CP subgenomic RNAs, respectively (Figure 8.1).

Comoviruses

Figure 8.2 shows the genome structure and translation strategy for a comovirus, cowpea mosaic virus (CPMV). The genome is segmented; it is split into two parts,

- Comoviruses have fragmented genomes that act as polycistronic mRNA.

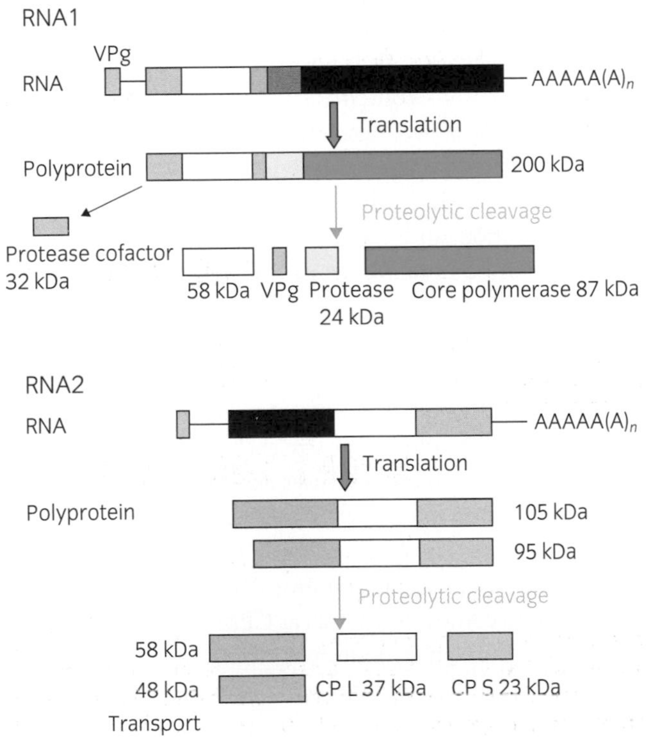

Figure 8.2 Translation of the cowpea mosaic virus (CPMV) genome. VPg, genome-linked viral protein. (Adapted with permission from Lomonossoff and Johnson (1991).)

RNA1 and RNA2. RNA1 has the coding sequences for the core polymerase, a protease, and a VPg (genome-linked viral protein). The VPg is a protein that is attached to the 5′ end of the molecule and fulfils a cap function. The coding sequence for, in this case, the two CP subunits, are found on RNA2 along with the MP. On entry into the cell, the genomes act as messages and can be translated directly. Large polyproteins are made that represent the entire coding sequence (Figure 8.2). These are then cut into the active proteins by specific proteases. Other members of the *Comoviridae*, such as nepoviruses, which will be discussed below, demonstrate this type of genome structure and mode of gene expression.

Entry and replication: points of inhibition

As with any virus infection, the virus has to enter the host cell to replicate. Individual viruses have specific mechanisms for their initial entry into the cell and subsequent movement through the plant. Many are transferred between plants by vectors such as nematodes, various arthropods (such as aphids, leaf hoppers, and whiteflies), and fungi.

Once inside the cell, the viruses uncoat, and the viral genetic material is then translated and replicated. Replication is carried out by the replicase complex encoded by the virus. The ultimate stage of the infection is the packaging of the replicated viral nucleic acid into capsids. One interesting point is that the nucleic acid molecules are packaged into the capsids separately: each RNA molecule is packaged into its own capsid. This has been shown for icosahedral particles of CPMV by gradient centrifugation, a technique whereby it is possible to size-fractionate separate populations on the basis of their RNA content. Also, electron microscopy has been used for rod-shaped viruses such as beet necrotic yellow vein virus (BNYVV), which has four RNAs in most isolates. In this case, it is possible to observe four different-sized populations of virus particles when viewed by electron microscopy. The mechanisms involving translation and replication will depend on the types of virus (further information can be obtained from the web-links listed at the end of the chapter).

● The mode of systemic movement may influence the success of PDR approaches.

For a systemic infection, the virus must move from cell to cell via the plasmodesmata and/or the vascular tissues. However, the exclusion limit of the plasmodesmata pore is normally too small to allow the passage of virus particles. The viral MPs may function by increasing the molecular size exclusion limit of the plasmodesmata, allowing passage of the virus. Different viruses exhibit different modes of movement; some are transferred as RNA, which may be coated with CPs or MPs, or they may be transferred as virus particles. For instance, with TMV the 30-kDa MP modifies the plasmodesmata to allow the passage of genomic RNA coated by the 30-kDa protein to pass from cell to cell. With comoviruses such as CPMV, the 58-/48-kDa proteins form tubular structures through the plasmodesmata that allow the passage of intact virus particles from one cell to another. Other mechanisms may exist for transport into the vascular tissue.

How has the agricultural community dealt with viruses?

Farmers and the agricultural industry have dealt with the problem of viruses in a number different ways, based on whether the viruses are normally present or not. Where viruses are endemic, the direct chemical control of viruses is not normally an option. However, some control can be exerted over the spread of the virus with chemical control methods if it is transmitted by an insect or another vector, but often this only has a limited impact.

Some plant systems have benefited from another approach, known as cross-protection. Since the 1920s, it has been known that some plants infected with a virus were apparently protected against infection by a related virus. One strategy therefore has been to deliberately infect plants with viruses that produce only mild symptoms but which generate resistance to infection by more-virulent strains. Mild strains have been obtained for some viruses from areas where plants apparently showed some resistance to infection. When analysed, the plants were found to harbour less-infective strains of the virus. These viruses were then used to protect other plants against infection. Cross-protection has been used with some success to combat disease caused by viruses such as the closterovirus citrus tristeza virus, which is the most important disease of citrus trees (a US$10 billion worldwide industry), and the potyviruses papaya ringspot virus (PRSV) and zucchini yellow mosaic virus (ZYMV). A similar approach has been used with strains of cucumber mosaic virus (CMV, a cucumovirus) that have satellite RNA associated with them. The presence of the satellite RNA reduces the pathogenicity of the initial virus infection, and it has been shown that it also reduces the symptoms caused by other CMV-related viruses that do not have the satellite.

- Cross-protection with other viruses is a form of biological control.

There is, though, some hesitance among the farming community to adopt the cross-protection strategy completely as there are some worries about deliberately infecting plants and controlling the properties of the mild viruses that could revert to a virulent form. However, these cross-protection strategies form the basis for the main genetic-modification (GM) strategy discussed later in this chapter.

Another approach that is used is the production of virus-free plant stocks. This has been particularly important with plants such as potato, sweet potato, and strawberry that are vegetatively propagated and are therefore clonal material. The symptoms of virus infections may not be readily apparent during vegetative growth, but they can have serious effects on yield and/or quality of the crop. Viruses such as potato leafroll virus (PLRV) and potato virus Y (PVY) can reduce tuber yield by 50–80%. Entire clonal populations of the potato stocks may become infected because of the mode of propagation, so great effort has been put into systems for the development of virus-free stocks. The most common systems that have been in use since the 1960s have been heat therapy and meristem tip culture (see Chapter 2). More recently, antiviral chemicals have also been used. In principle, combinations of these technologies can be used to produce virus-free stocks. These can then be used in the field, but they do not provide protection against subsequent infection.

CASE STUDY 8.1 Developments in the sugar beet industry

Many countries try to operate a quarantine system to prevent the import of virus diseases into non-endemic regions. This has been a major part of the strategy adopted in the UK, and is highlighted by the long-term struggle to keep the sugar beet disease, rhizomania, out of the country. It was first identified in Italy in the 1950s, since when the virus and its vector have been identified. This disease is caused by BNYVV, a positive-sense RNA benyvirus, that is carried in soil from one plant to another by spores of the fungus *Polymyxa betae*. Different pathotypes exist, A, B, and P, with different properties and genome organization. As mentioned in the previous section, this virus normally has four separate genome fragments (pathotypes A and B), but in some isolates, including P, a fifth genomic RNA has been detected. Some sequence differences exist between the various forms of RNA5, but versions of this pathotype seem to be more virulent, spreading through plants more rapidly. All the isolates can give serious infections, reducing sugar yield by quite significant amounts. The disease has moved slowly through the sugar beet-growing areas of the world, transmitted via contaminated soil on vehicles, tools, and unwashed crops.

● New molecular techniques are more sensitive and give faster results to help detect viruses in soil and infected plants and allow remedial measures to be taken.

Unfortunately, it has been difficult to identify contaminated locations before symptoms appear, thus allowing further spread to occur. One method of detection that is still being used is to fly over beet fields to look for changes in foliage appearance, a symptom of many virus infections. The major problem for detecting soil contaminated with BNYVV was that the assay used was a bait test, which is shown in Figure 8.3. Soil, potentially contaminated with BNYVV, was brought into the laboratory and used as a medium to grow non-contaminated test plants to see whether they developed the disease. The plants would then tested for the presence of the virus with an enzyme-linked immunosorbent assay (ELISA). The system was time-consuming and insensitive. To overcome these shortcomings, to improve sensitivity and shorten the test time, hand-held test systems such as lateral-flow devices—similar to pregnancy testing kits—have been developed that can be used in the field by farmers. Also molecular biology techniques such as hybridization and the polymerase chain reaction (PCR) have been used (see Figure 8.4) and microarray approaches are underway. PCR strategies have developed rapidly and now include real-time

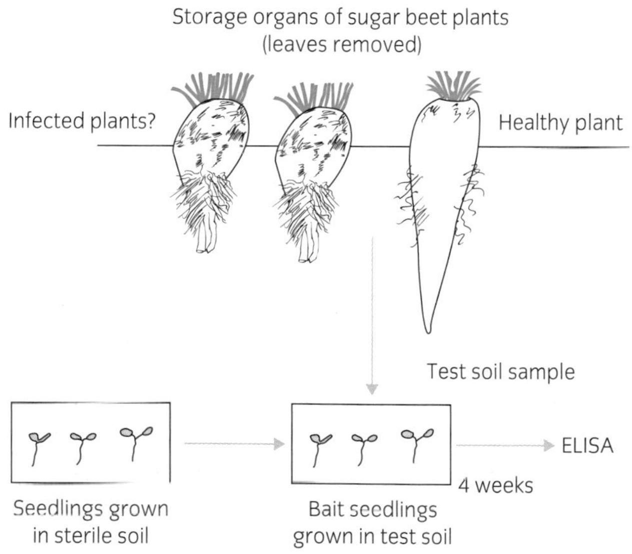

Figure 8.3 The soil bait test scheme.

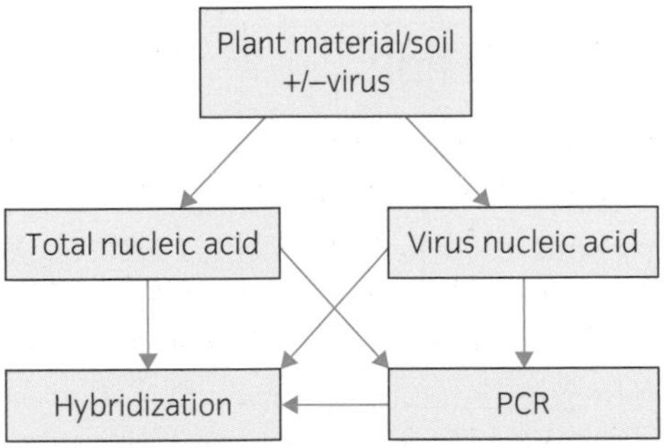

Figure 8.4 Molecular tests for viral nucleic acids.

PCR assays with *Taq*Man chemistry for the detection of *P. betae* in soil and multiplex reverse transcription PCR assays for the simultaneous detection of BNYVV, other beet viruses, and the fungal vector. The power of PCR and the availability of sequences also allows rapid identification of the different pathogens responsible for the disease. A quick examination of the literature shows these and other techniques being used for the detection of a whole range of viruses.

Unfortunately, these efforts have not worked and Great Britain is no longer a protected zone for rhizomania. To protect the sugar industry, sugar beet infected with the virus is allowed to be processed but is strictly managed to minimize any infected soil spreading the disease on to other land.

● The quarantine approach to combatting rhizomania failed, even on an island.

The most important approach to fighting BNYVV disease has been the development of crops with resistance. Previously, there had been some reticence about this strategy as in some cases the plants that were produced were tolerant, rather than resistant: the virus infected the plant, but the symptoms and viral titre were much reduced. While the crop was protected, it could act as a massive reservoir of virus that could infect non-resistant varieties. Recently, it has proved possible to produce varieties that resist rhizomania infection, which produce comparable yields of sugar in non-infected and infected trials and show low levels of BNYVV in the roots. This low level of BNYVV is important, as it will help to slow virus build-up in affected fields and help to restrict spread to uninfected areas and the processing factories. The production of such varieties has been possible because resistance genes exist within related beet species and it has been possible to breed them into elite lines. However, the main types of resistance are under threat. The main source of resistance in sugar beet has been a gene, designated *RZ1*, that maps close to a powdery mildew quantitative trait locus (QTL). It occurs in most of the commercialized resistant lines, giving protection against the A and B pathotypes. Another gene, *RZ2*, has also been identified and, in combination with *RZ1*, it is necessary for resistance against the P types. The requirement for two resistant genes is a cause for worry, as are the occasional reports of A-type viruses infecting resistant lines. Workers in the sugar beet industry are attempting to identify other sources of resistance to combat the disease.

The sugar beet story is repeated in a range of crops where resistance is available in a form that can be bred into elite lines. The pathogen eventually breaks through the barriers raised against them so now sources of resistance are always being looked for. Below, we will briefly consider which other genes have been identified that give resistance against viruses. In many crops, such resistance genes for specific diseases have not been identified. Therefore, other approaches need to be considered, in particular using transgenic strategies such as pathogen-derived resistance and the introduction of non-pathogen-derived genes.

The transgenic approach: PDR

PDR is a simple idea but with complex mechanisms of action.

The first, and still the main, antiviral transgenic approach used has been that of pathogen-derived resistance (PDR), originally known as *parasite*-derived resistance. In PDR, pathogen genomic sequences are deliberately engineered into the host plant's genome, on the non-scientific basis that the sequence may be expressed at an inappropriate time, in inappropriate amounts, or in an inappropriate form during the infection cycle and thereby induce some form of resistance in the plant. This type of strategy has, in part, developed from experiments such as cross-protection studies, where the introduction of a minor viral pathogen has induced resistance to a related but more virulent pathogen. However, many of the developments in this area have been made with no substantial knowledge base driving them forward. Indeed, it has been said that the reverse is true, and the use of PDR has led to a lot of basic information on virus biology and disease resistance being discovered. The mechanisms of PDR, as we will see, are varied and complex; the presence of the pathogen sequence may directly interfere with the replication of the pathogen or may induce some host defence mechanism. It has become clear that these interactions are not simple. As more is learnt about the underlying mechanisms and the resistance genes involved, new strategies involving plant resistance genes will be developed.

Potential types of interaction between transgenic sequences and virus life cycles can be categorized into two groups: those involving the synthesis of viral proteins and those involving viral RNA. In this section we will discuss some of these strategies.

Interactions involving viral proteins

Many protein genes have been used for PDR approaches but with varying success.

The first and most successful transgenic approach to viral resistance has been with the transgenic expression of the CP coding sequence. CP-mediated resistance was first reported with a TMV/tobacco model system in 1986. Subsequently, a large number of transgenic lines containing CP transgenes have been produced for a whole range of crop species and many different viruses; some examples are listed in Table 8.2. This is by no means a complete list, as there are far too many to include them all. Specific examples will be discussed in subsequent sections of this chapter. In many cases, some degree of resistance to virus infection has been reported; however, some variations in the level of resistance are found that can be related to transcriptional gene silencing, transgene position effects, and the relationship between the coding sequence and target virus. Some of these points are discussed below.

There are several important features of the CP strategy. First, there is some level of cross-protection against infection by related viruses. As will be discussed, resistance is not limited to viruses with the same nucleic acid sequence. Another important point is that, for TMV and several viruses from other groups, the level of resistance has been related to the level of transgenic CP produced. This is true for alfalfa mosaic virus (AlMV) and potato virus X (PVX) but not for PVY in which resistance was found in plants with no detectable level of transgenic CP. This suggests that several different

Table 8.2 Examples of CP-mediated resistance

Plant	Source of CP gene	Virus resistance exhibited to
Alfalfa	AlMV	AlMV
Citrus	CTV	CTV
Papaya	PRSV	PRSV
Peanut	TSWV (*N* gene)	TSWV
Potato	PVX	PVX, PVY
	PVY	PVY
	PLRV	PLRV
Rice	RTSV	RTSV
	RSV (*N* gene)	RSV
	RHBV (*N* gene)	RHBV
	RYMV	RYMV
Squash	CMV	CMV
	WMV 2	WMV 2
	ZYMV	ZYMV
Sugar beet	BNYVV	BNYVV
Tobacco	TMV	TMV, PVX, CMV, AlMV
	CMV	CMV
	AlMV	AlMV
Wheat	SBWMV	SBWMV
	BYDV	BYDV

List of virus names can be obtained from http://image.fs.uidaho.edu/vide. AlMV, alfalfa mosaic virus; BNYVV, beet necrotic yellow vein virus; BYDV, barley yellow dwarf virus; CMV, cucumber mosaic virus; CTV, citrus tristeza virus; PLRV, potato leafroll virus; PRSV, papaya ringspot virus; PVX, potato virus X; PVY, potato virus Y; RHBV, rice hoja blanca virus; RSV, rice stripe virus; RTSV, rice tungro spherical virus; RYMV, rice yellow mottle virus; SBWMV, soil-borne wheat mosaic virus; TMV, tobacco mosaic virus; TSWV, tomato spotted wilt virus; WMV 2, watermelon mosaic virus 2; ZYMV, zucchini yellow mosaic virus.

resistance mechanisms may be at work. Much of the work with TMV has focused on understanding the mechanism of TMV CP-mediated resistance (CPMR). Different mutant CP genes have been made that are either unable to form virus-like particles (VLPs), associate with other CPs, or have increased stability when introduced into tobacco. These studies suggested that CPMR TMV-like virus was due to an interaction between the transgenic CP and invading virus. With the non-VLP-forming mutant CPs, there was no resistance. With mutants that had increased CP stability in transgenic plants, there was an increase in CPMR when challenged with virus particles. The proposed explanation for TMV CPMR was that the CP was actively inhibiting the uncoating of the infecting virus, so preventing translation and replication. It was also suggested that the CP may interfere with the spread of some viruses. In addition, it was clear that transgenic TMV CP did not stop infection when the plants were challenged with naked TMV RNA. Transgenic plants expressing the CPs of other viruses, such as PVX and AlMV, did, however, display resistance to infection by their RNA. It is now becoming clear that there are three phenomena relating to the MP and CP. For some viruses the CP is not required but with others it is. In the third group, the viruses move as particles. The important point is that, as indicated above, other viruses have different properties to TMV, and that other mechanisms exist for CPMR at the protein level and, as we will see, at the RNA level.

There have been attempts at using other virus proteins. The genes for replication-related proteins (complete and partial sequences) have been introduced into plants with varying success. Potato plants containing Rep proteins have been commercialized (see below). Where resistance has been observed, it is generally specific to the virus from which the sequence was obtained. There is some debate over the mechanism of these effects; at the protein level, there may be some interactions between the transgenic replicase protein and that coded for by the virus. However, it has also been suggested that the resistance is an RNA effect. This will be discussed later in this chapter. Attempts to use the transgenic expression of mutant MPs to block the spread of viruses have met with varied success. As an example, transgenic plants containing a defective TMV MP show resistance to several tobamoviruses, AlMV, cauliflower mosaic virus (CaMV), and several other viruses.

Various expression strategies have been used with these genes to try and improve the resistance patterns obtained. The *CP* gene has been linked to inducible (wounding and chemical) and tissue-specific promoters. The gene has also been expressed in transgenic root stocks of non-transgenic grafts. An ecdysone agonist has been used as a soil drench for plants containing an ecdysone-inducible promoter linked to TMV CP. Drenching prior to infection with TMV brought about resistance. A PLRV complete-sense replicase gene has been linked to either a 35S promoter or a phloem-specific RolA (*Agrobacterium rhizogenes*) promoter and introduced into potato. The transgenic plants were challenged with aphids infected with PLRV. Two distinct patterns of resistance were observed. With the constitutive promoter, there was some initial evidence of infection but this was undetectable after 40 days. The RolA-expressed gene showed an opposite effect; initially no virus was observed but later the resistance was overcome. This example highlights the complexity of using the PDR approach: it is not a simple matter of expressing the virus-derived gene and obtaining resistance.

In the next section, we will look at a case study on the practical aspects of isolating a CP coding sequence and expressing it in a plant.

CASE STUDY 8.2 **Arabis mosaic virus**

Arabis mosaic virus (ArMV) is one of a group of economically important nepoviruses (which are transferred by nematodes) that also includes grapevine fanleaf virus and raspberry ringspot virus. These are very important in the wine industry, causing significant reductions in grape yield and quality. As the virus is transferred by nematodes it was felt that developing resistant root stocks would provide adequate protection against the virus. However, as no suitable resistance markers are available in the germplasm for breeding programmes, several groups have looked at transgenic approaches to produce resistant lines. In this case study, the production and laboratory testing of CP constructs in tobacco will be discussed. The aim of the project was to produce high levels of CP in the plant as there can be a positive correlation between CP expression levels and resistance. Therefore, a strong promoter, transcription enhancers, and a translation enhancer were used. The methods described in this case study were used for ArMV but they would be suitable for other viruses.

ArMV is a positive-sense RNA virus that has its genome on two separate molecules, RNA1 and RNA2. cDNA clones have been made to these sequences and it has been possible to locate the position of various protein-coding sequences in the genome. The CP is coded for on RNA2, but, as we have seen for the members of the *Comoviridae*, it is initially synthesized as part of a polyprotein that is processed in the cell to release the CP. The first step in producing the CP vector was to identify the precise location of the CP gene sequence in the genome. This was done, in part, by using computer analysis to look at sequence homologies. The capsid of CPMV has two CP subunits, but with the nepoviruses there is a single CP subunit (64 kDa) and this is to be found at the 3′ end of RNA2 (Figure 8.5). The 3′ end of the CP gene was defined by the stop codon (see Chapter 1) of the ORF. To identify the 5′ end of the protein sequence was more difficult as it is located in the region coding for the polyprotein. However, it was possible to obtain the N-terminal amino acid sequence of the CP by sequencing CP that had been purified by way of various phase separations and centrifugation in linear sucrose gradients. This could then be compared with the sequence obtained by computer (virtual) translation of the RNA sequence and the 5′ end identified.

PCR primers (oligonucleotides) were then designed to amplify, using PCR, the complete coding sequence of the CP. The primers were complementary to about 30 nucleotides of the virus and they had several restriction endonuclease sites, which did not occur in the sequence, incorporated at the 5′ end of the oligonucleotides. The primer for the N-terminal end of the sequence also contained an AUG start codon adjacent to and in frame with the CP coding sequence. PCR was carried out using RNA2 cDNA as the template. The amplified cDNA was then digested within the oligonucleotide restriction sites to allow the amplified CP construct to be directionally ligated into an *Escherichia coli* cloning vector. The ligation was carried out so that at the N terminus of the construct there was a CaMV promoter with several transcription enhancers and a TMV Ω sequence, to optimize the level of translation. At the C-terminal end, there was a nopaline synthase-derived transcription terminator sequence terminator (see Chapter 4). The complete sequence was then digested out of the intermediary plasmid and ligated into the binary vector pBIN19 (described in Chapter 4), which was used for *Agrobacterium*-based transformation. Many binary vectors are now available that make the intermediate step in the *E. coli* vector unnecessary. This general scheme is shown in Figure 8.5.

The CP construct was introduced into tobacco using the protocol described in Chapter 3, and transformed plant material was selected for by incubating the plant material on a medium containing kanamycin. A number of transgenic plant lines were obtained and these were screened for protein expression levels using ELISA.

Transgenic CP levels of the order of 20 µg g^{-1} fresh leaf tissue were found in various plants (a similar level of expression to the TMV subgenomic CP promoter levels of proteins discussed in Chapter 11). Interestingly, microscopic examination of the transgenic plants indicated that the CP was forming VLPs in the absence of the full viral RNA genome. These plants were tested for

● VLPs have the potential for transcapsidation or for use in molecular farming as vehicles for protein fusions.

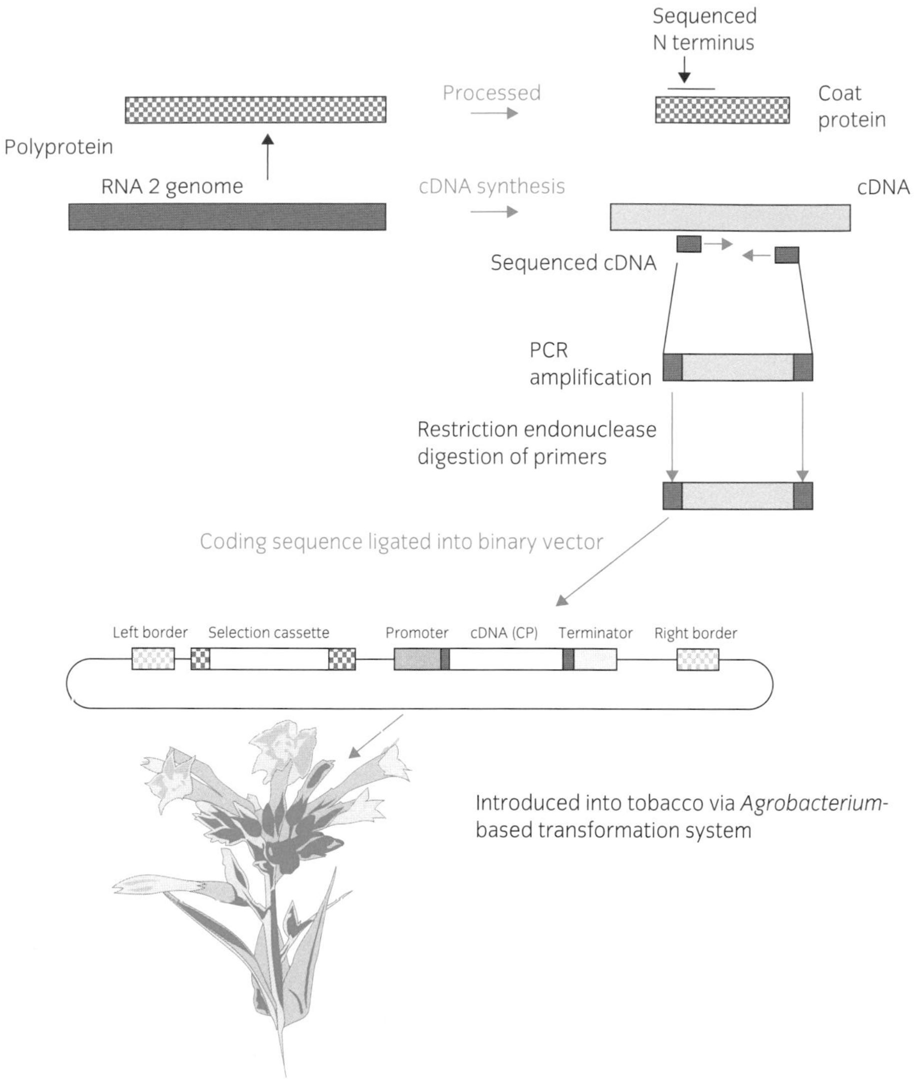

Figure 8.5 Strategy for the construction of ArMV transformation vectors and transgenic plants.

resistance to ArMV by mechanical inoculation of the leaves with whole virus, RNA, or water as a control. The local necrotic lesions induced by the virus on the inoculated leaves were counted, and it was found that the transgenic plants had much fewer than the control plants. ELISA analysis was also used to assay the inoculated leaves (local infection) and the tips of new leaves (systemic infection) for the quantity of virus antigen present. The transgenic plants were found to have much lower levels of CP antigen present in the new leaf material compared with the levels found in infected control plants. The levels of antigen detected are shown in Figure 8.6. The data indicate quite clearly that in the transgenic plants the presence of the ArMV CP is inhibiting the systemic infection process by both the virus particles and naked RNA (unlike the situation for TMV discussed above).

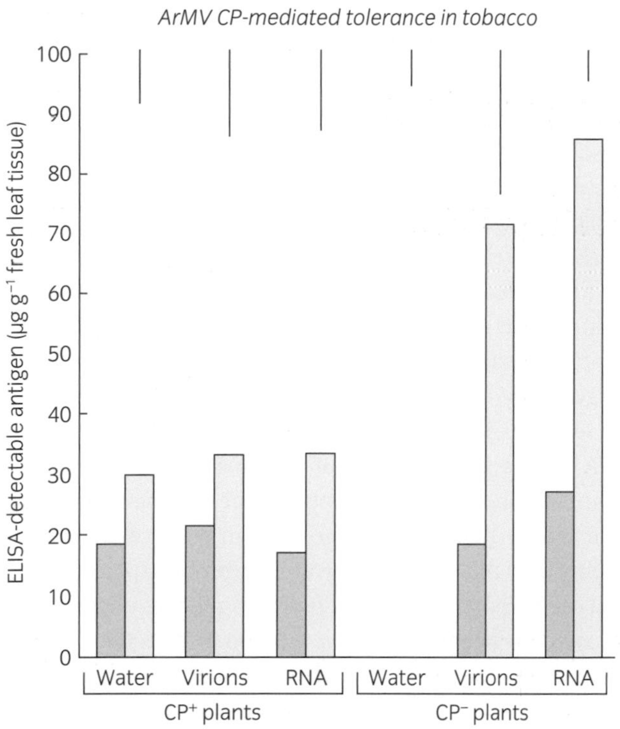

Figure 8.6 Quantities of ArMV CP, as detected by ELISA, in tip leaves (light-green bars) and inoculated leaves (dark-green bars) of plants expressing the ArMV CP (CP^+ plants) and control plants not expressing the ArMV CP (CP^- plants) inoculated with water, ArMV virions, or ArMV-derived RNA. The 95% confidence limits for the tip leaf data are given by the lines at the top of the graph. (Redrawn with permission from Bertioli *et al*. (1992).)

RNA effects

Some of the most interesting developments in the area of PDR relate to RNA effects. Several approaches have been developed that do not require protein synthesis, including satellite RNAs, antisense RNA, and ribozyme technologies. Perhaps the most important development in this area, though, is gene silencing. Studies where virus coding sequences were mutated and then introduced into plants led to the conclusion that high levels of RNA synthesis could lead to a post-transcriptional gene-silencing (co-suppression) phenomenon. This area of research has gone some way to explain some of the modes of resistance obtained with CP studies, but it has also led to the development of alternative strategies for PDR.

Satellite sequences

Plant viral satellite RNAs are small RNA molecules that are unable to multiply in host cells without the presence of a specific helper virus. The satellite is not required for virus replication but may affect disease symptoms. In many cases, the presence of a satellite RNA may reduce the effect of the virus, so one strategy has been to produce transgenic plants containing satellite sequences. In the section on cross-protection, it was noted that CMV symptoms were reduced when the virus was carrying a satellite and that this had been used as a commercial method for antivirus protection. Major studies have been carried out to see whether a transgenic approach could benefit this approach. Transgenic tobacco and tomato plants expressing CMV satellite RNA were tested in the field in China (1990–1992). Although some reduction in disease symptoms was found, it was not great enough to protect the plants. However, a strategy was developed in which satellite RNA PDR was developed in combination with CMV CPMR. It was found that the resistance was stronger than that of either CPMR or satellite PDR alone. This appears to be a good approach as combinations of genes are likely to generate more durable resistance, but some studies indicate that, whereas satellite RNAs generally have a neutral or attenuating effect on the helper virus, in some cases the severity of the disease can be increased. In one study, where the effect of satellite RNA on the pathogenicity of an ArMV isolate was investigated, of the 42 different plant species tested it was found that the presence of the satellite either had no effect or reduced the effect of the virus in most cases. However, the satellite worsened the effects of the virus in several leguminous species. The authors of this study comment on the potential risks of including the sequence in transgenic plants. A related approach has been the inclusion of small defective interfering RNAs and DNAs. These are small sequence fragments generated during the replication of some viruses. The presence of the sequences redirects the replication machinery towards the defective interfering RNAs and DNAs, reducing the amount of virus. This strategy has some relationship with co-suppression.

Antisense RNA and ribozymes

One simple strategy that has been attempted is the use of antisense RNA approaches (see Chapter 10). Ribozymes are also included in this broad category. To produce an antisense-mediated resistance to viral disease in transgenic plants, constructs have been designed that express a negative-sense RNA molecule that will hybridize with the infecting virus sequence (Figure 8.7a) The antisense approach was designed to specifically interfere with virus replication.

Ribozymes are antisense catalytic RNA molecules capable of catalysing the cleavage of the target sense RNA sequence. The constructs are similar to those used for the antisense approach, except that a short catalytic sequence is embedded within a specific target region (Figure 8.7b). The aim is to both block replication by the formation of a double-stranded RNA (dsRNA) hybrid and to cut a key region of the virus genome (such as within a replicase complex gene) before it is able to replicate. As both antisense and ribozyme technologies are based on sequence homology, they have limited cross-protection value unless the antisense sequence is designed against a very conserved region. These homology-based approaches have met with mixed success. There have been some reports that the antisense RNA does provide protection, but this may be due to gene silencing (which is described in the next section). The ribozyme approach has generally not been very successful in plants as an antiviral tool.

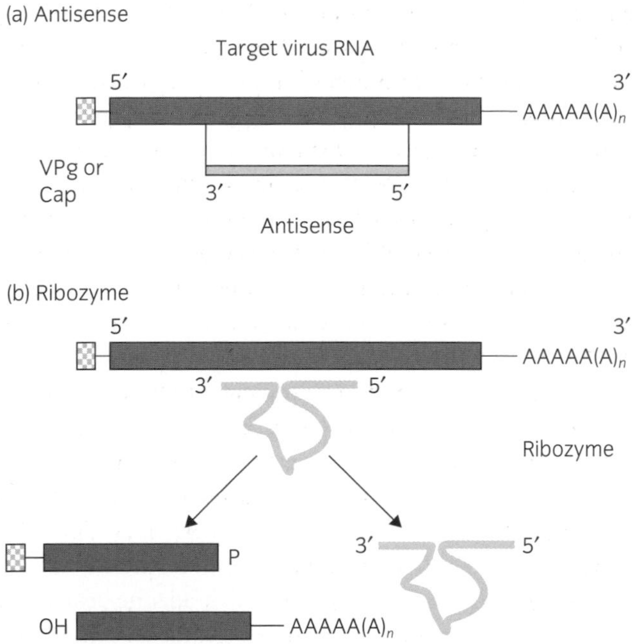

Figure 8.7 RNA approaches to PDR. (a) Antisense; (b) ribozyme.

Gene silencing/co-repression: the explanation for the RNA effects

The most important phenomenon associated with RNA-mediated PDR is post-transcriptional gene silencing (PTGS). PTGS was discussed in detail in Chapter 1 and briefly in Chapter 3 in the context of plant transformation. This phenomenon is found not just in plants but has been described in a wide range of organisms (fungi, nematode worms, humans) and seems to be an important system for preventing invasion by viruses and transposon sequences, as well as having some regulatory functions in the cell.

● A hot topic: the study of small RNA molecules has led from an understanding of unusual virus protein effects to an explanation for a whole new level of gene regulation.

Two separate avenues of research in plants uncovered the phenomenon of gene silencing. In studies with transgenic petunia plants expressing introduced chalcone synthase genes, under the control of strong promoters (which were designed to produce deep-purple flowers), it was found that pigment production in certain regions of the petals was often turned off and predominantly white variegated flowers were produced. This effect was termed co-suppression since both the introduced gene and the endogenous genes were turned off. With plant viruses, the first evidence for PTGS came from studies with tobacco etch virus (TEV). In these studies, virus CP transgenes, with in-frame stop codons so they were unable to produce full-length proteins, were found to give higher levels of protection against TEV infection than wild-type CP transgene constructs. Other related observations came from experiments performed with a mutated tomato spotted wilt virus (TMSV) nucleocapsid protein, PVY CP constructs, and various forms of the RNA-dependent RNA polymerase (RdRp) gene from PVX that were incapable of being translated into functional proteins. Initially it was thought this was related to an antisense effect, in which the positive-sense RNA from the transgene was binding to the negative-sense strand of the virus. Some experiments were also carried out with sense and antisense duplex constructs (hairpins) of the protease gene from PVY. It soon became clear that with the resistant lines there was often a high copy number of the transgene but there was no correlation between a high level of RNA from the transgene and the high level of resistance. On the contrary, the resistant plants often seemed to have low steady-state levels of the transgene RNA, although run-off experiments, where the actual level of gene expression is measured, indicated that transcription levels of the RNA seemed to be high. From these and other observations, a simple model for PTGS was developed. It was believed that the cell had a mechanism by which it could monitor RNA levels: it scanned for unusual amounts of double-stranded nucleic acids, such as in a viral infection where positive and negative hybrids may form during the replication process, and above a certain critical level the RNA was rapidly degraded. This model still holds in the context of PDR and virus infection: it is believed that plants showing good resistance due to the RNA effect are the ones in which the transgene RNA levels are very close to, or have exceeded, the critical level. The transgene RNA is being expressed at a high level in these plants, but is being rapidly degraded. So, on infection by the same or a very closely related virus, the cell is already primed to degrade the invasive RNA, which it does, therefore stopping the infection. While this critical-level model explains many of the results, there are some problems. It has been modified to allow for the possibility that it may not necessarily be the quantity of the dsRNA that is produced, but may, in part, be its quality that triggers destruction. It has been proposed that the cell can screen for dsRNA molecules that may contain aberrant RNA. How these are actually produced is still not clear, but they may result from cryptic transcription start sites (resulting in antisense RNA), transcription through inverted repeats, or the activity of an RdRp. This model could also be used to explain the co-repression phenomenon observed in transgenic plants.

Great progress has been made in our understanding of the mechanism of PTGS, partly due to the phenomenon being present in other eukaryotes. An overview of the current thinking is given in Chapter 1 and here summarized in Figure 8.8. The dsRNA is digested into 21–23 nucleotides long (19–21 base duplexes with two base overhangs at their 3′ ends) small interfering RNAs (siRNAs, which are also known as guide RNAs). This is carried out by an RNA III double-stranded-specific ribonuclease (in *Drosophila* this is known as Dicer). In plants, there are at least three Dicer-like proteins (DCLs): DCL2 cleaves dsRNAs from replicating viruses, and DCL3 cleaves dsRNAs derived from endogenous transcripts through the activity of RdRps 2 and 6. The third is DCL1, which is involved in the production of microRNAs (miRNAs; see below). The siRNA duplexes bind to a complex, which is believed to contain a different nuclease, to form the RNA-induced silencing complex (RISC). An associated ATP-dependent helicase then unwinds the duplexes and the RISC then targets homologous single-stranded RNA transcripts, via base-pairing interactions, and cleaves the RNA. It has been suggested that some of the short RNA molecules (possibly a different size range) act as primers for the synthesis of more double-stranded molecules, thus introducing an amplification step into the process. It is also believed

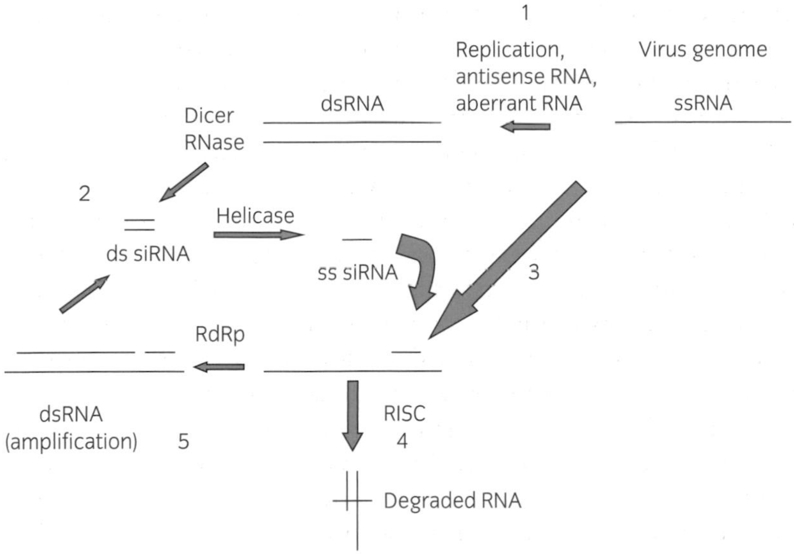

Figure 8.8 A model for the mechanism of post-transcriptional gene silencing (PTGS). (1) Double-stranded RNA (dsRNA) forms in the plant cell. (2) The dsRNA is recognized by the Dicer complex and is degraded to single-stranded small interfering RNA (ss siRNA). (3) Invading virus is detected entering the cell and is (4) degraded by the RNA-induced silencing complex (RISC). (5) Alternatively, the RNA is part of an amplification complex. (Adapted with permission from web-link 8.3.)

that the short siRNAs may act as systemic signals that are transported throughout the plant, thereby inducing systemic gene silencing.

miRNAs are important regulators of gene expression in plants and animals. In plants, they are involved in the control of development. miRNAs are single-stranded RNAs (20–24 nucleotides long) that are produced from longer pre-miRNA precursors by DCL1 (in *Arabidopsis*). These RNAs are recruited to the RISC complex and by base-pairing to specific target sequences, the complex down-regulates the expression of the target by one of two mechanisms. In animals, the base pairing results in translational repression. In plants, the base pairing results in cleavage of the target mRNAs.

One question that comes from these observations is that if plants contain effective defence mechanisms how do viruses manage to cause an infection? Part of the answer to this question is that virus proteins that can act as suppressers to gene silencing have been identified. In several potyviruses, the HcPro protein has been identified as having this function. Similarly, in CMV, the 2b protein has been shown to act as a suppresser of gene silencing. It is also interesting to note that these proteins function in different ways. HcPro blocks the maintenance of already established gene silencing. The 2b protein, on the other hand, suppresses the initiation of gene silencing. Other examples have been identified and it is becoming clear that successful virus infections can have serious effects on the host plant by interfering with both the siRNA and miRNA pathways.

It is likely that many of the effects seen with CP constructs are due to gene silencing. Recently, as the mechanism is more fully understood, there have been deliberate attempts to use PTGS as a mechanism to develop resistance. In particular, double-stranded sequences have been used as these may feed directly into the PTGS system, giving more effective resistance to the viruses. One group has inserted an inverted repeat of CMV cDNA into tobacco. On transcription, this generates an RNA sequence with the ability to form intramolecular double-stranded molecules. These have been shown to induce resistance to CMV by switching on the PTGS system. This was confirmed by the presence of the short siRNAs discussed above. In another study, the 3′ end of a PVY CP, which is conserved amongst different PVY isolates, was expressed as dsRNA in transgenic potatoes of the commercial variety Spunta. Twelve of 15 transgenic lines were found to be resistant to a number of different PVY lines. Another group have introduced four short tospovirus sequences, from TSWV, groundnut ringspot virus, tomato chlorotic spot virus, and watermelon silver mottle virus, as inverted repeats, into a single construct and introduced the unit into *Nicotiana benthamiana*. This approach using a single transgene gave resistance to the four viruses in transgenic plants. This sort of approach is not always successful. In another study where nepovirus sequences (coding for several proteins, including the MP) were introduced into grapevine root stocks as inverted-repeat constructs in order to induce PTGS, little effect was observed. Exactly why is not clear at this stage.

Even miRNAs have been considered as possible routes for blocking virus infections. Alteration of miRNA sequences to ones complementary to two plant viruses, turnip yellow mosaic virus and turnip mosaic virus, does not interfere with the synthesis of the miRNA so it has been possible to express the sequences in *Arabidopsis*, bringing about resistant phenotypes. This study also showed that using miRNAs may have advantages over PTGS approaches. Previously it had been observed that viral RNA-mediated resistance can be sensitive to reduced temperatures (15°C). The miRNA approach was durable for transgenic plants grown for 3 weeks at 15°C.

Some non-PDR approaches

There have been a number of attempts to induce virus resistance using non-PDR strategies. These have included antibodies to the virus or virus proteins and various mutants. One recent approach is the targeting of virus translation. There have been a number of genes introduced into plants to bring this about but perhaps the most interesting, if not surprising, is a human double-strand-dependent protein kinase. This enzyme is induced by interferon in humans and inhibits viral replication by the inactivation of translation initiation factor eIF-2α upon activation by dsRNA. The human gene has been fused to an *Arabidopsis* wound-inducible promoter—blue copper-binding protein gene—and introduced into tobacco plants. The plants expressing this protein were challenged with a range of viruses including CMV, TEV, and PVY and were found to show significant resistance. This seems to be an important development as naturally occurring resistance genes in crop plants have been shown to be due to mutant forms of another translation initiation protein eIF4E: rym4/5. These give resistance in barley to barley yellow mosaic virus and in melon to melon necrotic spot virus. This is clearly an area of some importance for further work.

CASE STUDY 8.3 **DNA viruses**

Although the RNA viruses seem to dominate the literature, there are some extremely important DNA viruses. As an example of these, we will consider the geminiviruses. The geminiviruses represent a family of viruses that is rapidly becoming a major problem in subtropical and tropical countries. There are four genera now defined within this group of viruses. The four genera are *Begomovirus*, *Topocuvirus*, *Curtovirus*, and *Mastrevirus* and they all require insect vectors (whitefly, treehopper, leafhopper, and leafhopper, respectively). The first three genera all have dicotyledonous hosts and have similar genome organization. The mastreviruses are distinct in genome organization and have monocotyledonous hosts. The begomoviruses are perhaps the most important group—comprising 117 of the 133 recognized viruses—and it is these that cause most of the economically important diseases. The majority of the viruses in this group have two single-stranded DNA components (bipartite). African cassava mosaic virus, which is discussed below, is an important example of these. A few of the begomoviruses have a single genome (single-stranded DNA, monopartite).

The two genomes of the bipartite begomoviruses code for a limited number of proteins, and both DNA molecules are required for infection. Genome A codes for DNA-replication proteins (replication-associated protein, Rep; replication enhancer protein, Ren), proteins with host-gene regulation and silencing activity (transcription activator protein, TrAP, AC4), and the CP. Genome B codes for two movement proteins (movement protein, MP; nuclear shuttle protein, NSP).

On being deposited into the host plant cell by the vector insect, the virus particles uncoat, giving circles of single-stranded DNA. These enter the nucleus, where they are made double-stranded. This molecule serves as a template for transcription (both strands are transcribed) or replication by a rolling-circle mechanism. The RNA moves out of the nucleus into the cytoplasm where it is translated to give the various proteins. The movement proteins are involved in moving the virus between cells and the CP packages the single-stranded DNA into virus particles.

Within the life cycle are a number of potential targets for engineering but, as with other viruses, there are a number of different factors that should be considered when designing a strategy if durable virus resistance is to be produced. For instance, these viruses often undergo recombination, which can lead to the emergence of new strains. Also, the influence of environmental conditions and the vector may influence the response of the plant to infection. Some breeding approaches have given some success, but a transgenic approach could allow direct introduction of resistance genes into virus-sensitive landraces or into breeding lines, or to stack resistance genes to improve the durability and spectrum of resistance.

The CP-based strategies have been considered for these DNA viruses. The introduction of transgenic CP into host plants may influence the life cycle of homologous viruses in a number of ways. For instance, with TMV it is believed to interfere with virus uncoating. With other viruses, the transgenic CP may interfere with virus transport or the sequence may lead to RNA effects (see above). With the monopartite geminiviruses, the CP is required for systemic spread of the particle. With these, the CP approach has had some success. Tomato plants expressing the CP of tomato yellow leaf curl virus show delayed symptom development when challenged with the virus. However, the bipartite viruses do not require the CP for systemic spread. It seems that the NSP can functionally replace the protein. It is not surprising that with these viruses the CP approach has not been successful. Other strategies have been considered where mutant virus proteins have been introduced into plants. One interesting idea is that mutant forms of the CP, that lead to reduced infectivity and vector interaction, could be introduced into plants to reduce the spread of the virus within and between plants. Other strategies that have been considered include the use of defective movement and replication proteins, using ribozyme-inhibitor proteins and RNases linked to virus regulatory systems (making them inducible) and DNA- and RNA-interference approaches. The latter strategies are interesting as they build on the complexity of the viruses' life cycle. African cassava mosaic virus (ACMV)-infected plants seem to contain small amounts of virus DNA, either twice or half the size of the genome, of unknown function. These sequences act as defective interfering DNA and plants transformed with a tandem repeat of subgenomic defective interfering ACMV DNA B showed reduced symptoms. This approach was specific for the parent virus. RNA interference approaches have included silencing the virus promoter. DNA constructs expressing dsRNA homologous to the bidirectional virus promoter have reduced expression levels in a transient expression system. Other RNA approaches have been used, such as silencing of coding sequences and antisense approaches. Up to now, these approaches have been the most successful. Both have been used to target virus protein genes. Resistance to African cassava mosaic disease has been induced in cassava plants transformed with a construct expressing antisense RNAs targeting the virus genes *Rep* (*AC1*), *TrAP* (*AC2*), and *REn* (*AC3*). An alternative approach has been to target the *AC1* gene. A copy of this gene from ACMV has been introduced into cassava where expression led to resistance to ACMV and two other related viruses. Analysis of the plants in both these systems revealed that siRNAs specific to the transgenes were produced, leading to PTGS and consequent protection against infection.

What has been commercialized in Western agriculture?

- Virus-resistant plants are available but have not been generally accepted by the industry. The exception is the papaya, but there has been a lot of hostility from pressure groups.

Yellow squash and zucchini

The Asgrow seeds company (Monsanto) markets several varieties of squash and zucchini under the names Independence II, Liberator III, Freedom III, and Destiny III. These lines are resistant to three important viral diseases: ZYMV, watermelon mosaic virus 2, and CMV. These are all serious pathogens of curcubits, causing leaf mottling, yellowing, dwarfing, and deformed fruit. Conventional resistance routes have been ineffective; the genetically engineered strains provide better levels of resistance, although they are susceptible to variations in the pathogen strain and the environment. There are two interesting points about these lines. The constructs used for some of the strains contain three separate CP coding sequences for the viruses. They also benefited from the serendipitous loss of linkage between the selectable marker and the CP constructs. This allowed marker-free plants to be produced by simple breeding strategies.

Papaya

Several different genetically engineered papaya lines have been produced and marketed commercially. The amazing thing about the production and commercialization of these is the rapid time scale in which it was brought about. Beginning in 1992, a very serious outbreak of PRSV had reduced Hawaii's papaya production by over 40% within 5 years. Workers at Cornell University and in Hawaii produced two genetically engineered lines, SunUp and Rainbow, that have been available to farmers since 1998. There are some interesting points to this story. The transgenic plants were able to be established in areas of quite bad virus infection, which allowed farms to develop the new lines without having to clear all the papaya. Both SunUp and Rainbow give durable resistance to PRSV but the range of virus isolates is different. SunUp, which is homozygous for the transgenic CP gene, is resistant to a range of viruses that includes isolates from outside Hawaii. This is believed to be due to a gene-dosage effect leading to enhanced PTGS. Rainbow, on the other hand, is hemizygous for the transgene, having only one copy. It is an F_1 hybrid between SunUp and a non-transgenic line, Kapoho. This dilution effect leads to it being more sensitive to non-Hawaii virus isolates and presumably it may be less durable to local isolates. To date, there is no evidence for this. Despite the potential weakness of Rainbow, it is the one preferred by the market and the growers.

At present, there is still a requirement for non-transgenic plants for some overseas markets. Transgenic plants are tagged with the β-glucuronidase (GUS) protein gene for ease of detection. Thus, there are considerable areas of non-transgenic planting. Interestingly, in areas where these plants are surrounded by transgenic varieties, there is the potential for a buffer zone composed of transgenic plants. Aphids carrying the virus would feed on the transgenic plants and be purged of the virus before arriving at the non-transgenic plants. The success of this crop in Hawaii has lead to significant examples of technology transfer to produce virus-resistant crops in countries such as Brazil and Jamaica.

Potato

The main examples here are plants that contain transgenic replication-related sequences. Monsanto marketed NewLeaf potato lines that had both *Bacillus thuringiensis (Bt)* resistance and resistance to several virus lines (such as PLRV and PVY). Due to a low take-up of these lines, they have been withdrawn from sale. There is still interest in creating new transgenic commercial plants but at present the market has not developed.

Risk

The commercialization of the CP lines has been a particular target for opponents of GM technology who have attacked them over the problems discussed in Box 8.1 (transcapsidation and recombination). They have also raised issues over the presence of the CP sequence giving the plants some growth advantage, in essence turning them

BOX 8.1 *The risk of producing a new virus*

It seems that CPMR works very well for many of the constructs, but deliberate expression of both RNA and protein, to bring about the CP effect, leads to two levels of interaction where problems could be encountered. There is a danger of RNA recombination (see panel a in the figure) and when expressing a CP, a serious consideration has been whether a phenomenon known as transcapsidation (which is also called heterologous encapsidation, transencapsidation, or genome

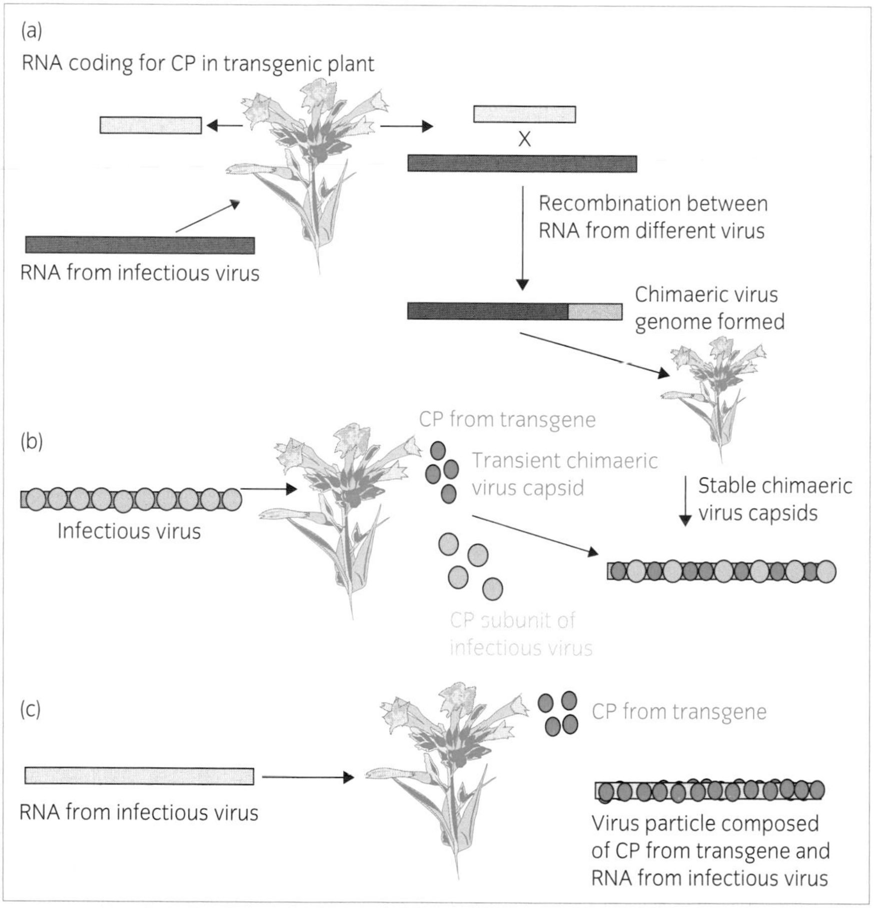

Recombination, transcapsidation, and chimaeric particles.

BOX 8.1 *Continued*

masking) could occur. The danger is that the CP protein produced from the transgene could form chimaeric particles with invading-virus CP (panel b in the figure) or package the nucleic acid of compatible viruses (panel c in the figure). This is important because CPs may affect a range of virus properties, for example protection of the nucleic acid, vector transmission and specificity, systemic invasion, and symptom expression.

Many studies have been carried out to investigate the possibility of transcapsidation. Chimaeric particles were first demonstrated in a study where a plant producing transgenic TMV CP was challenged with a mutant strain defective in CP production. The transgenically produced CP was able to package the defective RNA. Phenotypic mixing has also been detected in transgenic potato plants producing the CP of PVY strain N. When the plants were infected with PVY strain O, as many as 75% of the PVY-O particles were found to contain transgenic PVY-N CPs. Mixing of CPs has also been shown to occur between two unrelated viruses, as was the case when transgenic plants producing the CP of AlMV were infected with CMV.

The biological significance of these observations is still under debate as many other studies have been carried out that fail to show transcapsidation. However, perhaps the most important type of observation, as far as plant biotechnology is concerned, is that CP mixing has been demonstrated to restore aphid transmissibility to a non-aphid-transmissible isolate of ZYMV. This happened after phenotypic mixing with a transgenic CP of plum pox virus (PPV). Again, contradictory results have been obtained from field studies. Curcubit plants expressing the CP of a highly aphid-transmissible strain of CMV were inoculated with non-aphid-transmissible strains of the same virus. No aphid-mediated transmission of the second strain to control plants was detected. Apparently, in this case, the transgenically produced CP did not mediate transfer.

It is possible that the synthesis of CP leading to transcapsidation could alter virus transmissibility, but it could be argued that such phenomena already happen in nature. In many cases this is probably true, although genetic engineering can produce situations that do not normally occur in nature. Problems may arise if, for instance, CP is produced in tissues where the virus is not normally found. This is the case for PLRV in potato, which is normally expressed in vascular tissues during infections. However, with constitutive expression of a CP transgene in potato, the coat is expressed all over the plant and will then be exposed to other viruses it does not normally meet.

The frequency of transcapsidation is likely to be very low, and (as there is no exchange of genetic material) effects, such as changes of vector, will be transitory. Nevertheless, some effort has gone into removing the potential problem from more recent constructs. Several strategies have been adopted to reduce the likelihood of transcapsidation. In one approach, the CP sequences have been mutated, either to remove sequences that may be involved in transfer (such as the sequences in potyvirus CP required for aphid transfer) or to truncate the sequence to the minimum size required for protection. Another approach, PTGS, fully precludes protein expression; constructs are being generated that do not allow any protein expression, but rely on gene silencing to produce resistance.

(Continued overleaf)

BOX 8.1 *Continued*

RNA synthesis from the CP transgene and the reliance on the RNA-based approaches to viral resistance also have some potential problems; in particular, recombination could occur between different virus genomes. The most likely mechanism is one of template-switching. During replication of the virus genome, the viral RNA polymerase moves along the RNA sequence (the template), synthesizing the complementary RNA strand. It has been argued that, in the replication of some multipartite viruses, part of the process involves moving from one RNA fragment to others within the genome. In essence, the viral polymerase pauses during replication then switches to the different template and resumes synthesis. Potentially, this could be a very serious problem if the polymerase was to switch to the RNA of another virus, as this may result in the production of a recombinant virus. In nature, recombination is quite a common phenomenon within virus groups such as *Luteovirus* and *Tobravirus*, but it has also been shown to occur between totally unrelated viruses. Tests have been carried out with sugar beet plants containing the CP of an A-type BNYVV grown in soil containing *Polymyxa betae* carrying B-type virus isolates to see whether recombination occurred between the two virus CP sequences. To do this, PCR was used with primers designed to amplify part of the transgenic A-type sequence and part of the CP from the B type that is outside of the transgenic sequence. Despite the extreme sensitivity claimed for this assay, no evidence for recombination was found. Nevertheless, the risk of recombination should be minimized. Several possibilities have been suggested. It is possible to include stop codons in the sequence chosen to ensure that proteins cannot be expressed from the sequence. This would also make it unlikely that a viable recombinant virus could form as a result of recombination between the transgenic sequence and any wild-type virus. Other approaches, such as using short sequences that may prime the PTGS, may be considered.

into weeds. The same argument has been used for situations where the CP gene may be transferred to wild relatives. There have been a number of studies along this line with sugar beet and its wild relatives. Transfer has been shown to be possible and it may give some advantage to the hybrid when stressed with virus infections. As some virus-resistant plants are in the field, the big experiment has begun and time will tell as to the outcomes. The durability of disease resistance will also now be tested.

SUMMARY

- A combination of molecular breeding and transgenic approaches will produce the plants needed for agriculture during this century.

In this chapter, PDR has been considered in some detail. Some attempt has been made to explain the effects at the protein level by examining the work done with TMV and other related viruses. PTGS has also been raised as a mechanism of natural virus protection and not just an issue in transgenesis. Some transgenic plants have been commercialized with varying degrees of success. The transgenic papaya fruit lines have been welcomed in Hawaii as a saviour for the small farmer, whereas the NewLeaf transgenic potato lines currently seem to have been rejected or at least removed from sale in mainland USA.

The future of PDR-type resistance will depend on the public acceptance of these crops as much as whether they provide durable resistance. Other approaches to virus resistance are being considered. Enhancement of the PTGS system is an obvious strategy for virus resistance. Chapter 7 considered the possibility of activating generalized resistance pathways, and this potential route for virus resistance is being explored. It has been known for some time that tobacco strains containing the *N* gene are resistant to infection by TMV. It is now apparent that the product of the *N* gene belongs to the Toll-interleukin-1 receptor (TIR), nucleotide-binding (NB) domain, leucine-rich repeats (LRR) class of *R* genes (TIR-NBS–LRR). Many other *R* genes and interactions between the various signalling pathways are now known for viruses (see the Further reading section). A lot of effort is now going into the mapping and identification of further resistance genes for use in transgenic approaches or for the generation of molecular probes that can be used for marker-based breeding.

As with the bacterial pathogens, it would seem that combinations of approaches will provide the best strategy to keep ahead of virus-based diseases. In a recent study comparing CP- and replicase-based resistance in wheat to wheat streak mosaic virus with several non-transgenic resistant lines, the transgenic approach came out behind the others when actually tested in the field. The interaction between the pathogen and host defence systems discussed in Chapter 7 can influence the durability of virus resistance. There are examples where the resistance phenotype exhibited towards one virus can be switched off by another infecting virus. Despite these negative points for papaya, the GM strategy has been a great success to date and has potential for protecting susceptible lines in mixed growing.

FURTHER READING

Visit www.oxfordtextbooks.co.uk/orc/slaterplants2e/ for hyperlinked references that take you directly to online abstracts for the journal articles listed below:

PDR approaches

Bertioli, D. J., Harris, R. D., Edwards, M. L., Cooper, J. I., and Hawes, W. S. (1991) Transgenic plants and insect cells expressing the coat protein of arabis mosaic virus produce empty virus-like particles. *Journal of General Virology* **72**, 1201–1209.

Bertioli, D. J., Cooper, J. I., Edwards, M. L., and Hawes, W. S. (1992) Arabis mosaic nepovirus coat protein in transgenic tobacco lessens disease severity and virus replication. *Annals of Applied Biology* **120**, 47–54.

Virus groups and expression patterns

Lomonossoff, G. P. and Johnson, J. E. (1991) The synthesis and structure of comovirus capsids. *Progress in Biophysics and Molecular Biology* **55**, 107–137.

Non-GM approaches to crop protection

Fenby, N. S., Scott, N. W., Slater, A., and Elliott, M. C. (1995) PCR and non-isotopic labelling techniques for plant virus detection. *Cellular and Molecular Biology* **41**, 639–652.

Osman, F. and Rowhani, A. (2006) Application of a spotting sample preparation technique for the detection of pathogens in woody plants by RT-PCR and real-time PCR (TaqMan). *Journal of Virological Methods* **133**, 130–136.

Transgenic resistance

Beachy, R. N. (1997) Mechanisms and applications of pathogen-derived resistance in transgenic plants. *Current Opinion in Biotechnology* **8**, 215–220.

Dasgupta, I., Malathi, V. G., and Mukherjee, S. K. (2003) Genetic engineering for virus resistance. *Current Science* **84**, 341–354.

Gonsalves, D. and Slightom, J. L. (1993) Coat protein-mediated protection: analysis of transgenic plants for resistance in a variety of crops. *Seminars in Virology* **4**, 397–405.

Robaglia, C. and Caranta, C. (2006) Translation initiation factors: a weak link in plant RNA virus infection. *Trends in Plant Science* **11**, 40–45.

Gene silencing

Bucher, E., Lohuis, D., van Poppel, P. M. J. A., Geerts-Dimitriadou, C., Goldbach, R., and Prins, M. (2006) Multiple virus resistance at a frequency using a single transgene construct. *Journal of General Virology* **87**, 3697–3701.

Lecellier, C.-H. and Voinnet, O. (2004) RNA silencing: no mercy for viruses? *Immunological Reviews* **198**, 285–303.

MacDiarmid, R. (2005) RNA silencing in productive virus infections. *Annual Review of Phytopathology* **43**, 523–544.

Mlotshwa, S., Voinnet, O., Mette, M. F., Matzke, M., Vaicjeret, H., Ding, S. W., Pruss, G., and Vance, V. B. (2002) RNA silencing and the mobile silencing signal. *Plant Cell* (Suppl.), S289–S301.

Niu, Q.-W., Lin, S.-S., Reyes, J. L., Chen, K.-C., Wu, H.-W., Yeh, S.-D., and Chua, N.-H. (2006) Expression of artificial microRNAs in transgenic *Arabidopsis thaliana* confers virus resistance. *Nature Biotechnology* **24**, 1420–1428.

Scholthof, H. B. (2005) Plant virus transport: motions of functional equivalence. *Trends in Plant Science* **10**, 376–382.

Sontheimer, E. J. (2005) Assembly and function of RNA silencing complexes. *Nature Reviews: Molecular Cell Biology* **6**, 127–138.

Waterhouse, P. M., Wang, M. B., and Lough, T. (2001) Gene silencing as an adaptive defence against viruses. *Nature* **411**, 834–842.

DNA viruses

Vanderschuren, H., Stupak, M., Futterer, J., Gruissem, W., and Zhang, P. (2007) Engineering resistance to geminiviruses—review and perspectives. *Plant Biotechnology Journal* **5**, 207–220.

Vanitharani, R., Chellappan, P., and Fauquet, C. M. (2005) Geminiviruses and RNA silencing. *Trends in Plant Science* **10**, 144–151.

Virus-resistance genes

Kang, B.-C., Yeam, I., and Jahn, M. M. (2005) Genetics of plant virus resistance. *Annual Review of Phytopathology* **43**, 581–621

Ritzenthaler, C. (2005) resistance to plant viruses: old issue, news answers? *Current Opinion in Biotechnology* **16**, 118–122.

Risk

Falk, B. W. and Bruening, G. (1994) Will transgenic crops generate new viruses and new diseases? *Science* **263**, 1395–1396.

Greene, A. E. and Allison, R. F. (1994) Recombination between viral RNA and transgenic plant transcripts. *Science* **263**, 1423–1425.

Tepfer, M. (1993) Viral genes and transgenic plants. *Biotechnology* **11**, 1125–1132.

Commercialization

Fuchs, M. and Gonsalves, D. (1995) Resistance of transgenic hybrid squash ZW-20 expressing the coat protein genes of zucchini yellow mosaic virus and watermelon mosaic virus 2 to infections by both potyviruses. *Biotechnology* **13**, 1466–1473.

Tricoli, D. M., Carney, K. J., Russell, P. F., McMaster, J. R., Groff, D. W., Hadden, K. C., Himmel, P. T., Hubbard, J. P., Boeshore, M. L., Reynolds, J. F., and Quemada, H. D. (1995) Field evaluation of transgenic squash containing single or multiple virus coat protein gene constructs for resistance to cucumber mosaic virus, watermelon mosaic virus 2, and zucchini yellow mosaic virus. *Biotechnology* **13**, 1458–1465.

Web-links

Bustamante, P. I. and Hull, R. (1998) Plant virus gene expression strategies. *Electronic Journal of Biotechnology*, web-link 8.1: www.ejb.org/content/vol1/issue2/full/3/index.html

Fuchs, M., Ferreira, S., and Gonsalves, D. (1997) Management of virus diseases by classical and engineered protection. *Molecular Plant Pathology On-Line*, web-link 8.2: www.bspp.org.uk/mppol/1997/0116fuchs

Melcher, U., web-link 8.3: www.opbs.okstate.edu/~melcher/mg/MGW3/MG33322.html

Rybicki, E., web-link 8.4: www.uct.ac.za/microbiology/tutorial/RNA%20virus%20replication.htm

Universal Virus Database, ICTVdB, web-link 8.5: www.ncbi.nlm.nih.gov/ICTVdb/index.htm

9

Strategies for engineering stress tolerance

INTRODUCTION

● Plants are subject to a diverse range of abiotic stresses.

● Most stresses ultimately generate potentially damaging reactive oxygen species.

Crop plants are subject to a range of external factors that adversely affect their growth and development. In the previous four chapters, we have considered the effects of various biological factors (weeds, pests, and diseases) on crop yields, and biotechnological strategies to resist them. In this chapter, we will look at the effects of environmental conditions such as temperature, water availability, and salinity on crop plants. Figure 9.1 shows a number of physical factors that may impose an abiotic stress on plants and adversely affect their growth and development. These include a number that can be grouped together as temperature stresses (heat, chilling, and freezing), which in turn belong to a larger subgroup that can be categorized as stresses that result in water deficit. The figure also emphasizes the point that most abiotic and biotic stresses directly or indirectly lead to the production of free radicals and reactive oxygen species (ROS), creating oxidative stress.

The impact of abiotic stresses on crop yield compared with biotic stresses (weed, pest, and disease effects) is shown in Table 9.1. One of the first things to notice is the large difference between the average yields of crops and the record yields (an indicator of the maximum yields possible under ideal conditions). It is clear from these data that the major difference between record yield and average yield is accounted for by abiotic stress. Thus, the variation in environmental conditions from one year to the next produces such a variation in yield for wheat that the average yield is only 13% of the maximum. In contrast, as discussed at the beginning of Chapter 5, the control of biotic stresses in industrialized farming is such that they tend to reduce the annual yield by a fairly stable proportion, which is generally less than the most adverse abiotic stresses. Improving the tolerance of crops to abiotic stresses could therefore enable them to maintain growth and development during the normal fluctuations of adverse conditions, and consequently buffer crops against the large swings in yield experienced from one year to the next.

In the longer term, the predicted depletion of the ozone layer and climate changes associated with global warming are likely to add to the burden of environmental

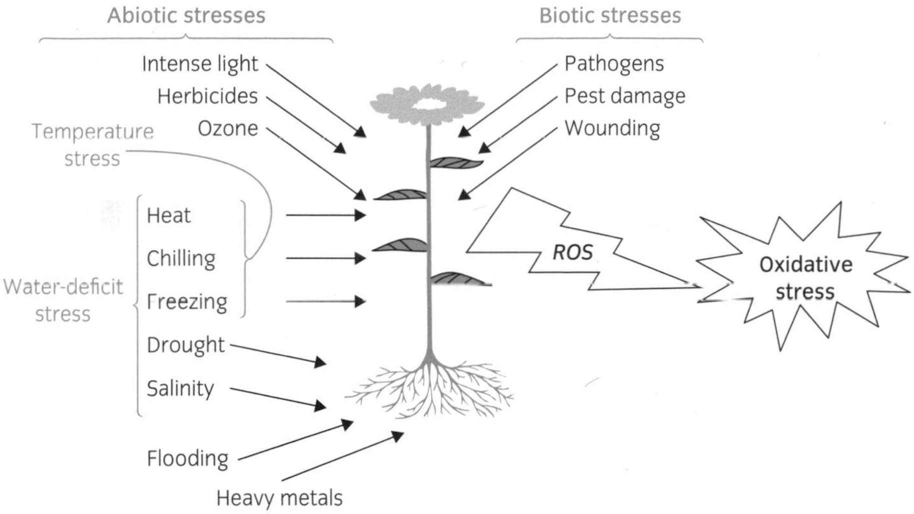

Figure 9.1 The different types of external stress that affect plant growth and development. The figure shows some of the external stresses impacting upon plant growth and development. The stresses are grouped according to common characteristics. Thus, abiotic and biotic stresses are defined, and the range of abiotic stresses resulting in water deficit is shown. The figure emphasizes the point that virtually all stresses result in the production of reactive oxygen species (ROS) and this creates oxidative stress.

Table 9.1 Average and record yields of some major crops

Crop	Record yield (kg ha^{-1})	Average yield (kg ha^{-1})	Average yield (% of record yield)	Average loss (% of record yield)	
				Biotic	Abiotic
Wheat	14 500	1 880	13.0	5.0	82.1
Barley	11 400	2 050	18.0	6.7	75.4
Soybean	7 390	1 610	21.8	9.0	69.3
Corn	19 300	4 600	23.8	10.1	65.8
Potato	94 100	28 300	30.1	18.9	54.1
Sugar beet	121 000	42 600	35.2	14.1	50.7

(Data from Bray *et al.* (2000).)

stresses on crop plants, and increase the imperative to develop stress-tolerant varieties. Furthermore, there is increasing pressure to extend the area of crop cultivation to environments that are not optimal for the growth of major crops (e.g. desert or high-salt conditions). The development of stress-tolerant plants is therefore a major goal of crop biotechnology, and one that is likely to become increasingly important.

The nature of abiotic stress

Stress is not easy to define: stressful conditions for one plant may be optimal for another.

When discussing the subject of stress tolerance, it is necessary first to try to define stress in relation to plant physiology. Plants are subject to many types of fluctuation in the physical environment. Many of the strategies used by animals to avoid the effects of these fluctuations are not available to plants, because of the sessile nature of their growth habit. Plants therefore depend largely upon internal mechanisms for tolerating variations in the external environment. Not all such fluctuations present a stress to plants, since they are able to cope with normal variation by virtue of their plasticity (the ability to alter metabolic profiles and developmental trajectories in response to external conditions; see Chapter 2). Thus, plants are adapted to function in a fluctuating environment, and normal external changes are countered by internal changes without detriment to growth and development. It is only acute or chronic extremes of environmental condition that lead to environmental stress that has the potential to cause physical damage to the plant. Note also that the extremes of environmental condition are defined by the normal environment of the particular plant species in question; desert plants do not prosper in the 'mild' climate of the temperate regions.

Given the range of abiotic stresses to which plants are exposed, it might be thought that a wide range of different strategies would be required to engineer particular types of stress tolerance. However, many different stresses cause similar types of damage to plants. This chapter will concentrate on the two major types of damage that result from a variety of different stresses. One of these is the damage that results from water deficit caused by a number of different environmental conditions, including drought, salinity, heat, and cold (see Figure 9.1). The other type of damage results from the production of ROS, causing chemical damage to the cellular constituents of plants. Most abiotic and biotic stresses cause the production of ROS, either directly or indirectly.

The nature of water-deficit stress

Several different abiotic stresses result in water deficit.

Several of the physical changes shown in Figure 9.1 result in a water deficit; that is, a situation in which the demand for water exceeds the supply. The supply is determined by the availability of water in the soil to the roots. The demand for water is determined mainly by the plant transpiration rate. Transpiration, the loss of water from the aerial parts of the plant via the leaf stomata, is required to dissipate the energy received from solar radiation and the ambient air temperature. Thus, transpiration causes leaves to cool relative to the ambient temperature when the environmental energy load on the plant is high. The rate of transpiration is also affected by the relative humidity and wind speed.

Two parameters are used to describe the water status of plant cells and tissues: water potential and relative water content. Water potential is measured in units of pressure (mega-Pascals or MPa) and is similar in concept to electrical potential:

current flows from a compartment of high potential to one of low potential if a connection is made between them. Thus, water is driven through the plant from the soil to the atmosphere by the difference in water potential between the atmosphere (very low potential) and the soil (relatively high potential when wet). As water transpires from the leaf, leaf water potential is reduced, but if water is available in the soil, water will flow into the leaf to replenish the loss. However, if the soil water potential starts to reduce (because the soil is drying out, for example), a deficit between the supply from the root and demand from the leaf is created. Clearly, a continued water deficit would lead to dehydration and death, but plants can respond to water deficit in a number of different ways. One way is to reduce the water potential in the leaf to create the necessary gradient to maintain the flow of water from soil.

● Osmotic adjustment by solute accumulation maintains turgor.

Two major factors determine water potential: the concentration of solutes that creates the solute or osmotic potential, and the physical pressure of the cell or tissue boundaries that forms the pressure or turgor potential. The osmotic potential is generally lower (more negative) than the water potential, and the turgor potential is the difference between them. Leaf turgor is lost (turgor potential = 0) when the leaf water potential drops to a value equal to the osmotic potential. As leaf turgor falls, the stomata close to reduce transpiration. When turgor is lost, the leaf cells collapse and the leaf wilts. These effects have the result of reducing water loss from the leaf, by closure of the stomata, and because wilting or rolling up of leaves reduces their exposure to sunlight.

However, the reduction in stomatal conductance also causes a reduction in photosynthetic assimilation, thus reducing growth. There may also be an increase in leaf temperature to a level that could cause heat damage to the leaf. The maintenance of turgor and transpiration are therefore important to maintain the growth of plants under drought stress. Turgor could be sustained either by keeping the leaf water potential high (by maintaining water uptake from the drying soil) or by reducing the osmotic potential. The latter could be achieved by solute accumulation (remember that an increase in solute concentration creates a more negative osmotic potential, i.e. a steeper gradient for the movement of water): a process called osmotic adjustment.

Different abiotic stresses create a water deficit

● Water deficit is damaging: it inhibits photosynthesis, increases the concentration of toxic ions, and removes the protective hydration shell around macromolecules.

Drought is, by definition, a water-deficit stress, because the environmental conditions either reduce the soil water potential and/or increase the leaf water potential due to hot, dry, or windy conditions. High-salt conditions can result in water deficit because the soil water potential is decreased (because the osmotic potential of a salt solution is lower than that of water), making it more difficult for roots to extract water from the environment. As stated above, high ambient temperatures cause increased water loss by evaporation, so heat stress creates a water deficit. However, freezing temperatures also cause osmotic stress because the formation of ice crystals in the extracellular space reduces the water potential and results in the efflux of intracellular water. In general, these various causes of water deficit result in the efflux of cellular water, leading to plasmolysis and eventually cell death. Water deficit is a

BOX 9.1 *The water shell around macromolecules*

The precise folding of soluble proteins is determined not only by intramolecular interactions between residues in the molecule, but by its interactions with the surrounding solution. For example, hydrophobic side chains are normally coiled within the interior of the molecule to minimize the interaction with the surrounding water, whereas polar and charged groups are often positioned around the external surface of the molecule. The water molecules around the protein form a highly ordered (and therefore low-entropy) structure called a **hydration shell** or **water shell**, which stabilizes the protein and serves to buffer it from interactions with polar solutes. Increases in the ionic concentration of the medium surrounding the protein will disrupt the structure of the protein, because Na^+ and Cl^- ions can effectively penetrate the hydration shell, and interfere with the non-covalent bonding that contributes to the stability of the protein. On the other hand, compatible solutes are excluded from the protein water shell, so the protein does not come into contact with the solute, and is not denatured.

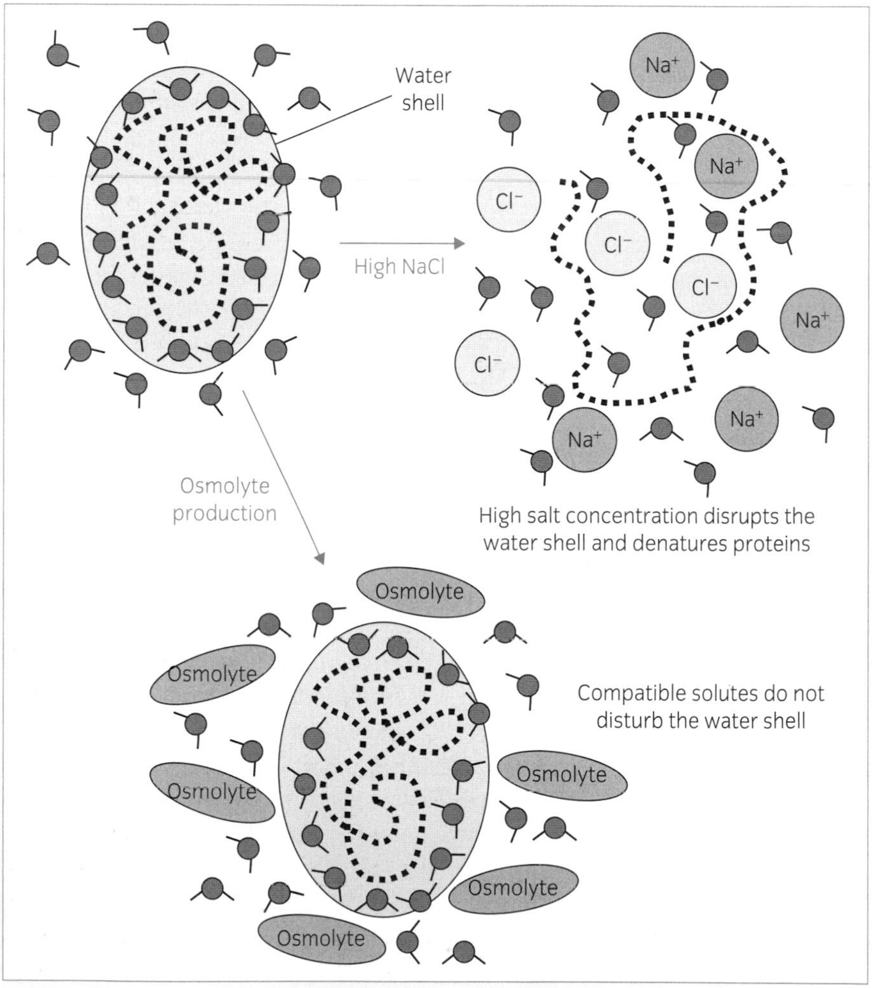

The water shell around a protein is disrupted when the concentration of ions is increased, and the protein is denatured. An equivalent concentration of a compatible solute does not disrupt the water shell and the protein is not denatured.

particular problem to plant cells because it inhibits photosynthesis via its effect on the thylakoid membranes. Water deficit is also potentially damaging to all cells because of the increase in the concentration of toxic ions and the loss of the protective hydration shell around vulnerable molecules (Box 9.1).

As described above, cells are able to accommodate some fluctuation in the water potential by osmotic adjustment, increasing the solute concentration and hence reducing the osmotic potential. However, as shown in Box 9.1, an increase in ions such as Na^+ and Cl^- would be counterproductive, given their effects on vulnerable molecules. Instead, plant cells respond to osmotic stress by producing non-toxic compounds called compatible solutes (or osmolytes, or osmoprotectants) to reduce the osmotic potential. It has been proposed that one reason for the non-toxicity of these compounds is that they can accumulate to high intracellular concentrations without disrupting the hydration shell around proteins and membranes (see Box 9.1).

● Plants may respond to water deficit by the accumulation of compatible solutes or osmoprotectants.

Figure 9.2 shows a range of compatible solutes produced by plants. These fall into two broad classes: sugars and sugar alcohols, and zwitterionic compounds (i.e. carrying a positive and negative charge). The former class includes sugar alcohols such as mannitol, sorbitol, pinitol, and D-ononitol, and oligosaccharides such as trehalose and fructans. (The production of some of these compounds in transgenic plants is described in more detail in Chapter 11). The latter class includes amino acids such as

● Production of osmoprotectants is a broad strategy for countering water deficit stress.

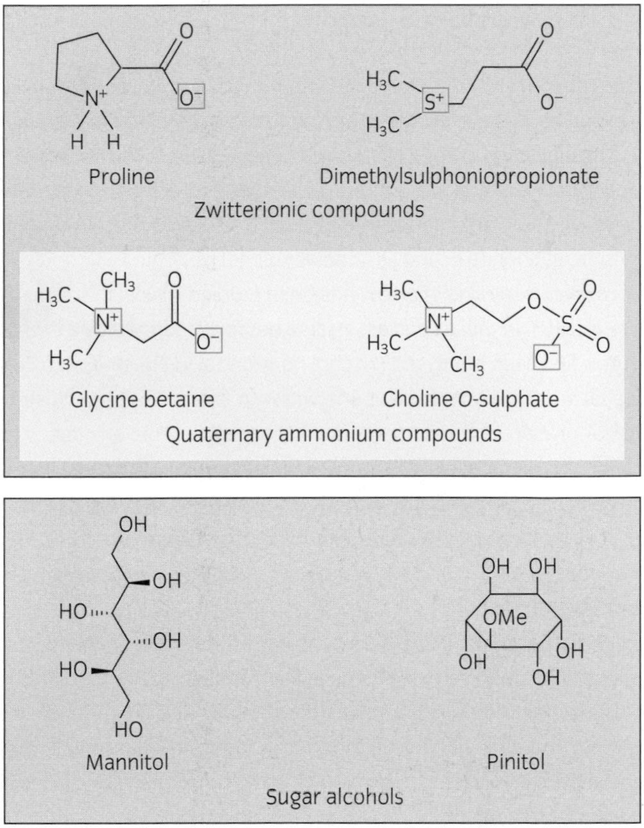

Figure 9.2 Compatible solutes. The chemical structures of a number of compatible solutes are shown, grouped together as zwitterionic or sugar-based compounds.

proline, and quaternary ammonium compounds such as glycine betaine. Different plant species tend to produce different compatible solutes. For example, mannitol is produced in large amounts by celery in response to salt stress, whereas spinach accumulates glycine betaine. In fact, spinach cells grown in a high-salt medium accumulate glycine betaine to a concentration of 300 mM in the cytosol. This maintains the osmotic potential of the cytoplasm relative to the apoplast (the connected extracellular space outside the cell; the symplast is the continuous intracellular compartment within protoplasts, connected by plasmodesmata) while keeping the cytoplasmic concentration of Na^+ and Cl^- ions low (<50 mM) and storing high concentrations of Na^+ (200 mM) and Cl^- (150 mM) ions in the vacuole.

● Some important crop plants lack osmoprotectants.

It is particularly significant that certain crop plants (e.g. rice and tobacco) lack significant amounts of any of these major classes of osmoprotectants. Basic strategies for engineering resistance to water-deficit stress have therefore focused on the production of osmoprotectants as a mechanism for overcoming the osmotic stress generated by water deficit. These strategies have generally required the determination of the biosynthetic pathways for various osmoprotectants, isolation of the relevant genes, and appropriate engineering of constructs to target gene expression and protein destination.

CASE STUDY 9.1 **Glycine betaine production**

Glycine betaine is a quaternary ammonium compound found in at least 10 families of flowering plants and also in marine algae. The charged groups mean it is extremely soluble in water, and is electrically neutral over a wide range of physiological pH values. The three methyl groups that comprise the quaternary ammonium group enable it to interact with hydrophobic as well as hydrophilic molecules. Thus, glycine betaine appears to stabilize proteins and membranes via these interactions, in addition to acting as a cellular osmolyte.

The synthesis of glycine betaine in plants has been studied mainly in species of Chenopodacieae (spinach, beet, etc.). In these species, glycine betaine is formed by a two-stage oxidation of choline, with the first step catalysed by choline monooxygenase (CMO; Figure 9.3). CMO encodes an unusual ferredoxin-dependent enzyme with an iron–sulphur cluster similar to that found in the Rieske protein of the cytochrome b_6/f complex. This enzyme is localized in the chloroplast stroma, where its activity is strongly light-dependent due to its requirement for reduced ferredoxin generated by photosynthetic electron transport. The second step in glycine betaine synthesis is catalysed by betaine aldehyde dehydrogenase (BADH), which is also stromal in Chenopodacieae. CMO and BADH are both induced by osmotic stress and have both been cloned from several plants.

The biosynthetic pathway of glycine betaine is different in *Escherichia coli* compared with plants, in that the production of betaine aldehyde from choline is catalysed by choline dehydrogenase, but the subsequent step still involves a BADH enzyme. In contrast, the synthesis of glycine betaine in *Arthrobacter globiformis* involves a single enzyme, choline oxidase (encoded by the *codA* gene), which catalyses the reaction directly from choline to glycine betaine (Figure 9.3), producing hydrogen peroxide.

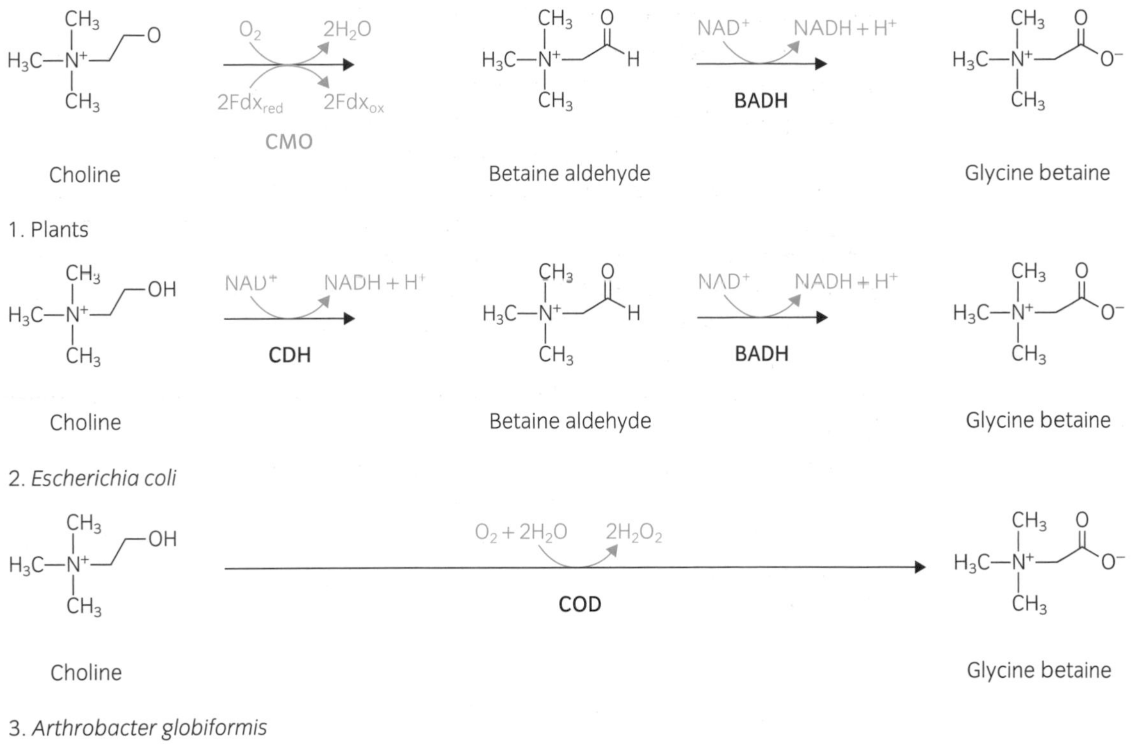

Figure 9.3 Pathways to glycine betaine. The different pathways from choline to glycine betaine in plants, *Escherichia coli*, and *Arthrobacter globiformis* are compared. BADH, betaine aldehyde dehydrogenase; CDH, choline dehydrogenase; CMO, choline monooxygenase; COD, choline oxidase; Fdx, ferredoxin; ox, oxidized; red, reduced.

Attempts have been made to manipulate glycine betaine production by transformation with the relevant plant, *E. coli*, or *A. globiformis* genes (Table 9.2). Transformation of plants with the betaine aldehyde dehydrogenase genes from plants or *E. coli* permitted the accumulation of glycine betaine when plants were supplied with betaine aldehyde, and conferred resistance to this toxic compound (see Chapter 4). However, plants did not accumulate glycine betaine in the absence of betaine aldehyde, presumably because the conversion of choline to betaine aldehyde does not occur in plants that do not normally accumulate glycine betaine.

Transformation of tobacco with the plant *cmo* gene targeted to the chloroplast resulted in very low levels of glycine betaine accumulation (Table 9.2). It has been suggested that this is a result of the very low pool of choline present in tobacco cells, or it could be attributable to the lack of betaine aldehyde dehydrogenase activity in the tobacco chloroplasts. Tobacco plants transformed with the *E. coli* choline dehydrogenase gene, *betA*, also failed to accumulate glycine betaine, even though this enzyme can also catalyse the conversion of betaine aldehyde to betaine. However, in this case, the gene was not targeted to the chloroplast, which might explain this failure. However, there was an observed increase in the salt tolerance of these transgenic plants that cannot be accounted for by the accumulation of glycine betaine to high concentrations.

Higher levels of glycine betaine in transgenic plants have been obtained using the choline oxidase gene from certain *Arthrobacter* species, particularly when transformed into plant

Table 9.2 Glycine betaine production in transgenic plants

Transgene	Host plant	Accumulation of glycine betaine	Stress tolerance tested
Barley *badh*	Tobacco peroxisome	Not tested	Not tested
Spinach *badh*	Tobacco chloroplast	20 μmol g^{-1} FW (in 5 mM betaine aldehyde)	Not tested
Spinach *cmo*	Tobacco chloroplast	<0.05 μmol g^{-1} FW	Not tested
E. coli betB (betaine aldehyde dehydrogenase)	Tobacco chloroplast or cytosol	Not tested	Not tested
E. coli betA (choline dehydrogenase)	Tobacco cytosol (membranes)	Not detected	Salt
E. coli betA/betB	Tobacco	0.035 μmol g^{-1} FW	Chilling, salt
E. coli betA	Rice	5.0 μmol g^{-1} FW	Drought, salt
A. globiformis codA (choline oxidase)	*Arabidopsis* chloroplast	1.2 μmol g^{-1} FW	Salt, chilling, freezing, heat, strong light
A. globiformis codA	Rice	5.3 μmol g^{-1} FW	Salt, chilling
Arthrobacter pascens cox (choline oxidase)	*Arabidopsis*	19 μmol g^{-1} DW	Freezing, salt
Arthrobacter pascens cox	*Brassica napus*	13 μmol g^{-1} DW	Drought, salt
Arthrobacter pascens cox	Tobacco	13 μmol g^{-1} DW	Salt

badh, betaine aldehyde dehydrogenase; *cmo*, choline monooxygenase; DW, dry weight; FW, fresh weight.

species (e.g. *Arabidopsis* and rice) that have significantly higher choline pools than does tobacco. In a number of cases where glycine betaine accumulation has been achieved, the tolerance to various water-deficit stresses has been determined, including salt, chilling, freezing, heat, and drought (see Table 9.2). It is apparent that production of glycine betaine in transgenic plants can improve tolerance to a range of osmotic stresses under the test conditions applied, validating the initial concept that osmoprotectants can enhance tolerance to a variety of stresses that cause water deficit. However, it is not entirely clear what the protective effect of the glycine betaine is. The levels of glycine betaine produced in most of these transgenic plants is relatively low (about 10% of the levels that accumulate in plants such as spinach) and is not sufficient to account for the improved osmotic stress tolerance by osmotic adjustment. It may be that the glycine betaine exerts a protective effect on vulnerable macromolecular structures at a lower concentration than its effect as an osmolyte.

In the same way that production of glycine betaine increases tolerance to a range of stresses that cause water deficit, other osmoprotectants have been produced in transgenic plants to test the feasibility of this approach. Table 9.3 shows the range of genes and products that have been used to test this strategy in a range of crop plants.

Table 9.3 Transgenic plants engineered to synthesize osmoprotectants other than glycine betaine

Osmoprotectant	Transgene	Crop plant	Accumulation	Stress tolerance
Polyamines	Arginine decarboxylase	Rice		Drought
	S-Adenosylmethionine decarboxylase	Rice		Salt
	Spermidine synthase	*Arabidopsis*		Cold, salt, drought
Proline	Mothbean *P5CS*	Tobacco		Salt
		Rice		Drought, salt, oxidative
		Soybean		Osmotic pressure, heat
	$P5CS^{F127A}$ (feedback-inhibition-insensitive)	Tobacco	4 mg g^{-1} FW	Salt
	Anti-ProDH	*Arabidopsis*	0.6 mg g^{-1} FW	Salt, freezing
Mannitol	*E. coli mt1D*	*Arabidopsis*	10 µg g^{-1} FW	Salt
		Tobacco	6 µmol g^{-1} FW	Salt
Sorbitol	Apple *s6pdh*	Tobacco		Oxidative stress
		Persimmon	61.5 µmol g^{-1} FW	Salt
Trehalose	Yeast *tps1*	Tobacco	3.2 µg g^{-1} DW	Drought
		Potato		Drought
	E. coli otsA + otsB	Tobacco	90 µg g^{-1} FW	Drought
D-Ononitol	Ice plant *imtl*	Tobacco	35 µmol g^{-1} FW	Drought, salt
Fructans	*Bacillus subtilis sacB*	Tobacco	0.35 mg g^{-1} FW	Drought
		Sugar beet	5 mg g^{-1} DW	Drought
Glutamine	GS2	Rice		Salt, chilling
Osmotin	*Osm1–Osm4* (protein accumulation)	Tobacco		Drought, salt

Anti-ProDH, proline dehydrogenase; DW, dry weight; FW, fresh weight; GS2, chloroplastic glutamine synthetase; *imtl*, myo-inositol-*o*-methyltransferase; mt1D, mannitol-1-phosphate dehydrogenase; *otsA + otsB*, trehalose-6-phosphate synthase and trehalose-6-phosphate phosphatase; *P5CS*, pyrroline carboxylase synthetase; *s6pdh*, sorbitol-6-phosphate dehydrogenase; *tps1*, trehalose-6-phosphate synthase.

What is clear from this table is the existence of a range of effective transgenes for the production of osmoprotectants in plants, and that these generally have a measurable effect on the tolerance of the plant to various stresses that cause water deficit. In some cases, tolerance to more than one stress has been demonstrated for the same transgenic plant, confirming the view that generic strategies to combat water-deficit stress could provide tolerance against a range of different stresses. However, several aspects of osmolyte accumulation still require further investigation and explanation. One point to note from Tables 9.2 and 9.3 is that the level of accumulation of most of the compatible solutes, apart from proline and D-ononitol, is considerably lower than could account for their protective effects by osmotic adjustment. The reasons for stress tolerance may be different in each case, with some compounds like glycine betaine acting primarily to stabilize macromolecular structures, and other compatible solutes such as mannitol, sorbitol, and proline being effective scavengers of ROS (see below). Another point to make is that the tests of stress tolerance of these transgenic plants are often performed under highly controlled laboratory conditions that present an acute osmotic shock rather than the chronic water-deficit stresses more generally experienced in the field. It should be noted that few of these transgenic

● Osmoprotectants have two roles: protection of vulnerable macromolecules and osmotic adjustment.

plants have been tested for stress tolerance in the field. Another consideration for developing these strategies further is that the accumulation of trehalose and sorbitol was found to inhibit plant growth.

● Drought tolerance can improve plant survival but may not increase yields to viable levels.

A final note of caution relates to the assumption that increasing osmolyte accumulation in transgenic plants to levels that can maintain cell and tissue turgor pressure under water-deficit conditions will improve crop yield under drought conditions. Under conditions of reduced soil water, the maintenance of leaf turgor and hence transpiration to sustain growth will reduce soil water further and hasten the dehydration of the leaves. What needs to be understood is that under conditions where the water deficit is severe enough to threaten plant survival, crop yields are so low that even large increases in fractional yield may be of little practical benefit to the grower.

Targeted approaches to manipulating tolerance to specific water-deficit stresses

The production of osmoprotectants is not the only possible strategy to engineer resistance to the water-deficit stresses shown in Figure 9.1. Individual stresses may be susceptible to specific mechanisms to improve tolerance. Some of these will be described in the following section.

Alternative approaches to salt stress

● Salt tolerance can be improved by preventing cytoplasmic accumulation of Na^+ and Cl^- ions.

Salt tolerance is increasingly becoming a major target for crop improvement as substantial areas of irrigated land are damaged by the accumulation of salt. Furthermore, the pressure for land has made it necessary to consider the possibility of growing crops in more-saline conditions, with poorer-quality water. As described above, saline conditions lead to osmotic stress by preventing water uptake by the roots and water efflux from the cells. However, the accumulation of Na^+ and Cl^- ions in the cytoplasm may also have direct toxic effects by inhibiting protein synthesis, photosynthesis, and susceptible enzymes. Thus, strategies for engineering water stress tolerance via the production of compatible solutes may provide protection against the osmotic effects of saline conditions, but not against ion toxicity. Additional approaches to minimize the toxic effects of specific ions may also be required. Strategies for engineering salt tolerance have been developed by comparing the response to salt stress of plants that are sensitive to high-salt conditions (glycophytes) with that of plants that can tolerate high-salt conditions (halophytes). As a general rule, glycophytes respond to salt stress by accumulating osmoprotectants whereas halophytes employ specific mechanisms to avoid the toxic effects of Na^+ and Cl^- ions. Some halophytes actually excrete salt via specialized glands on their leaf surfaces, but it is more common for plants to regulate Na^+ ion accumulation in the cytoplasm. In vulnerable cells with small vacuoles (e.g. meristematic cells), this involves transport of Na^+ ions out of the cell, whereas other cells with large vacuoles may act as sinks for the accumulation of excess sodium by transport into the vacuole.

CASE STUDY 9.2 Na^+/H^+ antiporters improve salt tolerance in transgenic plants

One approach to enhancing salt tolerance would be to copy the strategies used by halophytes to transport Na^+ ions out of the cytoplasm. In order to put this into practice, it is necessary to consider the mechanisms of ion transport out of the cytoplasm. The first transgenic experiments considered transport of Na^+ ions into the vacuole. Since this transport is working against a concentration gradient, it requires the input of energy. This is achieved by coupling the transport protein to a proton pump, transporting H^+ ions in the opposite direction (hence it is called an antiport protein). The vacuolar Na^+/H^+ antiport protein AtNHX1 of *Arabidopsis* has been studied extensively and is known to be coupled to proton pumps such as AVP1, a vacuolar H^+-translocating pyrophosphatase. An analogy that has been used is to compare AtNHX1 with a revolving door, and AVP1 as providing the energy for the door to spin. To increase the traffic through the membrane, one could therefore either increase the number of doors, or provide more energy for the existing doors to spin faster.

The first approach was successfully used to engineer salt tolerance in tomato plants by transformation with the *Arabidopsis* AtNHX1 antiport protein gene. These normally susceptible plants were able to grow in a high-salt environment to fertile maturity. Tomato plants grown in 200 mM NaCl accumulated salt in the leaves to a level 20-fold higher than normal, but, significantly, the fruits had a low sodium and chloride content. The same strategy produced transgenic canola (a variety of oilseed rape) plants that could accumulate up to 6% sodium, while the oil quality and yield of the seeds remained normal. The *Arabidopsis* gene has also been transformed into maize and wheat plants, leading to enhanced salt tolerance and improved grain yields in high-salt conditions. The potential of the *NHX1* gene from other plant sources to enhance salt tolerance has been investigated. For example, rice transformed with the *NHX1* gene from rice and the halophytic plant *Atriplex gmelini* showed improved growth and accumulation of biomass under salt-stress conditions. Similarly, transgenic tobacco containing the *NHX1* gene from cotton, oilseed rape, and wild barley has higher salt tolerance under experimental conditions.

Another approach to enhancing vacuolar transport has been to overexpress the gene encoding AVP1, initially in *Arabidopsis*, to increase the proton-pumping potential of the vacuole, and hence its ability to transport sodium. This has improved not only the salt tolerance of these experimental plants, but also the drought tolerance, since the altered ion balance has enabled the plants to retain more water.

More recently, the effects of introducing a plasma membrane Na^+/H^+ antiporter to extrude Na^+ ions from the plant cells have been investigated. Overexpression of the *Arabidopsis* plasma membrane Na^+/H^+ antiporter *AtSOS1* gene in transgenic *Arabidopsis* improved the salt tolerance of callus cultures and reduced the levels of Na^+ ions in the xylem of transgenic plants. Expression of the fission yeast *SOD2* gene in *Arabidopsis* improved seed germination and seedling salt tolerance. Transformation of rice with the *E. coli nhaA* plasma membrane Na^+/H^+ antiporter improved germination rate, growth, and yield under conditions of both salt and drought stress. Interestingly, expression of *nhaA* also increased the levels of proline in the transgenic rice, suggesting that enhanced osmoregulation might in part be due to activation of proline biosynthesis.

Flowers (2004) has sounded a valuable note of caution about claims of salt tolerance in transgenic plants. In a survey of 68 papers published between 1993 and 2003, he found 19 with some kind of quantitative estimate of plant growth. Of these, only four had measured growth under appropriate field conditions, and had compared the response of normal and transgenic plants to both saline and non-saline conditions. Many laboratory experiments were conducted under conditions with little or no transpiration, and conclusions about cellular tolerance could not necessarily be extrapolated to salt tolerance at the whole-plant level.

Alternative approaches to cold stress

Cold tolerance can be engineered by enhancing cold acclimation processes.

Different plants vary enormously in their ability to withstand cold and freezing temperatures. Most tropical plants have virtually no capacity to survive freezing conditions. On the other hand, many temperate plants can survive a range of freezing temperatures from −5 to −30°C depending upon the species. Plants from colder regions routinely withstand temperatures even lower than this. It is known that plants are better able to withstand cold or freezing stress if they first undergo a period of cold acclimation, at a low but non-freezing temperature. For example, wheat plants grown at normal warm temperatures are killed by freezing at −5°C, but after a period of cold acclimation when the plant grows at temperatures below 10°C, they can survive freezing temperatures down to −20°C.

Plants differ in their ability to withstand cold or freezing conditions and cold tolerance is one of the traits that plant breeders have selected for over many centuries. However, there has been little improvement in the cold tolerance of major crop species over the past two decades by conventional breeding, prompting the search for molecular solutions to this problem.

CASE STUDY 9.3 **The COR regulon**

One approach has been to study the mechanisms of freezing resistance that exist in some plant species. During the period of acclimation, plants produce a number of cold-induced proteins that are assumed to play a role in the subsequent cold resistance. About 50 cold-induced proteins have been identified in different plant species. These fall into a small number of groups, but they all share the property of being extremely hydrophilic. Many of them also have relatively simple amino acid compositions, with repeated motifs. Some of these groups had previously been identified as late embryogenesis abundant (LEA) proteins, which appear to play a protective role during seed desiccation. Other groups of proteins are encoded by a class of genes designated as cold-responsive (*COR*) genes according to their patterns of expression. The precise function of these cold-induced genes is not yet known, but it has been speculated that they might contribute directly to freezing tolerance by mitigating the potentially damaging effects of dehydration associated with freezing. Overexpression or ectopic expression of these cold-induced proteins could therefore be a possible route to the specific engineering of cold or freezing stress tolerance.

There are some examples of the expression of cold-induced proteins in transgenic plants. For example, constitutive expression of the small, hydrophilic, chloroplast-targeted COR protein COR15a in *Arabidopsis* improved the freezing tolerance of chloroplasts frozen *in situ* and of protoplasts frozen *in vitro*. However, COR15a expression has no discernible effect on the survival of frozen plants. One explanation for this observation is that the cold-induced proteins may be targeted to different vulnerable cell components, and that they are all required to provide complete protection to the cell. By implication, many *COR* genes would need to be transformed into a transgenic crop to obtain an appreciable improvement in cold tolerance. As discussed in Chapter 4, insertion of several genes into a transgenic crop on the same vector is technically difficult, and the process of pyramiding or stacking genes by crossing transgenic lines (see, for example, Chapters 6 and 11) is time-consuming. A different strategy is therefore needed.

- Regulons allow one transgene to activate several endogenous genes.

One solution to the problem of engineering a multigene trait has emerged following the recognition that several different cold-tolerance-related genes contain a similar regulatory element in their promoters: the C-repeat (CRT) element/dehydration-responsive element (DRE)/low-temperature-response element (LTRE). Furthermore, it has been found that the transcription factor CBF1 binds to the CRT/DRE/LTRE element and activates expression of this group of genes, which comprise the COR regulon (see Box 9.2). Therefore, the strategy is to overexpress the *CBF1* gene, leading to the induction of this entire group of *COR* cold-tolerance genes. Transgenic *Arabidopsis* plants carrying a 35S promoter::*CBF1* gene construct have been produced. These plants express a number of *COR* genes without cold acclimation and have been shown to be freezing-tolerant without prior cold acclimation. As a control, transgenic plants overexpressing a single COR protein, COR15a, were found to be less freezing-tolerant than the *CBF1* plants. The inter-related nature of different stress responses was demonstrated in a similar experiment. The expression of a CBF1 homologue, DRE binding protein 1A (DREB1A) under the control of a stress-induced promoter in transgenic *Arabidopsis* resulted in plants that had improved drought, salt, and freezing tolerance. Indeed, it is now apparent that there is extensive cross-talk between low-temperature and water-stress signalling pathways (see Box 9.3).

BOX 9.2 *COR and heat-shock regulons*

A number of the cold-induced *COR* genes have been characterized, and the sequences of their promoters compared. One of the features of several different *COR* genes is that their promoters share a common regulatory element termed the C-repeat (CRT) or low-temperature-response element (LTRE), which is five nucleotides long and has a consensus sequence of CCGAC. This element had already been linked to drought resistance and termed the dehydration-responsive element (DRE). The CRT/LTRE/DRE is bound by a CBF transcription factor, CBF1. The structure of CBF1 is shown in the figure, indicating the nuclear localization sequence, DNA-binding domain, and an acidic region that may be involved in interactions with other proteins.

(Continued overleaf)

BOX 9.2 *Continued*

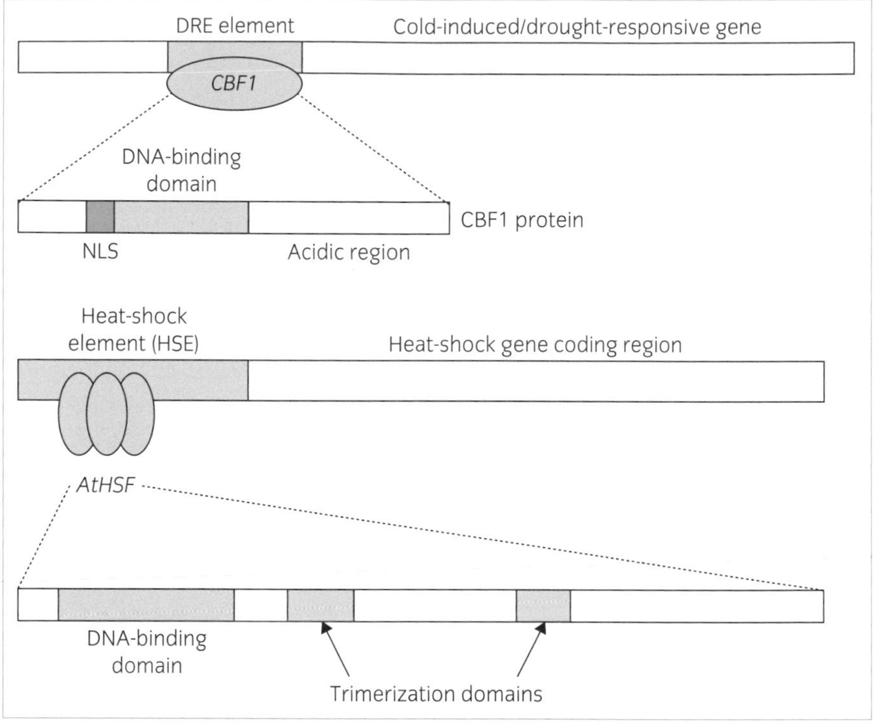

A number of cold-induced/drought-responsive genes contain the DRE sequence element in their promoters, which is bound by transcription factors of the CBF family, activating transcription. Overexpression of a single *CBF* gene therefore induces the expression of several cold-induced/drought-responsive genes. Heat-shock-induced genes also contain a heat-shock element (HSE) in their promoters. Heat-shock factor (HSF) proteins bind the HSE as a trimer. Overexpression of a HSF-encoding gene induces the expression of several heat-shock proteins. NLS, nuclear localization signal.

CBF1 expression is induced by cold acclimation, and leads to the expression of the *COR* genes. This group of genes, sharing a common regulatory mechanism, has been termed the COR regulon.

CBF1 is a member of a small gene family; *CBF2* and *CBF3* are also transcription factors, and expression of all three *CBF* genes is induced rapidly by low temperatures (but not by dehydration, abscisic acid (ABA), or high salt concentrations). In addition, *CBF3* overexpression results in several biochemical changes associated with cold acclimation, such as elevated levels of compatible osmolytes, proline, and soluble sugars.

Although low-temperature-induced gene expression, mediated by the CRT element, appears to be well conserved in plants, not all cold-induced genes (including the *CBF* genes) have the CCGAC element in their promoters. Other pathways of low-temperature-induced gene

BOX 9.2 *Continued*

expression, not mediated via CRT/CBF, appear to be present in plants and another sequence element, CCGAAA, has been identified as conferring low-temperature inducibility in some genes.

The heat-shock genes also comprise a regulon. They all contain the heat-shock element (HSE) consensus sequence in their promoter regions. The transcription factor HSF binds to the HSE as a trimer.

BOX 9.3 *Transcriptional regulation by low-temperature and dehydration signalling pathways*

The COR regulon described in Box 9.2 includes a number of proteins that are induced by low temperature and dehydration/osmotic stress. The reason for this can be seen in the figure, which shows a simplified signal transduction cascade regulating the promoter elements of an idealized stress-induced gene. What is clear from this diagram is that the CRT/DRE element is responsive to four signalling pathways: two low-temperature signalling pathways via ICE1 and ICE-like proteins, and two distinct dehydration signalling pathways: ABA and ABA-independent signal transduction. ABA-independent signalling involves the activation of the DREB2 transcription factor, while one of many effects of ABA signalling is the activation of CBF4. ABA also mediates the activity of several other groups of transcription factors, including the AREB/ABF basic zipper (bZIP) proteins that bind to ABA-response elements (ABREs) and MYC/MYB transcription factors that bind cooperatively to the MYCR/MYBR elements found in several ABA-responsive genes.

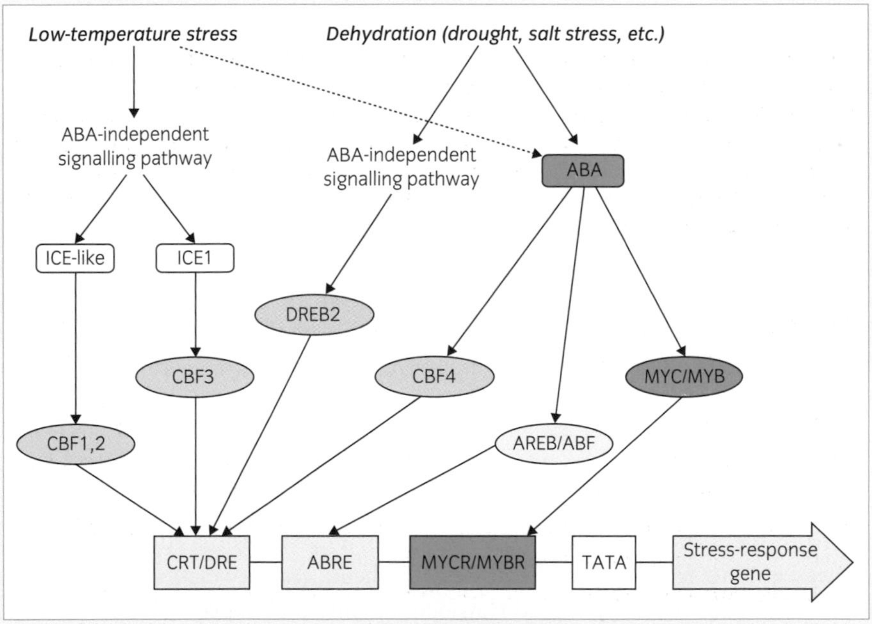

Signal transduction pathways involved in the regulation of stress-response gene expression.

Table 9.4 Transgenes used to manipulate heat tolerance

Gene	Protein	Transgenic plant
AtHSF1	Heat-shock transcription factor HSF1::GUS fusion	*Arabidopsis*
Hsp101	HSP100-class heat-shock protein	*Arabidopsis*
Hsp70	HSP70-class heat-shock protein	*Arabidopsis*
Hsp17.7	SmHSP small heat-shock protein family	Carrot
TLHS1	Class I smHSP	Tobacco

GUS, β-glucuronidase.

Tolerance to heat stress

For many years it has been known that heat stress applied to a wide range of organisms induces a specific set of heat-shock proteins (HSPs); they fall into five classes, four of which are highly conserved in prokaryotes and eukaryotes. These four are categorized according to size as the HSP100, HSP90, HSP70, and HSP60 classes whose members appear to function as molecular chaperones. Some of them are expressed constitutively and are involved in normal protein synthesis and folding. Those induced by heat appear to be involved in countering the effects of heat stress by protecting or refolding denatured proteins. Their expression is induced by heat treatment and, in some cases, can be correlated with the acquisition of thermotolerance. The fifth group of several classes of small HSPs are particularly abundant in plants, but their function is not yet clear.

● Activation of the heat shock regulon enhances heat tolerance.

In a way analogous to strategies for engineering cold/freezing tolerance, individual HSPs have been transformed into plants to enhance heat tolerance (Table 9.4). However, it is also known that the rapid heat-shock response is co-ordinated by a heat-shock factor (HSF). This protein is expressed constitutively, but in normal conditions exists as a monomer bound to one of the HSP70 proteins. Upon heat stress, the HSP70 dissociates and the HSF assembles into a trimer (Box 9.2), which binds to a heat-shock element (HSE) common to the promoters of HSP genes. The HSE is made up of 5-bp repeats in alternating orientations with the consensus sequence NGAAN; five to seven of these repeats occur in the promoter close to the TATA box.

When the *AtHSF1* gene was overexpressed in *Arabidopsis*, the transcription factor was not active, and there was no effect on thermotolerance. However, fusion of *AtHSF1* to the N or C terminus of the *gusA* reporter gene produced a fusion protein that was able to trimerize in the absence of heat. Transformation of this fusion protein into *Arabidopsis* produced transgenic plants that expressed HSPs constitutively and demonstrated enhanced thermotolerance without requiring prior heat treatment.

Secondary effects of abiotic stress: the production of ROS

Many, if not all, forms of environmental stress (abiotic and biotic) result in oxidative stress. This occurs most directly as a result of ozone pollution or ionizing radiation. However, oxidative stress is a secondary effect of many types of stress, from pathogen attack to water-deficit stress. The oxidative stress arises from the production of free radicals and the subsequent ensuing cascade of reactions. Much of the oxidative stress-induced damage to cells occurs as a result of the formation of ROS (see Box 9.4).

● ROS damage important macromolecules.

BOX 9.4 *ROS*

To understand the significance of oxidative stress and the production of ROS, it is necessary to review some of the basic chemistry of oxygen. Atmospheric oxygen in its ground state has two unpaired electrons. These two unpaired electrons have parallel spins and, in consequence, oxygen is usually non-reactive with most organic molecules, which have paired electrons with opposite spins. (Oxygen cannot react with a divalent reductant unless it has two unpaired electrons with a parallel spin opposite to the oxygen. This spin restriction means that the most common mechanisms of oxygen reduction in biochemical reactions are those involving transfer of only a single electron). This means that oxygen has to be activated before it can participate in reactions with organic molecules. The activation of oxygen may occur by two different mechanisms.

1. The first is the absorption of sufficient energy to reverse the spin on one of the unpaired electrons to form the singlet state, in which the two electrons have opposite spins (see Figure 9.4). Singlet oxygen can participate in reactions involving the simultaneous transfer of two electrons, and since paired electrons are common in organic molecules, singlet oxygen is much more reactive towards organic molecules than its triplet counterpart. In photosynthetic plants, singlet oxygen is often formed by the photosystems and plays a role in metabolic reactions such as the oxidation of xenobiotics and the polymerization of lignin. However, it may also be formed by the dysfunctioning of enzymes or electron transport systems, as a result of perturbations in metabolism caused by chemical or environmental stress.

2. The second mechanism is the stepwise reduction of oxygen to form ROS: superoxide ($\cdot O_2^-$), hydrogen peroxide (H_2O_2), and the hydroxyl radical ($\cdot OH$).

Superoxide is a powerful oxidant or a reductant. It can oxidize sulphur, ascorbic acid or NADPH, but it can also reduce cytochrome *c* and metal ions. The reaction leading to the formation of hydrogen peroxide and oxygen is catalysed by the enzyme superoxide dismutase (SOD), but it can also occur spontaneously (see Figure 9.4).

Hydrogen peroxide is a substrate in a number of oxidation reactions catalysed by peroxidases involving the synthesis of complex organic molecules. It readily permeates membranes and is therefore not compartmentalized in plant cells. The reactivity of hydrogen peroxide is the result of its reduction by metal ions to form the highly reactive hydroxyl radical.

The hydroxyl radical oxidizes organic molecules by two different reaction mechanisms: addition of the hydroxyl radical to form a hydroxylated compound, or the formation of an organic radical plus water. The organic radical can react with triplet oxygen to form a peroxyl radical, which can produce another organic radical. This can proceed to generate a chain reaction in which further organic radicals are produced.

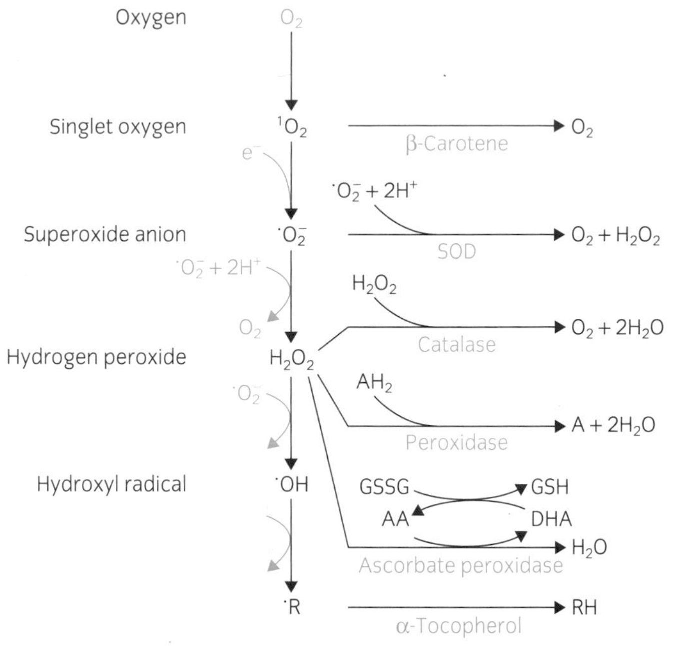

Figure 9.4 Formation of ROS and enzymes that remove ROS. The formation of ROS and the cascade of subsequent reactions that result are shown. The roles of the antioxidants and antioxidant enzymes are shown. AA, ascorbic acid; DHA, dehydroascorbate; GSH, glutathione; GSSG, oxidized glutathione; SOD, superoxide dismutase.

Figure 9.4 shows the cascade of successive reactions that form superoxide, hydrogen peroxide, and hydroxyl radical species. The ROS cause damage as a result of their reactions with cellular macromolecules. Damage to cellular membranes may result from a low concentration of hydroxyl radicals triggering a chain reaction of lipid peroxidation. Oxidative damage to proteins may involve specific amino acid modifications, fragmentation of the peptide chain, aggregation of crosslinked reaction products, altered electrical charge, and increased susceptibility to proteolysis. In DNA, both the sugar and bases are susceptible to oxidation by ROS, causing base degradation, single-strand breakage, and crosslinking to protein. These lesions in DNA can result in deletions, mutations, and other damaging genetic effects.

Antioxidants act as ROS scavengers.

Plants contain a number of enzymes that catalyse the cascade of ROS and convert them to less-reactive products (Figure 9.4). One key point to notice is the key role of certain enzymes in this cascade, notably superoxide dismutase (SOD), catalase, and peroxidases in neutralizing these reactive species. Also shown in the figure are a number of antioxidant compounds that react with the activated oxygen compounds to produce harmless, regenerable products. These include three different vitamin compounds: β-carotene (provitamin A), ascorbic acid (vitamin C), and α-tocopherol (vitamin E). Other important antioxidants include glutathione and zeaxanthin (see Box 9.5). Central to this antioxidant system is the ascorbate–glutathione cycle (Figure 9.4). The coupling of ascorbate and glutathione redox cycling has been most

BOX 9.5 *Antioxidants in plants*

L-Ascorbic acid

Ascorbate (vitamin C) has an important role in plant metabolism, growth, and development. Among other functions, it acts as a reductant for many free radicals to minimize the damage caused by oxidative stress. Ascorbate can directly scavenge ROS with and without enzyme catalysts, and can indirectly scavenge them by recycling α-tocopherol (see below) to the reduced form. Indeed, it reacts with ROS more readily than any other aqueous component. For example, ascorbate reacts with superoxide to form hydrogen peroxide and dehydroascorbate. It also reacts with hydrogen peroxide to form monodehydroascorbate and water, in a reaction catalysed by ascorbate peroxidase. It also regenerates membrane-bound antioxidants such as α-tocopherol that scavenge peroxyl radicals and singlet oxygen. The tocopheroxyl radical is converted to α-tocopherol via the oxidation of ascorbate to monodehydroascorbate.

Note that there are two different products of ascorbate oxidation—monodehydroascorbate and dehydroascorbate—which represent one- and two-electron transfers, respectively. Two molecules of monodehydroascorbate may spontaneously dismutate to form ascorbate and dehydroascorbate, or can be reduced to ascorbate by NAD(P)H monodehydroascorbate reductase. Dehydroascorbate is unstable at pH >6, forming tartrate and oxalate. To prevent this, dehydroascorbate is reduced rapidly to ascorbate by dehydroascorbate reductase using reducing equivalents from glutathione (see below).

Glutathione

Glutathione (GSH) is a tripeptide (Glu–Cys–Gly) that was met in Chapter 5 with reference to herbicide detoxification. The antioxidant function of glutathione involves the thiol group of the cysteine residue, which forms a thiyl radical on oxidation, and then reacts with a second oxidized GSH to form a disulphide bond (GSSG). The redox potential of GSH enables it to reduce dehydroascorbate to ascorbate or to reduce the disulphide bonds of proteins.

GSH functions directly as a free-radical scavenger by reacting chemically with singlet oxygen, superoxide, and hydroxyl radicals. It also stabilizes membranes by removing acyl peroxides formed by lipid peroxidation reactions. It functions to regenerate reduced ascorbic acid from dehydroascorbate.

The reduction of GSSG to GSH is catalysed by the enzyme glutathione reductase, which exists in plants in multiple forms associated with different subcellular compartments. Glutathione peroxidase provides an alternative means of detoxifying activated oxygen, by using GSH to reduce hydrogen peroxide, producing GSSG.

α-Tocopherol

α-Tocopherol (vitamin E; see Chapter 10) is a membrane-bound scavenger of oxygen free radicals, lipid peroxyl radicals, and singlet oxygen. It also serves as a membrane-stabilizing agent. The reaction with peroxyl radicals formed in the lipid bilayer appears to be particularly important. The tocopheroxyl radical produced is stable and therefore does not propagate further free radicals. α-Tocopherol can be regenerated by ascorbate, as described above.

(Continued overleaf)

BOX 9.5 *Continued*

Carotenoids

Carotenoids are C_{40} isoprenoids and tetraterpenes, and include the carotenes and xanthophylls (see Chapter 10). They are located in the plastids of both photosynthetic and non-photosynthetic plant tissues. In chloroplasts, they protect the photosystems by detoxifying ROS and triplet chlorophyll produced as a result of excitation of the photosynthetic complexes by light. Thus, they can perform a similar role to α-tocopherol in scavenging the products of lipid peroxidation. They can react with singlet oxygen and dissipate the energy as heat. They also dissipate excess excitation energy through the xanthophyll cycle, and they can quench triplet or excited chlorophyll molecules to prevent the formation of singlet oxygen.

extensively characterized in the chloroplast. Chloroplasts produce superoxide and hydrogen peroxide in the light, particularly from photosystem I. The superoxide generated is itself converted into hydrogen peroxide by either spontaneous dismutation or by the activity of SOD. The hydrogen peroxide produced is scavenged by ascorbate and the enzyme ascorbate peroxidase. The monodehydroascorbate produced from this reaction (see Box 9.5) may be regenerated in two ways. One is by reduction of NAD(P)H catalysed by monodehydroascorbate reductase. The other occurs via the dismutation of two monodehydroascorbate molecules to form ascorbate and dehydroascorbate, followed by the reduction of the dehydroascorbate by glutathione, catalysed by dehydroascorbate reductase.

● There are two broad strategies for engineering tolerance to oxidative stress.

From the outline of plant antioxidants and oxygen-scavenging activities given above, two general strategies for engineering tolerance to oxidative stresses are apparent: either to increase the level of enzymes that remove ROS, or to increase the level of antioxidant compounds that react with ROS.

Strategy 1: Expression of enzymes involved in scavenging ROS

SODs catalyse the reaction shown in Figure 9.4 in which the superoxide radical combines with two hydrogen ions to form oxygen and hydrogen peroxide. The activity of this enzyme therefore determines the concentration of the two substrates (superoxide and hydrogen peroxide) that react to form the hydroxyl radical. There is a broad family of SOD enzymes, requiring different metal ions for activity, and predominating in different cellular compartments. Thus, Mn, Fe, and Cu/Zn forms are found, specific to different organelles (Box 9.6).

Transgenic tobacco plants containing each class of enzyme have been produced and tested for their ability to withstand oxidative stress, as shown in Table 9.5. In each case, there has been some measurable increase in the ability to withstand oxidative stresses such as exposure to ozone. The Mn-SOD has also been transformed into alfalfa and targeted to chloroplasts or mitochondria. In this case, the results of a 3-year field trial indicated significant improvements in the yield and survival of the

BOX 9.6 *The SOD enzyme family*

SOD catalyses the dismutation of superoxide to hydrogen peroxide and oxygen. SOD is present in most (if not all) subcellular compartments of the plant cell that generate activated oxygen, and is assumed to play a central role in the defence against oxidative stress. There are three distinct types of SOD classified on the basis of the metal cofactor: the copper/zinc (Cu/Zn-SOD), the manganese (Mn-SOD), and the iron (Fe-SOD) isozymes. The Mn-SOD is found in the mitochondria, whereas Cu/Zn-SOD isozymes are found in the cytosol, or in the chloroplasts. The Fe-SOD isozymes, though not always detected in plants, are usually found in the chloroplast. All of the plant *SOD* genes are in the nuclear genome and are targeted to their respective subcellular compartments by an N-terminal targeting sequence. However, they are not regulated co-ordinately, but independently according to the degree of oxidative stress experienced in the respective subcellular compartments.

Table 9.5 Transgenes used to engineer tolerance to oxidative stress

Gene	Host	Stress tolerance
Mitochondrial Mn-SOD, tobacco	Alfalfa chloroplast	2× increase in SOD, increased field drought tolerance, increased freezing tolerance
Mitochondrial Mn-SOD, tobacco	Tobacco chloroplast	2–4× increase in SOD, increased ozone tolerance
Mitochondrial Mn-SOD, tobacco	Tobacco mitochondria	8× increase in SOD, no effect on ozone tolerance
Mn-SOD	Canola	Increased aluminium tolerance
Chloroplast Cu/Zn-SOD, pea	Tobacco chloroplast	3–15× increase in SOD, increased tolerance to high light and chilling
Cytosolic Cu/Zn-SOD	Tobacco cytosol	1.5–6× increase in SOD, reduced damage from acute ozone exposure
Arabidopsis Fe-SOD	Tobacco	Protected plants from ozone damage
Apx3	Tobacco	Increased protection against oxidative stress
Apx1	*Arabidopsis*	Heat tolerance
GST/GPX	Tobacco	Increased stress tolerance
Nt107 (glutathione *S*-transferase)	Tobacco	Sustained growth under cold and salinity stress
ParB (glutathione *S*-transferase)	*Arabidopsis*	Protects against aluminium toxicity and oxidative stress
NtPox (glutathione peroxidase)	*Arabidopsis*	Protects against aluminium toxicity and oxidative stress
Glutathione reductase, *E. coli*	Tobacco chloroplast	3–6× increase in foliar GR, increased tolerance to SO_2 and paraquat
Glutathione reductase, *E. coli*	Tobacco cytosol	1–35× increase in GR, increased tolerance to paraquat
Glutathione synthetase, *E. coli*	Poplar cytosol	100× increase in GS, GSH not increased, no effect on paraquat tolerance
MsFer, alfalfa ferritin	Tobacco	Increased tolerance to oxidative damage caused by excess iron

Apx, ascorbate peroxidase; GPX, glutathione peroxidase; GR, glutathione reductase; GS, glutathione synthetase; GSH, glutathione; GST, glutathione *S*-transferase; SOD, superoxide dismutase.

transgenic plants. However, it is not clear whether the protective effect is due to the removal of superoxide by the enzyme, or whether the hydrogen peroxide product enhances stress tolerance by inducing other stress-related genes. More recently, the introduction of the Cu/Zn-SOD, Mn-SOD, and Fe-SOD genes together into alfalfa has been shown to increase the winter hardiness of the crop.

Strategy 2: Production of antioxidants

Table 9.5 also shows the alternative strategy for oxidative stress tolerance in action. Thus, three of the enzymes shown in Figure 9.4—ascorbate peroxidase, glutathione peroxidase, and glutathione reductase—have been transformed into *Arabidopsis* and tobacco plants and shown to have some effect on tolerance to various abiotic stresses such as heat, cold, and salinity. Glutathione reductase also provided resistance to the oxidative stress resulting from paraquat treatment. On the other hand, simply increasing the cellular concentration of glutathione by expressing glutathione synthase had no effect on stress tolerance.

SUMMARY

This chapter has highlighted the importance of abiotic stress in determining the large annual fluctuations in crop yield. Although crops experience a number of different abiotic stresses, several of these cause the same two major problems: water-deficit and oxidative stress. Two general strategies for engineering tolerance to abiotic stresses in plants are therefore possible. Some measure of tolerance to water-deficit stress can be provided by the synthesis of compatible solutes. On the other hand, expressing enzymes involved in protection against ROS can combat oxidative stress.

In some cases, single-gene mechanisms for tolerating specific stresses can be deployed (e.g. salt stress and cold stress). However, the overriding theme from this chapter is that abiotic stresses induce complex reactions from plants, and that optimal protection may well involve several genes.

FURTHER READING

Visit www.oxfordtextbooks.co.uk/orc/slaterplants2e/ for hyperlinked references that take you directly to online abstracts for the journal articles listed below:

Plant stress—general

Bray, E. A., Bailey-Serres, J., and Weretilnyk, E. (2000) Responses to abiotic stresses. In *Biochemistry and Molecular Biology of Plants* (B. B. Buchanan, W. Gruissem, and R. L. Jones, eds), pp. 1158–1203. American Society of Plant Physiologists, Rockville, MD.

Fujita, M., Fujita, Y., Noutoshi, Y., Takahashi, F., Narusaka, Y., Yamaguchi-Shinozaki, K., and Shinozaki, K. (2006) Crosstalk between abiotic and biotic stress responses: a current view from the points of convergence in the stress signaling networks. *Current Opinion in Plant Biology* **9**, 436–442.

Pastori, G. M. and Foyer, C. H. (2002) Common components, networks, and pathways of cross-tolerance to stress. The central role of "redox" and abscisic acid-mediated controls. *Plant Physiology* **129**, 460–468.

Stress tolerance—general

Holmberg, N. and Bulow, L. (1998) Improving stress tolerance in plants by gene transfer. *Trends in Plant Science* **3**, 61–66.

Wang, W., Vinocour, B., and Altman, A. (2003) Plant responses to drought, salinity and extreme temperatures: towards genetic engineering for stress tolerance. *Planta* **218**, 1–14.

Zhang, J., Kueva N. Y., Wang, Z., Wu, R., Ho, T.-H., and Nguyen, H. T. (2000) Genetic engineering for abiotic stress resistance in crop plants. *In vitro Cellular and Developmental Biology—Plant* **36**, 108–114.

Zhang, J., Creelman, R. A., and Zhu, J.-K. (2004) From laboratory to field. Using information from *Arabidopsis* to engineer salt, cold, and drought tolerance in crops. *Plant Physiology* **135**, 615–621.

Osmoprotectants and stress tolerance

Chen, T. H. H. and Murata, N. (2002) Enhancement of tolerance of abiotic stress by metabolic engineering of betaines and other compatible solutes. *Current Opinion in Plant Biology* **5**, 250–257.

Nuccio, M. L., Rhodes, D., McNeil, S. D., and Hanson, A. D. (1999) Metabolic engineering of plants for osmotic stress resistance. *Current Opinion in Plant Biology* **2**, 128–134.

Sakamoto, A. and Murata, N. (2000) Genetic engineering of glycine betaine synthesis in plants: current status and implications for enhancement of stress tolerance. *Journal of Experimental Botany* **51**, 81–88.

Sakamoto, A. and Murata, N. (2002) The role of glycine betaine in the protection of plants from stress: clues from transgenic plants. *Plant, Cell and Environment* **25**, 163–171.

Serraj, R. and Sinclair, T. R. (2002) Osmolyte accumulation: can it really help increase crop yield under drought conditions? *Plant, Cell and Environment* **25**, 333–341.

Salt tolerance

Apse, M. P., Aharon, G. S., Snedden, W. A., and Blumwald, E. (1999) Salt tolerance conferred by overexpression of a vacuolar Na^+/H^+ antiport in *Arabidopsis*. *Science* **285**, 1256–1258.

Chinnusamy, V., Jagendorf, A., and Zhu, J.-K. (2005) Understanding and improving salt tolerance in plants. *Crop Science* **45**, 437–448.

Flowers, T. J. (2004) Improving crop salt tolerance. *Journal of Experimental Botany* **55**, 307–319.

Frommer, W. B., Ludewig, U., and Rentsch, D. (1999) Taking transgenic plants with a pinch of salt. *Science* **285**, 1222–1223.

Yamaguchi, T. and Blumwald, E. (2005) Developing salt-tolerant crop plants: challenges and opportunities. *Trends in Plant Science* **10**, 615–620.

Cold tolerance

Jaglo-Ottosen, K. R., Gilmour, S. J., Zarka, D. G., Schabenberger, O., and Thomashow, M. F. (1998) *Arabidopsis* CBF1 overexpression induces COR genes and enhances freezing tolerance. *Science* **280**, 104–106.

Kasuga, M., Liu, Q., Miura, S., Yamaguchi-Shinozaki, K., and Shinozaki, K. (1999) Improving plant drought, salt and freezing tolerance by gene transfer of a single stress-inducible transcription factor. *Nature Biotechnology* **17**, 287–291.

Thomashow, M. F. (1998) Role of cold-responsive genes in plant freezing tolerance. *Plant Physiology* **118**, 1–7.

Heat stress

Lee, J. H., Hubel, A., and Schoffl, F. (1995) Derepression of the activity of genetically engineered heat shock factor causes constitutive synthesis of heat-shock protein and increased thermotolerance in transgenic *Arabidopsis*. *Plant Journal* **8**, 603–612.

Oxidative stress

Mittler, R. (2002) Oxidative stress, antioxidants and stress tolerance. *Trends in Plant Science* **7**, 405–410.

Suzuki, N. and Mittler, R. (2006) Reactive oxygen species and temperature stresses: a delicate balance between signaling and destruction. *Physiologia Plantarum* **126**, 45–51.

Web-links

McKersie, B. Tutorial on oxidative stress, web-link 9.1: www.agronomy.psu.edu/Courses/AGRO518/oxygen.htm

Plant stress, web-link 9.2: www.plantstress.com

10

The improvement of crop yield and quality

INTRODUCTION

The previous five chapters have considered a variety of ways in which the productivity of crop plants can be improved by enhancing their ability to resist or tolerate biotic and abiotic stresses. These strategies can all contribute to an improvement in crop yield by allowing the plants to better withstand external factors that reduce the amount and quality of harvestable plant material. However, the performance of crop plants is also determined by endogenous factors that affect the yield and quality of the harvestable material produced. This chapter will look at a number of examples of how crop productivity can be genetically enhanced by increasing the amount, or improving the quality, of material produced by the crop.

The yield of a crop is ultimately determined by the amount of solar radiation intercepted by the crop canopy, the photosynthetic efficiency (i.e. the efficiency of conversion of radiant to chemical potential energy), and the harvest index (the fraction of dry matter allocated to the harvested part of the crop). Thus, manipulation of crop yield requires some understanding of the fundamental processes that determine photosynthetic efficiency and dry-matter partitioning. On the other hand, the quality of a crop is determined by a wide range of desirable characteristics such as nutritional quality, flavour, processing quality, and shelf-life. The nature of yield and quality are therefore much more varied and wide-ranging than the other traits described to this point, and the structure of this chapter is therefore somewhat different to reflect this. Rather than try to cover the entire range of targets for different crops, a small number of yield and quality traits have been selected to exemplify the types of approach that have been adopted in this area. The chapter therefore starts with an extended discussion of one system—tomato ripening—because it is possible to draw several general lessons from the techniques and concepts that were deployed in this project.

- Crop yield is determined by intercepted solar radiation, photosynthetic efficiency, and harvest index.

The genetic manipulation of fruit ripening

The predominance of herbicide and pest resistance amongst the commercially developed genetically modified (GM) traits has been discussed previously (Chapters 5 and 6). However, among the first GM products to reach the market, around 1994, were Calgene's FlavrSavr fresh tomatoes, and processed tomato products containing delayed-softening tomato fruit developed in the UK by Zeneca and the University of Nottingham. This may seem rather puzzling, because the tomato crop is relatively small compared with the major GM crops of North America (maize, soybean, oilseed rape, and cotton). Furthermore, it is not immediately clear why a large biotechnology company like Zeneca would be involved, when they had no direct interest in tomato seeds or processed products when the project was started (when the company was still ICI). Part of the answer lies in the early recognition by ICI and Calgene of the value of

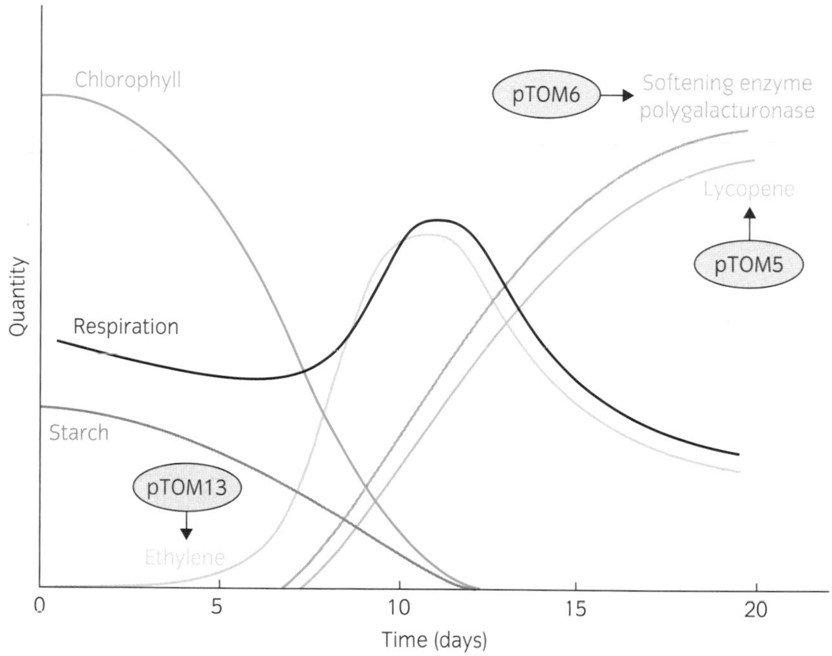

Figure 10.1 Biochemical changes during tomato ripening. The graph shows the changes that occur as tomato fruits progress from the mature green stage to fully ripe fruit. The onset of ripening is characterized by a burst of respiration, the respiratory climacteric, which is accompanied by a rapid increase in ethylene synthesis. The green coloration of the fruit is lost as chlorophyll is degraded, and the sweetness of the fruit is enhanced as starch is converted to sugars. The red pigment, lycopene, is synthesized, and the fruit softens as a result of the appearance of cell-wall-degrading enzymes such as polygalacturonase (PG). The graph also indicates the role of three ripening-related clones in the ripening process. The clone pTOM6 encodes PG, pTOM5 encodes phytoene synthase, and pTOM13 encodes the ethylene-forming enzyme, 1-amino-cyclopropane-1-carboxylic acid (ACC) oxidase. (Redrawn with permission from Grierson and Covey (1988).)

using tomato ripening as a model system on which to develop expertise and test the potential of plant biotechnology to modify crop quality. In order to understand why tomato ripening was seen as such a good model system, it is necessary to look at some of the basic biology of fruit ripening.

Fruit ripening is an active process triggered by ethylene.

Fruit ripening is an active process that, in climacteric fruit such as tomatoes, is characterized by a burst of respiration (the respiratory climacteric), ethylene production, softening, and changes to colour and flavour. Figure 10.1 shows the transient peaks in respiration and ethylene production at the start of tomato ripening, which are accompanied by softening, a change in colour from green to red, and enhanced flavour. The peak of ethylene production is significant, because ethylene is known to be the phytohormone that triggers ripening in climacteric fruit. The colour change results from the degradation of chlorophyll and the production of the red pigment, lycopene. Flavour changes occur as starch is broken down and sugars accumulate. A large number of secondary products that improve the smell and taste of the fruit are also produced. The softening of the fruit is largely the result of the cell-wall-degrading activity of the enzymes polygalacturonase (PG) and pectin methylesterase (PME). The PG enzyme is synthesized *de novo* during ripening and acts to break down the polygalacturonic acid chains that form the pectin glue of the middle lamella, which sticks neighbouring cells together. The realization that the ripening process involved the activation of specific ripening-related genes, such as the one encoding PG, enabled these genes to be cloned, as shown in Box 10.1.

BOX 10.1 *cDNA cloning of ripening-related genes*

The three ripening genes that will be discussed in this chapter were isolated from the same cDNA library made from ripe tomato fruit. Ripening-related clones were identified by differential screening of the cDNA library with cDNA sequences prepared from mRNA populations from mature green fruit and ripe fruit. Clones that hybridized to the ripe cDNA but not to the mature green cDNA were classed as ripening-related. Over 150 clones were isolated, grouped, and characterized. Three groups of clones, represented by pTOM5, pTOM6, and pTOM13, were confirmed as ripening-related and were also shown to be induced by ethylene. Their identities and roles in the ripening process were subsequently determined, as shown in the table.

Clone	Gene product	Function	Role in ripening
pTOM5	Phytoene synthase	Lycopene synthesis	Red coloration
pTOM6	Polygalacturonase	Cell wall degradation	Fruit softening
pTOM13	ACC oxidase	Ethylene formation	Ripening trigger

ACC, 1-amino-cyclopropane-1-carboxylic acid.

CASE STUDY 10.1 **The genetic manipulation of fruit softening**

Sequencing of the pTOM6 clone showed that the sequence matched the N-terminal sequence of the PG protein. In other words, one of the genes implicated in a particular part of the ripening process—softening—had been cloned. Northern blot analysis confirmed that expression of the *PG* gene was ripening-specific, and was induced by ethylene. The availability of the *PG* gene led to two types of question: (1) was it possible to use the cloned gene to investigate the precise role of PG in ripening and (2) was it possible to modify the expression of endogenous PG to manipulate the ripening process? This led to one of the first examples of using antisense techniques to alter plant gene expression (see Chapter 4).

The initial antisense PG construct comprised a partial cDNA fragment of pTOM6 inserted in reverse orientation downstream of the 35S promoter (Figure 10.2). Note that expression of the antisense gene was therefore constitutive, whereas expression of the endogenous gene is, of course, ripening-specific.

Antisense PG expression reduces both sense and antisense mRNA levels.

The effect of transforming this construct into tomato plants was quite instructive. Table 10.1 summarizes the results of Northern blots of RNA from mature green and ripe fruit from both transgenic and non-transgenic tomatoes, hybridized with probes to sense and antisense mRNA. There are a number of things to note from this blot. PG mRNA is not present in mature

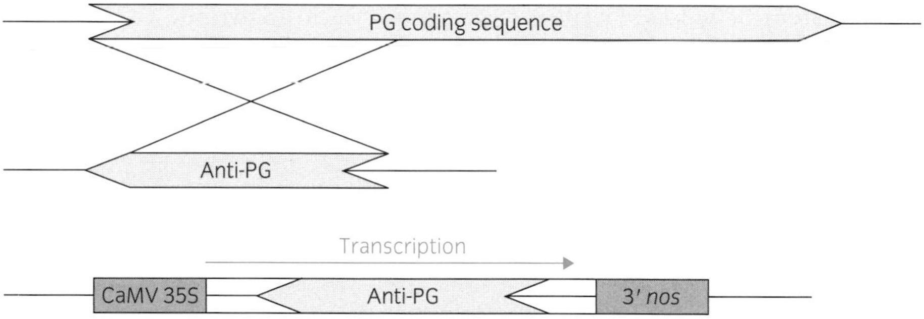

Figure 10.2 Construction of an antisense PG vector. The diagram shows the antisense concept in operation. Basically, all or part of a gene sequence is inserted into a vector in the wrong orientation, so that the opposite strand from normal is transcribed to produce an antisense transcript. This will be complementary to the endogenous mRNA, allowing the formation of double-stranded RNA.

Table 10.1 The expression of pTOM6 in transgenic tomato plants

	Non-transgenic fruit				Transgenic fruit			
	Mature green		Ripe		Mature green		Ripe	
Probe	Sense	AS	Sense	AS	Sense	AS	Sense	AS
Signal	ND	ND	Intense band	ND	ND	Intense bands	Faint band	Faint bands
Size of RNA			PG mRNA			<PG mRNA	PG mRNA	<PG mRNA

The results of Northern blot hybridization of samples from transgenic tomato fruit with sense and antisense pTOM6 probes are shown. AS, antisense; ND, not detected.

green wild-type fruit, but is induced during ripening. In ripe transgenic fruit, the level of PG mRNA is much reduced. The antisense RNA probe reveals that the antisense RNA accumulates to high levels in mature green fruit, but its level is also reduced in ripe fruit. The conclusion is that the interaction of sense and antisense mRNA leads to a reduction in both transcripts.

In these transgenic fruit, it could be shown that PG activity was much reduced during ripening. Furthermore, this caused a specific delay in softening, but not in other ripening events. For example, the red lycopene pigment still accumulated during the ripening of these transgenic fruit, while production of the softening enzyme was inhibited. It was possible to modulate the antisense effect by analysing heterozygote (one antisense gene) and homozygote (two antisense genes) plants. Two copies of the antisense gene were found to reduce the levels of PG to virtually undetectable levels during ripening. Thus, tomato fruit in which the activity of one specific enzyme had been effectively knocked out had been generated. It subsequently proved possible to reduce PG activity by co-suppression (see Chapter 4) using sense PG constructs.

- Antisense PG inhibits fruit softening.

The effect of reducing PG activity in low-PG transgenic fruit was to inhibit pectin degradation, as indicated by the large size of polyuronides remaining in the cell wall. The resulting changes to food processing properties were determined by Zeneca and are shown in Figure 10.3. The

- Low-PG and low-PME fruit have improved processing properties.

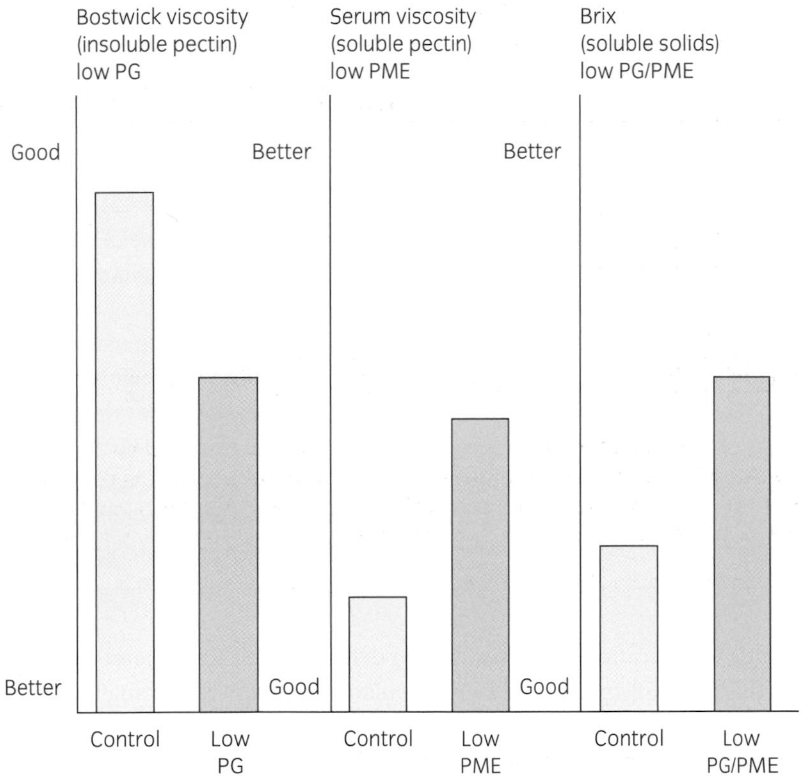

Figure 10.3 Relative processing performance of low-PG and low-PME fruit. Three different assays of processing quality were used to assess the effects of reducing two cell-wall-degrading enzymes: PG and PME. The low-PG fruit show an improved Bostwick viscosity (determined by insoluble pectin), whilst low-PME fruit have a better serum viscosity (related to soluble pectin). The total soluble solids (Brix) is improved in fruit with combined low-PG and low-PME traits. (Redrawn with permission from Grierson and Schuch (1993).)

Bostwick viscosity is influenced largely by the insoluble cell-wall polymers and is a measure of the potential paste yield of tomato fruit. The Bostwick viscosity in low-PG fruit (PG activity reduced by 99%) was enhanced by more than 80% as a result of the increase in size of the cell-wall pectins, indicating a significant increase in the yield of paste from these tomatoes. This characteristic enabled Zeneca to exploit these transgenic tomatoes for the production of paste for processed foods such as sauces and pizzas.

● The low-PG FlavrSavr fruit delayed softening on the vine.

Calgene in the USA exploited the delayed softening characteristics of low-PG fruit in quite a different way. Their low-PG varieties were found to remain firmer on the vine and during storage than unmodified fruit. Fresh tomatoes are generally harvested before they are fully ripe, to ensure they are firm enough to withstand handling during harvest, processing, and storage. Ripening may then be induced by ethylene prior to sale. Although this ensures the fruit are sold intact, at a uniform size, colour, and stage of ripeness, it is a common perception of consumers that tomatoes processed in this way have lost much of the flavour and aroma of those eaten straight from the vine. In contrast, the Calgene FlavrSavr tomatoes were allowed to ripen and develop their full flavour on the vine, but the ripe fruit were still firm enough to withstand damage from handling and also from post-harvest fungal infections.

Figure 10.3 also shows the effect of reducing PME activity in tomato fruit. This enzyme is present in unripe as well as ripe fruit, and removes methyl groups from cell-wall pectin. Low-PME fruit were found to have an enhanced serum viscosity, reflecting an increase in the proportion of soluble pectin. Serum viscosity contributes to the glossy appearance of tomato paste. When the low-PG and low-PME traits were combined (either by crossing parent tomatoes with a single trait, or by re-transformation with both genes in tandem), the fruit had the quality characteristics of both single gene lines, plus a significant increase in soluble solids as indicated by a high Brix value. The soluble solids are measured on the Brix scale, a value obtained using refractometry. The soluble solids are an important determinant of tomato-processing characteristics, with a high Brix value indicating good processing quality.

Despite their initial success, neither the Zeneca nor the Calgene low-PG tomato products can now be found, though for quite different reasons. The Zeneca tomato pastes were clearly labelled as GM products yet sold well when they were first introduced. However, they were cleared from the shelves of UK supermarkets following the anti-GM backlash catalysed by the Pusztai affair (see Chapter 6), despite the fact there has been no evidence of any risk to human health from this product. The FlavrSavr tomato met a different fate, in that the trait was not originally introduced into a suitable background and the tomato did not perform well in commercial production.

The genetic modification of fruit softening demonstrated the general validity of the concept that a specific trait could be manipulated by down-regulating the expression of an endogenous gene using antisense and co-suppression techniques. The specific example of reducing cell-wall degradation by targeting endogenous PG and PME has been applied to a number of other fruit, such as mango, peach, and pear. Furthermore, it is now apparent that pectin depolymerization is not the only determinant of texture, and the roles of other enzymes such as pectate lyases, cellulases, and xyloglucan hydrolases are being investigated. Plant cell-wall structure and metabolism are still poorly understood, but research in these areas holds out the prospect of being able to manipulate many aspects of plant growth and development.

CASE STUDY 10.2 The genetic modification of ethylene biosynthesis

The use of pTOM6 indicated that, although the ripening process is co-ordinated, specific elements of the process could be manipulated without affecting others. Thus, antisense PG fruit still change colour and accumulate sugars and flavour compounds. However, the cloning of several ripening-related sequences led to the proposal that some of these might be involved in the transient burst of ethylene production, and that the study of these clones could provide some insight into the regulation of the entire ripening process. Investigation of the expression of ripening-related clones in wounded leaves (which also produce ethylene) showed that the expression of pTOM13 was related to ethylene production rather than to ripening per se. Sequence analysis led to the preliminary identification of pTOM13 as the second ethylene-forming enzyme in the two-step pathway from *S*-adenosylmethionine to ethylene (Figure 10.4). The protein

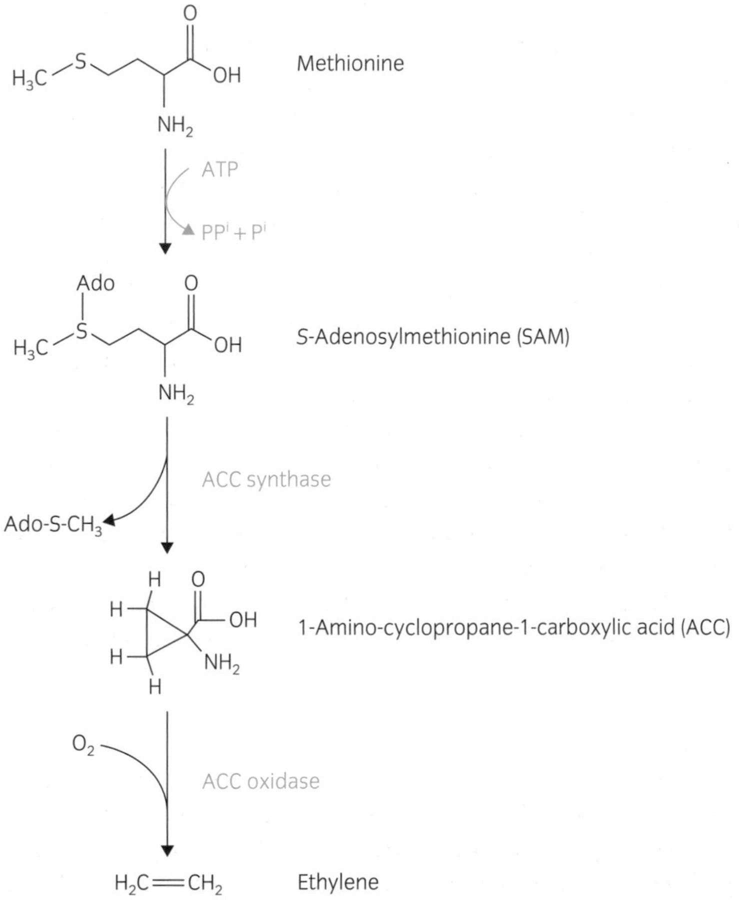

Figure 10.4 Ethylene biosynthesis and its regulation. The pathway to ethylene from methionine is shown. Only two committed steps from S-adenosylmethionine are involved. The first is catalysed by ACC synthase, generating the unusual three-sided ring compound, 1-amino-cyclopropane-1-carboxylic acid (ACC). This is converted to ethylene by ACC oxidase, producing CO_2, HCN, and H_2O in the process. Ascorbate is a co-substrate and is oxidized to dehydroascorbate (see Chapter 9).

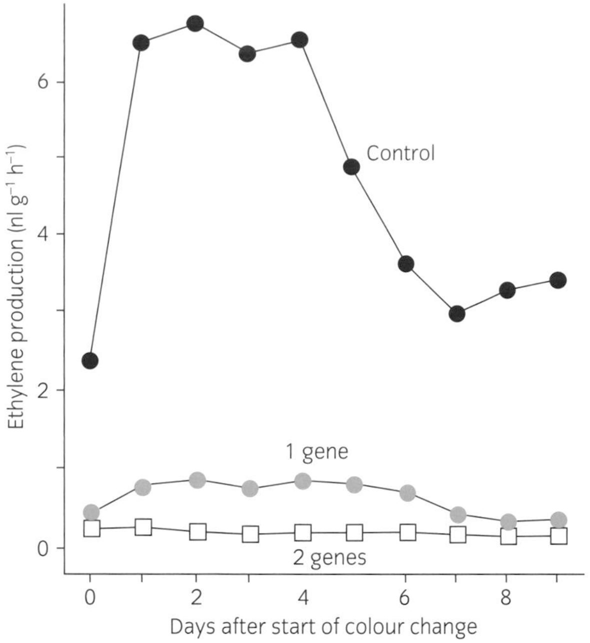

Figure 10.5 Ethylene production in tomato fruit transformed with an antisense ACC oxidase construct. The climacteric burst of ethylene production seen in control fruit is greatly decreased in fruit carrying one antisense gene, and effectively abolished in homozygotes carrying two antisense genes. (Adapted with permission from Hamilton *et al.* (1990).)

sequence, plus the availability of the clone for protein expression studies, also allowed the classification of the previously uncharacterized ethylene-forming enzyme as 1-amino-cyclopropane-1-carboxylic acid (ACC) oxidase.

The role of pTOM13 in ethylene biosynthesis was identified by antisense experiments.

The antisense strategy was initially used to confirm the role of pTOM13 in ethylene synthesis. Transgenic tomato plants carrying the antisense pTOM13 construct produced much less ethylene, either in wounded leaves or during ripening (Figure 10.5), than did control plants.

The effect of the much-reduced levels of ethylene produced during ripening was to delay the entire ripening process. Thus, lycopene production and softening (Figure 10.6) occur much more slowly than in normal plants. These ripening phenomena can be induced in the antisense plants by applying exogenous ethylene. The increased storage capacity of these fruit has tremendous potential for reducing spoilage of the fruit during storage and transport.

Low-ethylene technology has wide applications.

Antisense ethylene technology is applicable to all climacteric fruits, and to other systems triggered by ethylene, such as the spoilage of vegetables and the senescence of picked flowers. For example, it has been known for some time that the vase-life of cut flowers can be extended by adding silver salts to the water, which blocks the

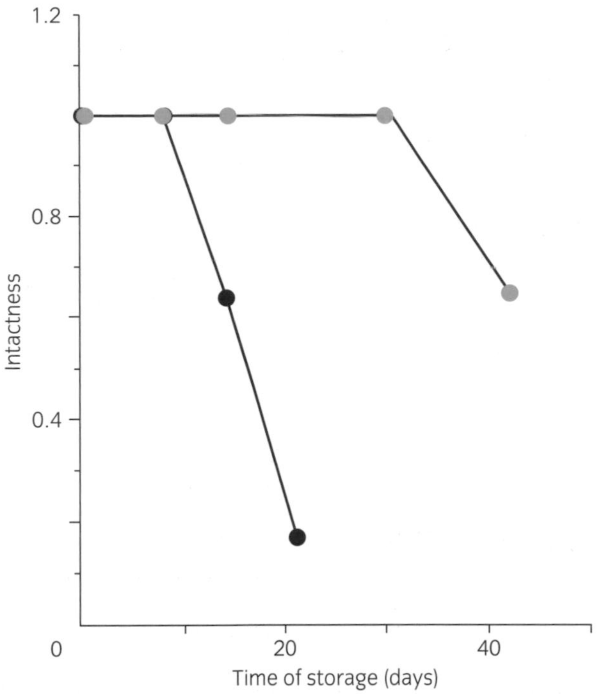

Figure 10.6 Increased storage of low ethylene fruit. The reduction of ethylene formation in antisense ACC oxidase fruit slows down the ripening process. One qualitative measure of this effect is on the intactness of the fruit: in other words, it takes a lot longer for the antisense fruit to become a soft pulp. Green circles, ACC oxidase fruit; black circles, control fruit. (Redrawn with permission from Grierson and Schuch (1993).)

response to ethylene. The introduction of antisense ACC synthase or ACC oxidase would have a similar effect of extending vase-life by suppressing ethylene synthesis. More importantly, the post-harvest spoilage of fruit and vegetables is often very rapid in tropical countries, and this can be a major barrier to the efficient distribution of food, particularly where the transport and food storage infrastructure is not well developed. The ability to delay ripening, senescence, and spoilage could make a significant contribution to the problems of food distribution in developing countries.

In addition, the successful manipulation of one plant growth regulator indicates the potential for engineering plant development by manipulating other plant growth regulators. The early availability of auxin and cytokinin biosynthetic genes from the *Agrobacterium tumefaciens* Ti plasmid (see Chapter 3) has meant that both phytohormones have been synthesized in a range of plants under the control of various promoters. The more recent cloning of genes involved in gibberellin and abscisic acid (ABA) synthesis/response has also permitted the genetic manipulation of important developmental processes (see below).

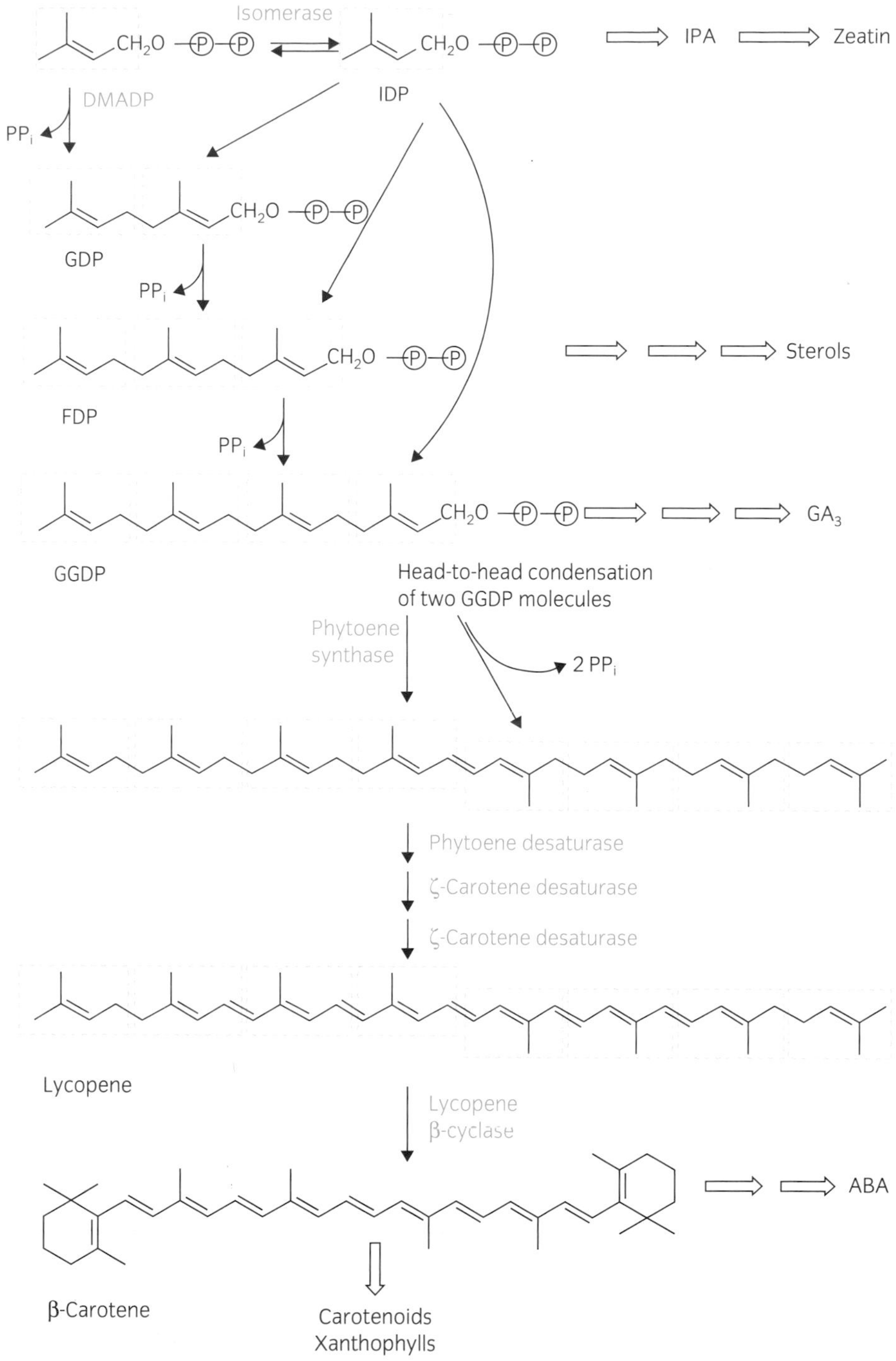

Figure 10.7 The carotenoid biosynthesis pathway. As shown, the synthesis of lycopene and β-carotene originates from the isoprenoid pathway. The sequential addition of five-carbon isoprenoid units (isopentenyl diphosphate, IDP; dimethylallyl diphosphate, DMADP) to form 10-carbon (geranyl diphosphate; GDP), 15-carbon (farnesyl diphosphate; FDP) and 20-carbon (geranyl geranyl diphosphate; GGDP) units is an important starting point for the synthesis of a range of compounds that include steroids, carotenoids, xanthophylls, and anthocyanins. Note that gibberellins and ABA are synthesized from isoprenoid precursors, whereas cytokinins are formed via the addition of an isopentenyl unit to the N6 position of purines.

CASE STUDY 10.3 **Modification of colour**

The third ripening-related clone, pTOM5, highlighted at the beginning of this chapter (Box 10.1) encodes phytoene synthase, a key enzyme in the biosynthesis of the red pigment, lycopene (Figure 10.7). The position of this step in the isoprenoid pathways of plants is shown. Many classes of compound are synthesized by building up molecules from the five-carbon isoprenoid unit. In particular, this route forms the basis of the biosynthetic pathways of three classical plant hormones (Chapter 2). Cytokinins such as zeatin are produced by the addition of a single isoprenoid unit to adenosine. The biosynthesis of gibberellins starts from the 20-carbon unit geranyl geranyl diphosphate (GGDP). ABA is formed from β-carotene, as are a wide range of carotenoids and xanthophylls.

As might be expected, antisense pTOM5 tomato plants produce yellow fruit during ripening. On the other hand, plants overexpressing pTOM5 under the control of the 35S promoter were found not only to produce lycopene ectopically, but also to have a shorter height. This latter finding was unexpected, but can be explained with reference to Figure 10.7. Analysis of gibberellin synthesis in these plants showed that the ectopic phytoene synthase activity was channelling 20-carbon units into the lycopene pathway, leading to a reduction in gibberellin synthesis. Gibberellins play a role in stem elongation, so the reduction in gibberellin synthesis caused the dwarf phenotype.

● Overexpression of pTOM5 has unforeseen consequences.

The manipulation of lycopene synthesis, and its effect on gibberellin synthesis, leads on to three different examples of the manipulation of yield and quality. The first general area to consider is the modification of colour. There are many examples of this, particularly among the ornamental flowers (see Box 10.2).

Another area that relates to the genetic manipulation of tomato colour is the effect that overexpression of phytoene synthase had on the gibberellin pathway, and the resultant dwarfing. This is of interest because dwarfing is an important agronomic trait. Dwarf plants require less of the plant's resources to be committed to the growth of the stem, allowing more dry matter to be partitioned to the grain (the main commercial product) rather than the straw. This increase in harvest index directly improves yield, and, in addition, the dwarf crops are less prone to damage by wind and rain. The dwarf character was a major feature of the high-yielding varieties of wheat and maize that gave rise to the Green Revolution (see Box 10.3).

The Green Revolution strains of wheat (Box 10.3) were shown to be short because they respond abnormally to gibberellin. This trait is conferred by mutant dwarfing alleles at two reduced height-1 loci (*Rht-B1* and *Rht-D1*). Recent work has shown that *Rht-B1*/*Rht-D1* and the maize dwarf-8 genes are orthologues of the *Arabidopsis* gibberellin-insensitive (*gai*) gene. The *gai* gene encodes one of a family of DELLA proteins that act as nuclear regulators of gibberellin action, and appear to be integrators of auxin, ethylene, and gibberellin growth-regulatory signals. The mutant *gai* protein lacks a region of 17 amino acids called the DELLA domain, which renders it insensitive to gibberellin signalling, thereby acting as a constitutive repressor of

BOX 10.2 *Flower colour*

Flower colours are due mainly to flavonoids, carotenoids, and betalains. The most important flavonoids in this respect are the anthocyanins, which contribute to red and blue flowers. Part of the anthocyanin pathway is shown below.

The first enzyme indicated is chalcone synthase, which was discussed in Chapter 8 in the context of gene silencing. The anthocyanin synthesis pathway is quite complex; at various stages the compounds can be modified by enzymes such as dihydroflavonol-

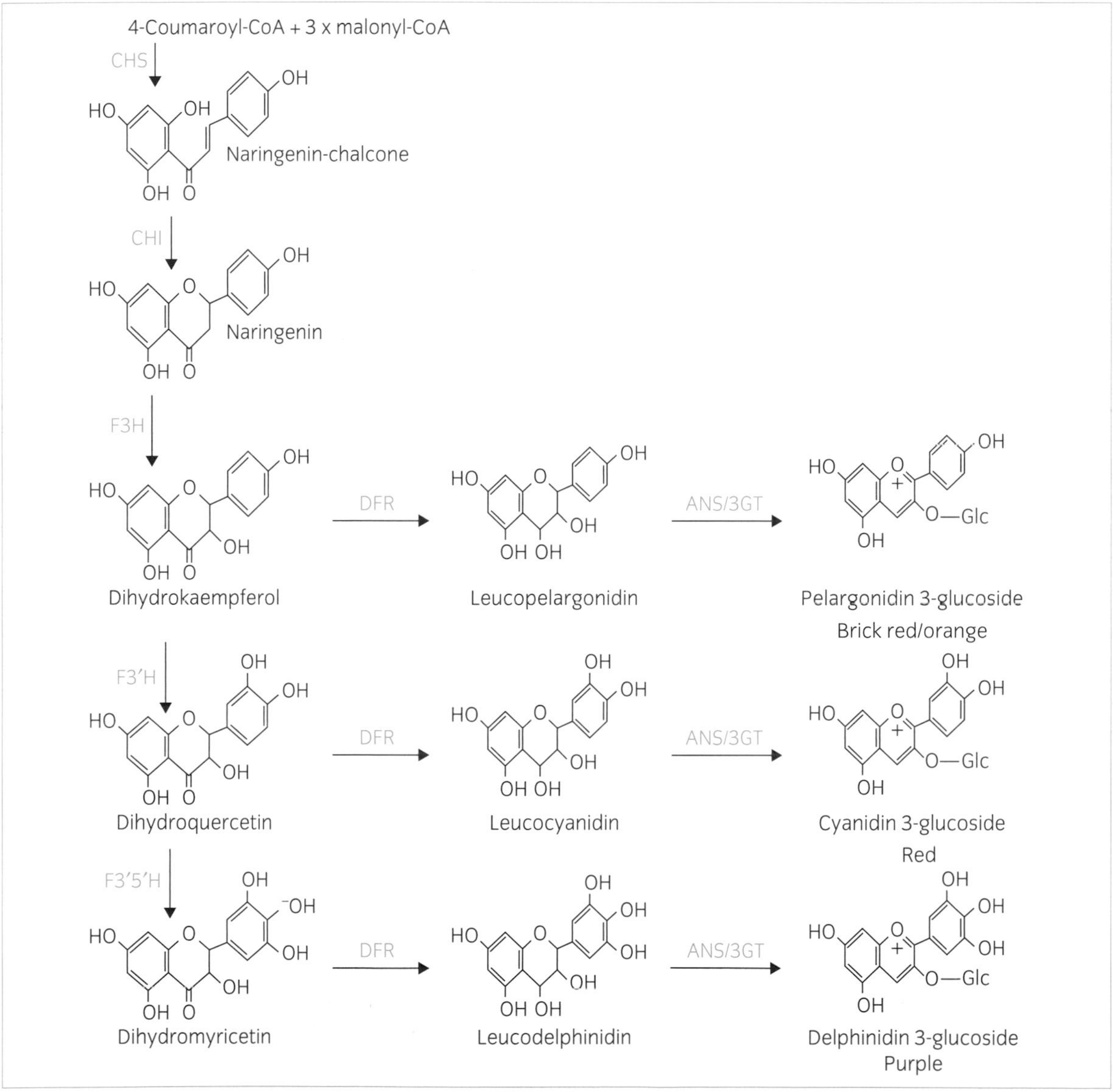

Anthocyanin synthesis pathway. CHS, chalcone synthase; CHI, chalcone flavanone isomerase; F3H, flavanone-3-hydroxylase; F3′H, flavonoid-3′-hydroxylase; F3′5′H, flavonoid-3′,5′-hydroxylase; DFR, dihydroflavonol-4-reductase; ANS, anthocyanidin synthase; 3GT, UDP-flavonol 3-*O*-glucosyltransferase.

BOX 10.2 *Continued*

4-reductase (DFR) to produce different coloured pigments. However, not all plants have all the enzymes of the pathway, so it is impossible to get some colours of flowers in some species. A combination of mutation and genetic manipulation of the anthocyanin pathway has the potential to generate colours in plants that previously were not possible. For example, petunias are unable to produce the pelargonidin-related pigments (brick red/orange) as their DFR is unable to use dihydrokaempferol. However, a line of petunias has been isolated with mutations in the flavonoid-3′-hydroxylase (*F3′H*) and the flavonoid-3′,5′-hydroxylase (*F3′5′H*) genes. The plants are unable to use dihydrokaempferol in the other parts of the pathway, so there is a build-up of the substrate. This plant line has been transformed with a maize *DFR* gene. The expression of the maize protein allows the dihydrokaempferol to be utilized to produce the brick-red pigment pelargonidin 3-glucoside, so producing brick-red petunia flowers.

Companies such as Florigene investigated the genetics of the basic pathways involved in the flower colour and they have patented the use of the genes for colour modification. They have engineered the anthocyanin biosynthetic pathway in plants such as roses, carnations, chrysanthemums, and gerberas to produce flowers in the blue spectrum (mauve–violet–blue). These flowers normally lack the anthocyanin delphinidin, due to the absence of F3′5′H. The enzyme has been cloned and introduced into a range of flower types to produce blue flowers. Florigene introduced its first mauve-coloured carnation on to the market in 1996—called Moondust™—it was the world's first GM flower on sale. Subsequently, other flower products have been marketed in various countries around the world, but not the European Union (EU). As recently as the autumn of 2006, the EU was debating whether to allow cut flowers to be imported into, not grown in, the EU. As well as patents, the companies maintain plant breeders' rights on the lines produced.

One interesting twist to this line of commercialization, particularly for the regulators, is one of cultural practice. In the West, cut flowers would seem to be a relatively safe product for marketing; they are often grown in greenhouses and do not represent a threat to human health or the environment. However, in countries such as Japan, many flower petals are used as decorations for foods and are often eaten. This is actually an important point, which indicates that care must be taken when considering what risks are involved when marketing/releasing transgenic plants. However, the converse to this is that some anthocyanin molecules are seen as nutraceuticals and eating anthocyanin-enhanced plant material may actually be a good thing.

growth. Transgenic rice and wheat plants into which a mutant *gai* gene was introduced were shown to have a reduced response to gibberellin, and to have a dwarf phenotype. This result is particularly significant because, although the dwarf character has been found in rice, attempts to breed a dwarf strain of basmati rice have failed to date because the resulting short plants have lost their characteristic flavour. The success of the *gai* gene experiments suggests that the dwarf character could be introduced into a wide range of crop species to improve crop yield, without having to shuffle all the other genes that contribute to the desirable characteristics of the current elite lines. An important lesson here is that transgenic approaches to crop improvement, particularly those that use well-characterized genes, can be much more precise and predictable than conventional breeding. However, as we look at effects of modifying phytochromes in a later section, it will become clear that this is not always the case.

The final example of crop quality enhancement that flows from the study of carotenoid biosynthesis is that of provitamin A production in rice grains: so-called Golden Rice.

BOX 10.3 *The Green Revolution*

● Dwarf-phenotype genes could be used to engineer Green Revolution traits.

The development of dwarf, high-yielding varieties of wheat and rice by conventional breeding during the 1960s enabled several developing countries, most notably in the Indian subcontinent, to move from a position of food scarcity to become net exporters of these cereals. This step-change in crop yield was termed the Green Revolution, and the key wheat breeder involved, Norman Borlaug, was awarded the Nobel Prize for Peace for his contribution to global food security. The key to his success lay in the development of semi-dwarf, high-yielding, winter wheat varieties adapted to heavy rain conditions in the USA after the Second World War. This wheat was unsuitable for growing in the tropics or semi-tropical climates because, as a winter wheat, it required a period of cold during early growth. It was also found to be susceptible to rust fungus. Using the semi-dwarf wheat lines as a starting point, a breeding programme was established in Mexico to develop wheat strains that were adapted to the growing conditions found in poorer countries. This was done by crossing a number of desirable traits, from wheat varieties collected from around the world, into the semi-dwarf lines. The result was the production of spring wheat varieties that matured quickly and were insensitive to photoperiod, allowing them to be grown more than once per year. They were also rust-resistant and adapted to a variety of warm climates. These characteristics meant that yields could be doubled compared with those of traditional varieties, given adequate fertilizer.

At about the same time, similar breeding programmes were implemented for rice at the International Rice Research Institute (IRRI) in The Philippines. Some of the most successful crosses were between semi-dwarf japonica and relatively high-yielding indica lines, and rice varieties that were semi-dwarf, high yielding, early maturing, photoperiod-independent, and blast-fungus-resistant were produced. At IRRI, the new varieties could yield three- to fourfold more than traditional varieties under appropriate experimental conditions.

The Green Revolution required that farmers not only adopted the new cereal seeds, but also signed up to a high-input method of agriculture—including the use of nitrogen fertilizers, herbicides, and pesticides as well as equipment for tilling and irrigation—to get the maximum yields from these varieties. There were significant cereal yield improvements, though often not as much as predicted from the experimental studies. Many small farmers were unable to afford to get the best out of the new seeds, whereas the richer farmers were able to produce more crops for less input of labour and thus started to acquire more land. The result was that smaller farmers were displaced from the land. Thus, the impressive increase in global food production was achieved at some cost in terms of higher agrochemical inputs, the loss of locally adapted crop varieties, and increased poverty for some.

Analyses of current trends in population growth indicate the need for another step-change in crop yields comparable to the Green Revolution. It is clear that the genetic modification of crops could make an important contribution to raising the so-called yield ceiling reached by conventional breeding efforts. However, it is also important that the implementation of this second Green Revolution takes account of the negative aspects of the first.

CASE STUDY 10.4 **Golden Rice**

Rice is the most important food crop in the world, and is eaten by some 3.8 billion people. In some regions of the world where rice forms a staple component of the diet, vitamin A deficiency is a major nutritional problem. Deficiency of this vitamin can cause symptoms ranging from night blindness to total blindness as a result of xerophthalmia and keratomalacia. It has been estimated that around 124 million children are vitamin A deficient, causing about 500 000 children to go blind each year. Vitamin A deficiency also exacerbates other health problems, for instance diarrhoea, respiratory diseases, and childhood diseases such as measles. In consequence, it is estimated that improved vitamin A nutrition could prevent 1–2 million childhood deaths per year. One of the causes of vitamin A deficiency in regions where the majority of calories consumed comes from rice is that milled rice contains no β-carotene (provitamin A). One of the solutions proposed for this problem is to engineer rice to produce provitamin A in the rice endosperm. Since the successful manipulation of β-carotene synthesis in the rice grains gives them a characteristic yellow/orange colour, rice which is genetically enriched in provitamin A has been described as Golden Rice.

● Golden Rice produces provitamin A in the rice grain.

Figure 10.7 shows that the biosynthetic pathway of provitamin A is a continuation of the lycopene pathway, discussed in the previous section on tomato ripening. Immature rice endosperm is capable of synthesizing GGDP, but subsequent stages of the pathway are not expressed in this tissue. Early transformation experiments with a phytoene synthase (*psy*) gene from daffodil fused to a rice endosperm-specific promoter indicated that phytoene could be synthesized from GGDP in the rice grain. However, three subsequent steps are required to convert phytoene to β-carotene: phytoene desaturase and ζ-carotene desaturase to introduce the double bonds to form lycopene, and lycopene β-cyclase to form the rings in β-carotene. Fortunately, a bacterial carotene desaturase gene capable of introducing all four double bonds can be substituted for the phytoene desaturase and ζ-carotene desaturase. Nevertheless, initial studies of the pathway indicated that manipulation of Golden Rice would require the introduction of three genes: phytoene synthase, carotene desaturase, and lycopene β-cyclase.

● The vitamin A pathway is an extension of the lycopene pathway.

One of the initial strategies used to insert three transgenes plus a selectable marker gene was by co-transformation, and the two constructs used are shown in Figure 10.8. The daffodil *psy* gene::rice glutelin promoter construct was inserted into the vector pZPsC, along with the bacterial carotene desaturase gene (*crt*1) from *Erwinia uredovora* controlled by the 35S promoter. Both enzymes were targeted to the plastid (the site of GGDP synthesis): the *psy* gene by its own transit peptide, and the *crt*1 gene by fusion to a pea ribulose-1,5-bisphosphate carboxylase/oxygenase small subunit (rbcS) transit peptide sequence. The lycopene β-cyclase gene from daffodil with a functional transit peptide was inserted into the vector pZLcyH under the control of the rice endosperm-specific glutelin promoter, along with a hygromycin-resistance selectable marker gene.

● Three genes are required to manipulate provitamin A synthesis in rice.

Rice immature embryos were inoculated with a mixture of *Agrobacterium* LBA4404 containing each of the two plasmids (see Chapter 3). A total of 60 hygromycin-resistant regenerated lines were selected at random, all of which were shown to contain the pZLcyH construct. Of these, 12 were also found to contain the pZPsC cassette. Most of the seeds from these

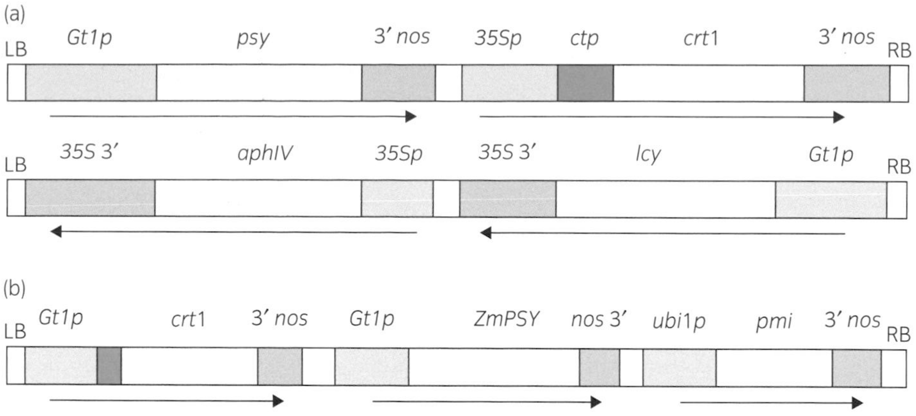

Figure 10.8 Constructs for the production of Golden Rice. (a) The most successful initial strategy for the production of Golden Rice involved transformation with two independent constructs. One of these, pZPsC, contains a daffodil phytoene synthase (*psy*) gene fused to the endosperm-specific rice glutelin promoter (*Gt1p*) in tandem with a bacterial carotene desaturase gene (*crt*1) driven by the 35S promoter. The other, pZLcyH, contains the daffodil lycopene β-cyclase gene (*lcy*) also fused to *Gt1p* in tandem with the hygromycin-resistance selectable marker gene *aphIV*. (b) Golden Rice 2 is produced by transformation with the improved construct pSYN12424. This contains the *crt*1 and maize phytoene synthase (*ZmPSY*) genes both driven by *Gt1p*, and the phosphomannomutatase (*pmi*) selectable marker gene fused to the ubiquitin gene promoter (*ubi*1*p*). *ctp*, Pea *rbcS* chloroplast transit peptide sequence.

transgenic lines containing both constructs were found to be yellow, indicating carotenoid synthesis. A range of carotenoids was found in some of these lines, whereas β-carotene was the only carotenoid in others. The highest-producing line of this type was found to contain 1.6 μg β-carotene g^{-1} endosperm from a mixed population of segregating grains. It was calculated that the homozygous grains of this line would produce about 2 μg g^{-1} provitamin A, which corresponds to a daily intake of 100 μg retinol equivalents, assuming that 300 g of rice is consumed per day. This would make a significant contribution to the daily intake of vitamin A, though it is not sufficient to provide the complete dietary requirement of the vitamin (see Box 10.4).

Since the initial demonstration of the scientific feasibility of this approach, a number of technical improvements have been made to Golden Rice. These include the following.

- The *pmi* (phosphomannose isomerase; also known as mannose-6-phosphate isomerase; see Table 4.1) mannose-selection gene was substituted to avoid antibiotic selection using the hygromycin-resistance gene.
- The Golden Rice trait was genetically engineered into indica rice cultivars. Indica rice is consumed by 90% of the Asian population, whereas the original Golden Rice was produced using the japonica variety Taipei 309.
- Subsequent research confirmed results obtained during the initial phase of development, indicating that the lycopene β-cyclase transgene was not required to produce β carotene in the endosperm. Further work by Syngenta to optimize β-carotene production showed that the daffodil phytoene synthase was rate-limiting, and that a *psy* gene from maize was much more effective. Transformation of rice with the construct pSYN12424 (Figure 10.8) resulted in

BOX 10.4 *The anti-GM opposition to Golden Rice*

One of the surprising features of the Golden Rice programme has been the opposition it has generated from opponents of GM food, because Golden Rice appears to meet all the objections to GM crops that have been presented by anti-GM groups. Thus, as its originator, Ingo Potrykus, has pointed out:

1. it benefits primarily the poor and disadvantaged;
2. it will be given free of charge and restrictions to subsistence farmers;
3. it can be re-sown every year from the saved harvest;
4. it does not create advantages for rich landowners;
5. it was not developed by or for the biotechnology industry, and industry does not benefit from it;
6. it fulfils an urgent need by complementing (rather than displacing) traditional interventions;
7. it is a sustainable, cost-free solution to vitamin A deficiency, not requiring other resources for extraction, manufacture, and distribution;
8. it avoids the unfortunate negative side effects of the Green Revolution (see Box 10.3);
9. it creates no new dependencies;
10. it can be grown without any additional inputs;
11. it does not reduce agricultural biodiversity;
12. it does not affect natural biodiversity;
13. there is no negative environmental impact envisaged;
14. there is no conceivable risk to consumer health;
15. it was impossible to develop the trait by traditional methods.

However, Golden Rice has been attacked by groups such as Greenpeace, mainly on the grounds that it is a Trojan horse for other GM crops. The nature of this opposition derives from a fundamental belief that all GM crops are undesirable, and that any concession that Golden Rice might be acceptable could give a green light to other GM crops. Thus, a number of arguments have been deployed against Golden Rice. One of the original criticisms was that the level of provitamin A in the rice was only sufficient to meet about 20% of the daily requirement for vitamin A. This point has now been overtaken by the development of Golden Rice 2. Another criticism was that vitamin A deficiency would be better met by encouraging a better, more varied diet of local vegetables (many of which used to be grown more widely, but have been replaced by rice). Although there would obviously be a number of benefits in encouraging local vegetables, there are problems related to the seasonal nature and rapid spoilage of certain crops, along with the economic cost of replacing the current rice crop. In any case, stating that there are alternative approaches to improving vitamin A nutrition (not exactly a major priority of the environmental groups of the developed world prior to Golden Rice) does not add up to an argument against Golden Rice.

a 23-fold increase (up to 37 μg g^{-1}) in carotenoids compared with the original Golden Rice, and has been named Golden Rice 2.

When the first laboratory-scale results were obtained, it was recognized that the scientific optimization was only one of a series of hurdles that would have to be negotiated before provitamin A-enriched rice reached those who could benefit from it. These challenges included the clarification of intellectual property rights (IPRs) and the mechanisms for technology transfer.

Golden Rice, IPRs, and technology transfer

● Golden Rice was developed using procedures protected by 70 IPRs and technology protection rights.

One of the major obstacles to the delivery of Golden Rice to its potential beneficiaries proved to be the protection of IPRs. Since Golden Rice was developed in public laboratories with public funding for humanitarian purposes, the aim was to patent-protect the process to fulfil the goal of supplying it to subsistence farmers free of charge and restriction. However, in compiling the international patent application, it was discovered that the research underpinning the development of Golden Rice had involved the use of procedures and technology protected by 70 IPRs and technology protection rights belonging to 32 different companies and universities! (One way of protecting technology rights is by the use of material transfer agreements, which require that certain conditions are met when DNA constructs are requested from the originating laboratory. These conditions typically require that the DNA is used for research purposes only.) In order to be able to grant so-called freedom-to-operate licences to public research institutions in developing countries so that they could introduce the trait into local varieties, it was necessary to obtain free licences from all of the patent holders.

The size and complexity of the task of drawing up the patent application and obtaining these licences led the public scientists to involve a commercial partner (Syngenta) in the patent application, in return for an exclusive licence for commercial exploitation of the technology (commercial exploitation is defined as income from Golden Rice exceeding US$10 000). Fortunately, free licences for humanitarian purposes have been largely granted by the IPR holders, at least in part because of the enlightened self-interest of the biotechnology companies involved. Golden Rice is seen by the biotechnology industry as a valuable *flagship project* to improve the public perception of GM technology (see Box 10.4). Conversely, there would be a lot of negative publicity for a company that was seen to be exploiting or standing in the way of the humanitarian aims of this project.

The next challenge faced by the developers of Golden Rice was that of technology transfer. Golden Rice originated in Switzerland as a series of laboratory lines of japonica rice. These lines would be poorly adapted to the range of conditions found in different rice-growing countries, for which locally adapted varieties and ecotypes have been developed. There is therefore an enormous task involved in transferring the Golden Rice trait into local varieties, which has only just started with the transfer of the trait into indica lines. To aid this technology-transfer process, a Golden Rice Humanitarian Board has been established. This board ensures that, for each region, there is a proper assessment of needs and a framework for the optimal use of Golden Rice. It also co-ordinates the assessments of bioavailability, food safety, and allergenicity required before the rice is made available for human consumption. The Golden Rice material is only given to public research institutions that can ensure proper handling and use of GM plants according to local rules and regulations. These institutions will transfer the Golden Rice trait into the best locally adapted lines by conventional breeding methods and/or *de novo* transformation of elite varieties. Socio-economic and environmental impact studies will also be conducted to ensure that the technology does reach the poor without damage to the environment. India has taken a leading role in this technology transfer process and serves as a model for other countries to follow.

From Golden Rice to biofortified rice

Those involved in the Golden Rice project and its humanitarian board have looked beyond Golden Rice to the development of a biofortified rice in which the levels of other vitamins and micronutrients are enhanced. One target is to develop rice varieties with high iron and zinc concentrations in the grain by conventional breeding and/or GM approaches. Iron deficiency is a major global nutritional disorder, estimated to affect about 30% of the world population, and is prevalent in areas where rice is a staple crop. One problem is the practice of polishing rice (removing the outer layer of the endosperm), which is required to prevent spoilage due to oxidation of the oils that accumulate in this layer. Unfortunately, this is also the location of micronutrient and vitamin accumulation, and iron and zinc levels are low in polished grains. Another reason for concentrating on iron and zinc is that these elements assist the absorption of provitamin A from the gut. Strategies for the genetic manipulation of iron and zinc accumulation in the rice grain include endosperm-specific expression of the soybean ferritin gene, which binds iron and stores it in a bioavailable, non-toxic form. This strategy has already been shown to have some success in transgenic rice. Another approach that could be deployed in tandem is to express high levels of phytase sufficient to degrade phytic acid, a major inhibitor of iron absorption in the diet. Expression of cysteine-rich proteins has also been proposed as a strategy to improve the bioavailability of iron in rice grains.

Another target is to increase the levels of vitamin E, a complex of amphipathic molecules (tocochromanols) comprising two groups: tocopherols and tocotrienols (see Chapter 9). Of these, α-tocopherol has the highest vitamin E activity due to its preferential retention in the liver. The tocochromanols are formed by addition of a hydrophobic prenyl tail group (synthesized via the isoprenoid pathway; see Figure 10.7) to an aromatic homogentisic acid head group (synthesized from tyrosine via the shikimate/prephenate pathway; see Box 5.2). The core enzymes involved in tocopherol synthesis have all been identified in *Arabidopsis* and transgenic approaches to the improvement of tocopherol production have been tested in this model plant. One approach has been to increase flux to the aromatic head group, homogentisic acid. Two enzymes are involved in the conversion of tyrosine to homogentisic acid: tyrosine amino transferase, which produces *p*-hydroxyphenylpyruvate (HPP), and HPP dioxygenase (HPPD). HPPD is predicted to be a key enzyme in regulating flux to tocopherols, but transformation with HPPD alone or with tyrosine amino transferase had little effect on tocopherol levels. However, insertion of the HPPD gene along with the yeast *TyrA* gene resulted in an eightfold increase in tobacco leaves and a two- to threefold increase in *Arabidopsis* seeds. *TyrA* encodes prephenate dehydrogenase, which converts prephenate directly to HPP, bypassing the feedback inhibition of arogenate dehydrogenase by tyrosine (see Box 5.2). The transgenic seeds were black due to oxidative polymerization of the very high levels of homogentisic acid, and germination was impaired. Another strategy has been to target the first committed step of tocochromanol synthesis by homogentisate prenyltransferases. Overexpression of the *Arabidopsis VTE2* gene encoding a homogentisate prenyltransferases that transfers phytyl diphosphate to homogentisic acid increased overall tocopherol levels in transgenic *Arabidopsis*. Co-expression of *VTE2* with *VTE4* (encoding γ-tocopherol methyl transferase) in *Arabidopsis* seed converted all the γ-tocopherol to α-tocopherol, resulting in an approximately 12-fold increase in vitamin E activity. These strategies have been applied to the enhancement of tocopherol in soybean seeds, with the most successful being co-expression of TyrA, HPPD, and VTE2, resulting in an 11-fold enhancement in vitamin E activity in the seed oil.

Attempts to enhance the levels of other vitamins have been reported. For example, the vitamin C levels in tobacco leaves and maize kernels were increased two- to fourfold by expression of a wheat dehydroascorbate reductase, with glutathione levels also increased (see Chapter 9). Folate levels have been improved in *Arabidopsis* by expression of the *Escherichia coli folE* gene encoding GTP cyclohydrase-1. This enzyme catalyses the first committed step in pterin synthesis and expression of the unregulated bacterial enzyme resulted in a 1250-fold and two- to fourfold enhancement of pterins and folates, respectively. These examples of nutritional enhancement fall within the objectives of HarvestPlus, a global alliance of research institutions and implementing agencies with a common aim to breed and disseminate biofortified crops for better nutrition.

Engineering plant protein composition for improved nutrition

Despite our view of plants as being healthy to eat, some lack essential nutrients.

The example of Golden Rice has dealt with one aspect of nutritional quality. Another that has become a target for genetic engineering is the amino acid content of plant foods. Humans are only capable of synthesizing 10 of the 20 naturally occurring amino acids. The other essential amino acids are obtained from the diet. Obtaining balanced quantities of the amino acids can be problematic if certain foods predominate in a diet. This is the case with cereal grains, which are commonly used as the principal energy source, and with legume seeds, which are important sources of proteins in the diet of humans and livestock. Cereal grains are often limiting for lysine, while the legume seeds have an adequate level of lysine but are limiting for the sulphur-containing amino acids, methionine and cysteine. It is important to note that animals can convert methionine into cysteine, but not the reverse, so it is possible to make up any short-fall in the sulphur-containing amino acids by increasing the methionine level. Transgenic routes have been used to increase the content of several different amino acids and some of the approaches will be discussed here.

One approach used has been to isolate the gene for a sulphur-poor protein and modify and enrich its nucleic acid sequence for the sulphur-containing amino acids. This has been attempted for proteins such as β-phaseolin from *Phaseolus vulgaris* and vicilin from *Vicia faba*, but the modified proteins were either unstable or contained too little methionine to make them useful. A more successful approach has been to construct totally artificial genes that code for proteins containing a high sulphuric amino acid content. One such totally synthetic protein, containing 13% methionine residues, has been expressed successfully in sweet potato (*Ipomoea batatas*). Several methionine-rich proteins have been identified in maize (21-kDa zein, 28% methionine;

10-kDa zein, 23% methionine), rice (10-kDa prolamin, 20% methionine), sunflower (2S sunflower seed albumin, 16% methionine and 8% cysteine), and Brazil nut (Brazil nut albumin, 18% methionine and 8% cysteine). The genes for these proteins have been introduced into a number of crops (maize, soybean, lupin, canola) to increase the level of sulphur-containing amino acids in blended stock feeds containing cereal and legume grains. One major problem associated with moving proteins between species has become apparent with the Brazil nut albumin; that is, the protein responsible for the potent allergenicity of Brazil nuts. This property is maintained in the seeds of Brazil nut albumin-containing transgenic plants, which makes them unacceptable for human consumption and has been a timely warning that safety issues should be paramount when dealing with food crops.

Some of the problems associated with these studies relate to the stability of the proteins and their localization within the plant. This point can be illustrated by considering the expression of high-lysine proteins in the seed endosperm of cereals such as rice. These studies raise important issues about the types of protein that can be used and how they are localized in the grain, issues that should be considered when ectopically expressing proteins for whatever reason. In cereal endosperm, storage proteins accumulate primarily in protein-storage vacuoles of terminally differentiated cells and as protein bodies assembled directly within the endoplasmic reticulum (ER). Specific storage proteins are stored in these structures. In rice for example, there are two main types of storage protein: prolamins, which form accretions within the lumen of the ER, and glutelins (related to 11S globulins). Glutelins comprising up to 80% of the total seed protein are synthesized on rough ER and then transported to protein-storage vacuoles. It has recently been found that mRNAs for both types of protein are found in rough ER polysomes, but that the prolamin transcripts are preferentially located on membranes surrounding prolamin-containing protein bodies, whereas glutelin mRNAs are associated predominantly with polysomes on the cisternal ER.

One interesting study has been carried out with rice, where a lysine-rich protein from soybean has been expressed in the endosperm. To optimize expression and localization, the soybean protein chosen was a glycinin, which is a member of the 11S globulin family. The important feature of the glycinin is that it is processed and stored in soybean by a process very similar to that found for the rice glutelins. When the transgenic rice plants were analysed, it was found that both globulins (glutelin and glycinin) were expressed and formed complexes of discrete sizes that were targeted correctly. If this had not been the case, then it would not be possible to obtain optimum levels of proteins. The other interesting point of glycinin proteins in soybean is that they are associated with a hyporcholesterolaemic effect (reduction in cholesterol) if the dietary intake of these proteins is in excess of 6 g day^{-1}. Therefore, expressing the soybean gene in rice potentially has two beneficial effects: it increases the amount of lysine in the endosperm protein, and it would contribute to the dietary levels of soybean globulins that lead to a reduction in the level of cholesterol.

● Several approaches will be needed in combination to improve amino acid content.

As more proteomic information becomes available, other approaches are becoming more feasible. To influence the amount of sulphur-containing amino acids, manipulation of the synthetic pathways has been considered. For example, to increase methionine levels in the forage crop alfalfa (*Medicago sativa*) an *Arabidopsis* cystathionine γ-synthase gene (*AtCGS*/*AtCγS*), which codes for the first enzyme in the methionine pathway, has been introduced linked to the *Arabidopsis* ribulose-1,5-bisphosphate carboxylase/oxygenase (Rubisco) small subunit promoter, giving expression in the leaves. This has given some interesting results: the amount of methionine incorporated into the water-soluble protein fraction was increased as well as the amount of protein-bound cysteine. Other key enzymes are targets for manipulation. Aspartate is a common precursor for lysine, threonine, methionine, and isoleucine, all of which are essential amino acids. Manipulation of the early stages of these pathways could improve the content of a range of amino acids. It has been shown that the introduction of a high-methionine protein, such as a maize β-zein, into alfalfa does not necessarily lead to higher levels of sulphur-rich protein synthesis. Rather, the pool of methionine might be limiting for the hoped for increase in expression. The dual approach of protein plus the incorporation of the *AtCGS* gene was shown to improve incorporation.

The genetic manipulation of crop yield by enhancement of photosynthesis

● Photosynthesis is a complex system that will not be easy to manipulate.

Crop yield is totally dependent upon light and its conversion (light harvesting and electron transport) into the usable energy (ATP, NADPH) that drives the dark reactions of photosynthesis (carbon dioxide conversion into carbohydrates; Figure 10.9). These reactions usually take place in chloroplasts within the leaves, the source, but the efficiency of the process is also dependent upon the capacity of sink tissues and organs to assimilate this fixed carbon. These are complex processes and outside the scope of this book, but some studies have been carried out to investigate the potential for enhancing photosynthesis by biotechnological means. It is clear that the introduction of a single gene will not lead to a durable increase in yield. However, several examples will be discussed briefly to describe the potential for manipulating complex physiological interactions. Direct manipulation of light harvesting, electron transfer, or the processes of photoprotection and photoacclimation will not be considered here.

Manipulation of light harvesting and the assimilate distribution: phytochromes

When grown in the field or in their natural environments, plants are not always under optimum light conditions, even during daylight hours. They are faced with changes in the amount, direction, duration, and quality of incident light radiation. Plants have evolved mechanisms that optimize the acquisition of available light energy for photosynthesis. In dense populations or in shady conditions, plants display the shade response. The plants use available resources to increase stem and petiole elongation

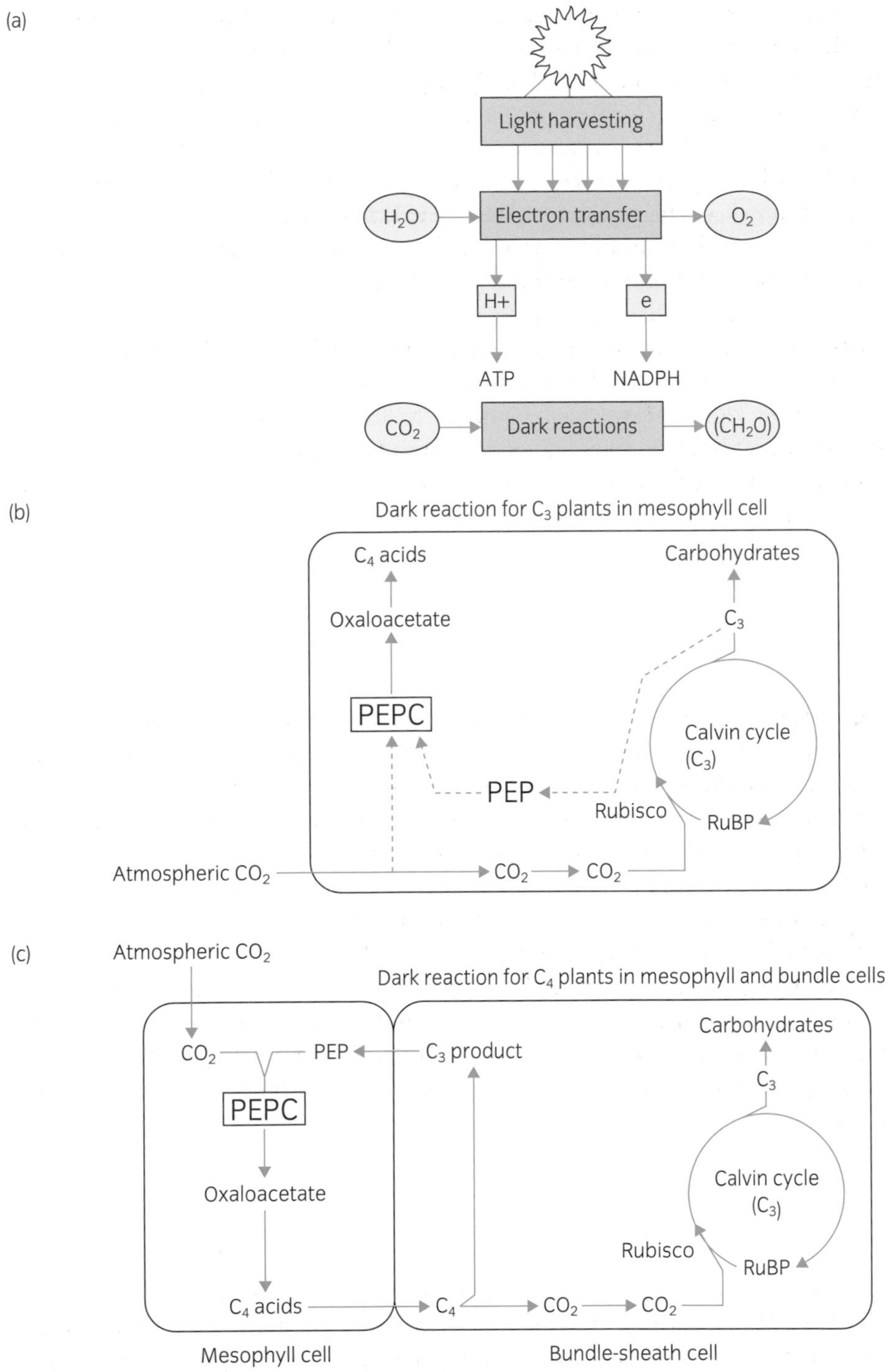

Figure 10.9
Photosynthesis. (a) Photosynthesis can be divided into several separate steps: light harvesting and electron transfer, in which light energy is converted into useful chemical energy (ATP, NADPH); and the dark reaction where CO_2 is fixed. (b) In C_3 plants, the fixation is carried out by Rubisco (ribulose-1,5-bisphosphate carboxylase/oxygenase), as part of the Calvin cycle (C_3 cycle), in the chloroplasts of mesophyll cells (solid arrows). (c) In C_4 plants, CO_2 fixation occurs in mesophyll cells, via phosphoenolpyruvate (PEP) carboxylase (PEPC), to form C_4 acids such as malate and aspartate. These are transferred to adjacent bundle-sheath cells where metabolism of the C_4 acids generates high concentrations of CO_2 that are used by Rubisco in a normal C_3 cycle. In C_3 transgenic plants containing PEPC (b), there is a conflict between the existing C_3 cycle and the presence of PEPC, which is likely to drain C_3 precursors from the Calvin cycle, thus limiting fixation into carbohydrates (dashed arrows). RuBP, ribulose 1,5-bisphosphate. (In part redrawn with permission from Edwards, G. (1999) Tuning up crop photosynthesis. *Nature Biotechnology* **17**, 22–23.)

to outgrow any shading plant; there is also a reduction in chlorophyll synthesis and leaf thickening and an increase in apical dominance. From an agricultural point of view, this shade response has important implications for yield, as assimilates are used to re-establish optimal light conditions for photosynthesis rather than being stored. The shade response is regulated through a series of photoreceptors. The phytochrome (Phy) family of proteins are the best characterized of these proteins (there are at least five Phy proteins in dicotyledons). They are able to detect the level and quality of the incident light and then control growth and development via a series of signal transduction pathways that regulate gene expression. Phy proteins are photochromic proteins; they have a protein moiety connected to a tetrapyrrole chromophore, and they can exist in two forms: physiologically inactive (P_R, red-light absorbing) and active (P_{FR}, far-red-light absorbing). These forms are interconvertible by red or far-red light, respectively. The photosynthetic pigments in leaves absorb most visible radiation (400–700 nm) but reflect light beyond 700 nm (far red, 700–800 nm). This leads to a high proportion of far-red radiation being found in dense plant stands. Plants are able to perceive this radiation and the proximity of other plants because the equilibrium of the phytochromes is shifted towards the inactive form. Two of the major phytochromes involved in this response are PhyA and PhyB. PhyA accumulates in the dark and is degraded rapidly upon conversion, by the absorption of red light, into the labile P_{FR} form. Despite this, it is responsible for the detection of continuous far-red light and dampens the shade-avoidance response. PhyB is responsible for the detection of red light, so under high light conditions (high relative proportions of red light), it is converted to the active form, in which form it suppresses the shade-avoidance response.

The manipulation of phytochromes is a general approach that can have pleiotropic effects on photosynthesis and assimilate distribution. Analysis of these systems suggests that phytochrome manipulation is likely to have important consequences on photosynthesis (the phytochromes co-ordinately regulate many of the genes involved in photosynthesis) and the sink–source relationship. Thus, a number of different transgenic plant types, ectopically expressing either PhyA or PhyB, have been constructed and some interesting effects have been observed. Plants expressing the phytochromes at very high levels were seriously affected in their growth, but it was possible to identify plants in which the phytochromes improved the harvest index under high densities in the field. Several examples will be discussed in more detail below.

In one set of experiments, tobacco plants, engineered to overexpress an oat PHYA cDNA, were grown at five different planting densities. At the higher planting density, the wild-type tobacco demonstrated the shade response, in that the plants were normally taller than those planted at low densities. With the transgenic lines, they elongated at lower rates; thus, at the high planting densities, relative to the lower planting densities, the result was dwarfed plants. The ectopically expressed PhyA was not only damping the shade response in response to far-red light, but was actually reducing stem elongation to below the normal levels found in plants grown in optimal light conditions. Overall yield was not affected, but assimilates showed an enhanced allocation to leaves with increases in the harvest index approaching 20%. Thus, the beneficial outcome of this approach was that at high planting densities assimilates were not being utilized in stem growth but were being utilized in leaf production or were being stored.

Similar dwarfing effects were found in an experiment with transgenic potato plants containing an ectopically expressed *Arabidopsis PHYB* sequence. However, overexpression of the phytochrome also led to an increase in chloroplast number and an overall increase in photosynthesis per leaf area. Photosynthesis was also less sensitive to photoinactivation under prolonged light stress. Although leaf senescence was not delayed, chlorophyll degradation was, leading to a prolonged period of photosynthetic activity. The increase in the rate of photosynthesis and the prolonged period of photosynthetic activity resulted in an increase in biomass. This was demonstrated by extended underground organs (roots/shoots) and with a large number of small tubers. In fact, tuber yields were increased by as much as 50% in weight.

The pleiotropic nature of this approach has some overlap with strategies to prolong the photosynthetic period of plants by delaying leaf senescence. One such strategy has involved the production of transgenic plants containing the senescence-related *SAG*12 promoter (*SAG*, senescence-activated gene) driving expression of the *Agrobacterium* cytokinin biosynthetic gene *ipt* (see Chapter 3). As the leaves begin to senesce, the expression of the *ipt* gene is switched on, thereby increasing the cytokinin level in the leaf. This delays senescence and maintains chlorophyll content. The prolonged period of photosynthetic activity is expected to increase assimilate production. However, it is not clear whether this strategy will be successful as there may be problems with nitrogen metabolism. As the nitrogen released from senescing leaves is required for other developmental processes, such as grain filling in cereals, its maintenance in the leaf is unlikely to be beneficial. In other studies, the *ipt* gene has been expressed in developing endosperm with the aim of increasing the sink size and strength. This has been shown to lead to an increase in seed size and assimilate content.

Direct manipulation of photosynthesis: enhancement of dark reactions

The sheer complexity of photosynthesis and assimilate storage suggests that the task of increasing yield will require whole suites of genes—nuclear and chloroplast—to be enhanced. The phytochrome and the *ipt* approaches allow for this, in that they have pleiotropic effects. It is likely that different crops will require different strategies to be adopted. This is the case for C_3 and C_4 plants (this refers to whether the plant uses a three-carbon or a four-carbon route in the fixation of CO_2), which will be discussed in the next section. However, identifying limiting steps in pathways can allow increases in yield to be obtained. In C_4 plants, it has been shown that carbonic anhydrase, the enzyme that catalyses the first reaction in the C_4 photosynthetic pathway, the conversion of atmospheric CO_2 to bicarbonate in the mesophyll cytosol, is essential for the functioning of the pathway. Also it has recently been found with tobacco that two enzymes in the Calvin cycle (C_3 cycle; Figure 10.9b) are limiting. Most of the enzymes in the Calvin cycle are present at levels in excess of those required to sustain a continued rate of CO_2 fixation. However, levels of fructose-1,6-bisphosphatase (FBPase) and sedoheptulose-1,7-bisphosphatase (SBPase) are extremely low compared with those of other enzymes in the Calvin cycle. Transgenic tobacco plants expressing a single gene for a chloroplast-targeted, dual-function cyanobacterial fructose-1,6/sedoheptulose-1,7-bisphosphatase show enhanced photosynthetic efficiency and growth characteristics. Dry matter and photosynthetic CO_2 fixation were 1.5- and 1.24-fold higher in the transgenic plants compared with wild-type tobacco. It was also found that Rubisco activity was increased, and that various Calvin cycle intermediates and the accumulation of carbohydrates were also higher. These results seem to relate to an increased metabolic flux, and support the notion that the enzymes were limiting the photosynthetic pathway in tobacco.

There have been a number of studies to investigate possible approaches to increase the efficiency of the Rubisco enzyme, looking at the enzyme or by protecting the system from environmental stresses, but the most challenging approach attempted has been to introduce various enzymes from the C_4 photosynthesis pathway into C_3 crop species. The C_3 plants—which include wheat, oats, and soybean—have chloroplast-containing mesophyll cells in the leaves. CO_2 is assimilated in these chloroplasts, via Rubisco, as part of the Calvin cycle (Figure 10.9b), and the first product from this is the C_3 molecule 3-phosphoglyceric acid. This is not a very efficient system as the CO_2 concentration in C_3 plants can be a rate-limiting factor. Also, Rubisco can use O_2 as an alternative substrate for reactions with ribulose 1,5-bisphosphate, which ultimately leads to photorespiration (metabolism of C_3 molecules leading to the release CO_2 in mitochondria). In essence, O_2 is a competitive inhibitor with respect to CO_2 fixation by Rubisco. In addition, high temperature decreases the availability of CO_2 to Rubisco, and water stress (by closing stomata) increases the resistance to CO_2 diffusion into the leaf. In C_4 plants, such as maize and sugar cane, two different types of chloroplast-containing cells are involved in CO_2 assimilation. CO_2 is first fixed in mesophyll cells by phosphoenolpyruvate carboxylase (PEPC; Figure 10.9c). C_4 acids then migrate into associated bundle-sheath cells where the C_4 pathway generates a high concentration of available CO_2, which is then used by Rubisco. This system is very efficient under all the conditions where the C_3 system fails.

Recently, a complete maize PEPC gene (maize is a C_4 plant) has been introduced into rice (a C_3 plant) with surprising results. Many of the plants obtained showed high levels of expression, two or three times that found in maize, and the enzyme accounted for up to 12% of the total leaf protein. The most important observation was that the O_2 sensitivity of photosynthesis in rice was decreased in the transformed plants. Why this should be the case was not clear, but it suggested that carbon fixation by the maize enzyme increased the supply of CO_2 to Rubisco. This experiment raises the possibility of manipulating other stages of the photosynthetic pathways. However, it is also likely to be a very difficult process, as even the expression of the *PEPC* gene had drawbacks. Expression of PEPC in the absence of the other enzymes of the C_4 pathway led to a drain of carbon into C_4 acids rather than the carbohydrates, so reducing plant yield (Figure 10.9b).

SUMMARY

This chapter has reviewed a number of specific examples of the genetic modification of yield and quality traits. The manipulation of tomato ripening has been considered in some detail because it is possible to demonstrate a number of techniques and approaches with this topic. Tomato ripening was one of the first systems in which antisense technology was used to manipulate endogenous gene expression, such that levels of enzyme activity that were less than 1% of the wild type were achieved. Antisense techniques were exploited to confirm the role of specific genes in the ripening process (PG) and identify ripening-related genes (ACC oxidase). The technique was also applied to the modification of ripening for commercial targets. Subsequent work on this system led to the recognition that endogenous gene expression was also inhibited by sense constructs, and that the phenomenon of co-suppression could also be used for genetic manipulation purposes. Three of the important genes that were identified and exploited exemplify three distinct target areas for genetic modification. The manipulation of PG demonstrates the possibility of manipulating the expression of a single gene that has very little detectable effect on related processes in the plant: softening can be inhibited without affecting the other ripening events. In contrast, the manipulation of a regulatory pathway (ethylene production) has profound effects on the entire ripening process. The third example, phytoene synthase, demonstrates the potential for, and pitfalls of, manipulating complex biochemical pathways in plants. This potential has been exploited most impressively in the engineering of Golden Rice. Enhancement of nutritional quality has also been discussed in respect of amino acid content. This area is likely to become more accessible to manipulation as metabolomic studies provide information on plant pathways.

Several preliminary studies on the enhancement of photosynthetic pathways were presented in this chapter. These indicate that it might be possible increase yield levels by manipulating the dark reactions. However, it is clear that this will not be a rapid process, for a lot more fundamental work is required in this area. Some work is also underway in understanding how it may be possible to manipulate the light reactions of photosynthesis. Again these are complex areas, and a lot more basic research is required before it will be possible to optimize yield production through the manipulation of light harvesting or the energy-production aspects of photosynthesis.

FURTHER READING

Visit www.oxfordtextbooks.co.uk/orc/slaterplants2e/ for hyperlinked references that take you directly to online abstracts for the journal articles listed below:

Crop yield and quality

Chapple, C. and Carpita, N. (1998) Plant cell walls as targets for biotechnology. *Current Opinion in Plant Biology* **1**, 179–185.

Mazur, B., Krebbers, E., and Tingey, S. (1999) Gene discovery and product development for grain quality traits. *Science* **285**, 372–375.

Tomato ripening

Bouzayen, M., Hamilton, A., Picton, S., Barton, S., and Grierson, D. (1992) Identification of genes for the ethylene-forming enzyme and inhibition of ethylene synthesis in transgenic plants using antisense genes. *Biochemical Society Transactions* **20**, 76–79.

Grierson, D. and Covey, S. (1988) *Plant Molecular Biology*, 2nd edn. Blackie, Glasgow.

Grierson, D. and Schuch, W. (1993) Control of ripening. *Philosophical Transactions of the Royal Society London Series B* **342**, 241–250.

Hamilton, A. J., Lycett, G. W., and Grierson, D. (1990) Antisense gene that inhibits synthesis of the hormone ethylene in transgenic plants. *Nature* **346**, 284–287.

Smith, C. J. S., Watson, C. F., Ray, J., Bird, C. R., Schuch, W., and Grierson, D. (1988) Antisense inhibition of polygalacturonase expression in transgenic tomatoes. *Nature* **334**, 724–726.

White, P. J. (2002) Recent advances in fruit development and ripening: an overview. *Journal of Experimental Botany* **53**, 1995–2000.

The Green Revolution

Ait-ali, T., Rands, C., and Harberd, N. P. (2003) Flexible control of plant architecture and yield via switchable expression of *Arabidopsis gai*. *Plant Biotechnology Journal* **1**, 337–343.

Alvey, L. and Harberd, N. P. (2005) DELLA proteins: integrators of multiple plant growth regulatory inputs? *Physiologia Plantarum* **123**, 153–160.

Chrispeels, M. J. and Sadava, D. (2003) *Plants, Genes and Crop Biotechnology*, 2nd edn. Jones and Bartlett Publishers, London.

Peng, J. R., Richards, D. E., Hartley, M. N., Murphy, G. P., Devos, K. M., Flintham, J. E., Beales, J., Fish, L. J., Worland, A. J., Pelica, F., Sudhakar, D., Christou, P., Snape, J. W., Gale, M. D., and Harberd, N. P. (1999) 'Green Revolution' genes encode mutant gibberellin response modulators. *Nature* **400**, 256–261.

Golden Rice

Al-Babili, S. and Beyer, P. (2005) Golden Rice—five years on the road—five years to go? *Trends in Plant Science* **10**, 565–573.

Mayer, J. E. (2005) The Golden Rice Controversy: useless science or unfounded criticism? *BioScience* **55**, 726–727.

Paine, J. A., Shipton, C. A., Chaggar, S., Howells, R. M., Kennedy, M. J., Vernon, G., Wright., S. Y., Hinchcliffe, E., Adams, J. L., Silverstone, A. L., and Drake, R. (2005) Improving the nutritional value of Golden Rice through increased pro-vitamin A content. *Nature Biotechnology* **23**, 482–487.

Potrykus, I. (2001) Golden Rice and beyond. *Plant Physiology* **125**, 1157–1161.

Ye, X., Al-Babili, S., Kloti, A., Zhang, J., Lucca, P., Beyer, P., and Potrykus, I. (2000) Engineering the provitamin A (β-carotene) biosynthetic pathway into (carotenoid-free) rice endosperm. *Science* **287**, 303–305.

Biofortification

Bouis, H. E., Chassy, B. M., and Ochanda, J. O. (2003) Genetically modified food crops and their contribution to human nutrition and food quality. *Trends in Food Science and Technology* **14**, 191–209.

Chen, Z., Young, T. E., Ling, J., Chang, S.-C., and Gallie, D. R. (2003) Increasing vitamin C content of plants through enhanced ascorbate recycling. *Proceedings of the National Academy of Sciences USA* **100**, 3525–3530.

DellaPenna, D. (2005) Progress in the dissection and manipulation of vitamin E synthesis. *Trends in Plant Science* **10**, 574–579.

Hossain, T., Rosenberg, I., Selhub, J., Kishore, G., Beachy, R., and Scheubert, K. (2004) Enhancement of folates in plants through metabolic engineering. *Proceedings of the National Academy of Sciences USA* **101**, 5158–5163.

Karunanandaa, B., Qi, Q., Hao, M., Baszis, S. R., Jensen, P. K., Wong, Y.-H. H., Jiang, J., Venkatramesh, M., Gruys, K. J., Moshiri, F., Post-Beittenmiller, D., Weiss, J. D., and Valentin, H. E. (2005) Metabolically engineered oilseed crops with enhanced seed tocopherol. *Metabolic Engineering* **7**, 384–400.

Lucca, P., Hurrell, R., and Potrykus, I. (2001) Genetic engineering approaches to improve the bioavailability and the level of iron in rice grains. *Theoretical and Applied Genetics* **102**, 392–397.

Welch, R. M. and Graham, R. D. (2004) Breeding for micronutrients in staple food crops from a human nutrition perspective. *Journal of Experimental Botany* **55**, 353–364.

Enhancement of photosynthesis

Ku, M. S. B., Agarie, S., Nomura, M., Fukayama, H., Tsuchida, H., Ono, K., Hirose, S., Toki, S., Miyao, M., and Matsuoka, M. (1999) High-level expression of maize phosphoenolpyruvate carboxylase in transgenic rice plants. *Nature Biotechnology* **17**, 76–80.

Miyagawa, Y., Tamoi, M., and Shigeoka, S. (2001) Overexpression of a cyanobacterial fructose-1,6-sedoheptulose-1,7-bisphosphatase in tobacco enhances photosynthesis and growth. *Nature Biotechnology* **19**, 965–969.

Raines, C. R. (2006) Transgenic approaches to manipulate the environmental responses of the C3 carbon fixation cycle. *Plant, Cell and Environment* **29**, 331–339.

Robson, R. H., McCormac, C., Irvine, S., and Smith, H. (1996) Genetic engineering of harvest index in tobacco through overexpression of a phytochrome gene. *Nature Biotechnology* **14**, 995–998.

Thiele, A., Herold, M., Lenk, I., Quail, H., and Gatz, C. (1999) Heterologous expression of *Arabidopsis* phytochrome B in transgenic potato influences photosynthetic performance and tuber development. *Plant Physiology* **120**, 73–81.

Enhancement of protein content

Bagga, S., Potenza, C., Ross, J., Martin, M. N., Leustek, T., and Sengupta-Gopalan, C. (2005) A transgene for high methionine protein is posttranscriptionally regulated by methionine. *In vitro Cellular and Developmental Biology—Plant* **41**, 731–741.

Herbers, K. and Sonnewald, U. (1999) Production of new/modified proteins in transgenic plants. *Current Opinion in Biotechnology* **10**, 163–168.

Herman, E. M. and Larkins, B. A. (1999) Protein storage bodies and vacuoles. *Plant Cell* **11**, 601–614.

Katsube, T., Kurisaka, N., Ogawa, M., Maruyama, N., Ohtsuka, R., Utsumi, S., and Takaiwa, F. (1999) Accumulation of soybean glycinin and its assembly with glutelins in rice. *Plant Physiology* **120**, 1063–1073.

Tabe, L. and Higgins, T. J. V. (1998) Engineering plant protein composition for improved nutrition. *Trends in Plant Science* **3**, 282–286.

Web-links

Florigene website, web-link 10.1: www.florigene.com

Golden Rice Humanitarian Board, web-link 10.2: www.goldenrice.org/index.html

HarvestPlus—Breeding Crops for Better Nutrition, web-link 10.3: www.harvestplus.org/index.html

Norman Borlaug, a listing of the best and most comprehensive online resources pertaining to Norman Borlaug, web-link 10.4: www.ideachannel.com/Borlaug.htm

11

Molecular farming

INTRODUCTION

Molecular farming (sometimes known as molecular *pharming*) is a term coined to describe the application of molecular-biological techniques to the synthesis of commercial products in plants. The term can be applied to a broad spectrum of activities, from the enhanced production of products that are already extracted from plants through to the manufacture of compounds that are completely novel to plants. A wide range of products have already been identified as likely targets for molecular farming, including a variety of carbohydrates, fats, and proteins, as well as secondary products. This chapter will consider examples of some of the major classes of molecules produced in transgenic plants. The first part of this chapter will look at some examples of molecular farming as applied to carbohydrates and oils. In the second part of the chapter, examples of molecular farming in which it is the transgenic protein itself that is the product will be looked at. Many of the proteins being 'farmed' in plants are antibodies, vaccines, or biopharmaceuticals aimed at improving human and animal health, and it is in the area of molecular farming that many of the most exciting and potentially beneficial developments in plant biotechnology are taking place. At the end of the chapter, we will also review some of the economic reasons why molecular farming might, in some cases, be an attractive alternative to current forms of manufacture of these compounds.

Carbohydrates and lipids

In this first section, we will look at some examples (see Table 11.1 for a summary) of the production of novel or modified carbohydrates and oils (and their derivatives) in plants. Many of these examples have multiple applications (as modified food- or feedstuffs, and industrially) that will be highlighted here.

Carbohydrate production

Plants produce a range of commercially valuable carbohydrates. The two most abundant carbohydrates are cellulose (used for textile fibres from crops such as cotton and flax, for paper-making from trees, and for a range of industrial products such as paints and polymers) and starch (used for food, feed, and industrial purposes). Some biotechnological effort is directed towards improving the yield and quality of these

Table 11.1 **Major targets for carbohydrate and lipid molecular farming**

Compound	Origin of genes	Application	Transgenic plant
Carbohydrates			
Amylose-free starch	Potato	Food, industrial	Potato
High-amylose starch	Potato	Food, industrial	Potato
Cyclodextrins	*Klebsiella pneumoniae*	Food, pharmaceutical	Potato
Fructans	*Bacillus subtilis*	Industrial	Tobacco, potato, maize
	Jerusalem artichoke	Food	Sugar beet
Trehalose	*Escherichia coli*	Food	Tobacco
Lipids			
Medium-chain fatty acids	California bay tree	Food, detergent, industrial	Oilseed rape
Saturated fatty acids	*Brassica rapa*	Food	Oilseed rape
Mono-unsaturated fatty acids	Rat	Food	Tobacco
Polyhydroxybutyrate	*Alcaligenes eutrophus*	Biodegradable plastics	*Arabidopsis*, soybean, oilseed rape, cotton

bulk carbohydrates. However, there are a number of other carbohydrates that it could be attractive to produce in transgenic plants, including oligofructans, cyclodextrins, and trehalose (Figure 11.1).

● Carbohydrates are synthesized and stored in a variety of cellular compartments.

Figure 11.2 indicates the biosynthetic pathways and cellular locations of these compounds in plants. One obvious point is that carbohydrates are synthesized and stored in a number of different cellular compartments. Thus, starch and its derivatives are synthesized in plastids, whereas the fructans, another important class of storage carbohydrates in plants, are synthesized and stored in the vacuole. Sugars and sugar alcohols are synthesized in the cytosol and accumulate throughout the cell, often in response to abiotic stress (see Chapter 9). Note also that Figure 11.2 shows a stylized cell, and that in practice different carbohydrates tend to accumulate in different tissues and organs. For example, starch is stored transiently in the leaves of cereals, and starch reserves accumulate in the amyloplasts of the grain endosperm. However, the fructans are also a significant storage carbohydrate in some cereals, and accumulate in the leaves and stems.

CASE STUDY 11.1 **Starch**

The major starch-producing crops—cereals and potatoes—are already widely grown to produce starch as a chemical feedstock. Approximately 70% of the starch produced in Europe and the USA is used for a variety of industrial purposes, with only 30% used for human consumption and animal feed. In Figure 11.2, the pathway for starch synthesis in chloroplasts is shown, starting with triose phosphate, which is generated in the Calvin cycle. This is subsequently converted to hexose phosphate and then ADP–glucose, which is the substrate for starch synthesis. In the case of amyloplasts, which are plastids specialized for starch accumulation, the hexose molecules are transported directly into the amyloplast. In the cereal endosperm, ADP–glucose may be synthesized in the cytosol and transported directly into the plastids. Starch synthesis

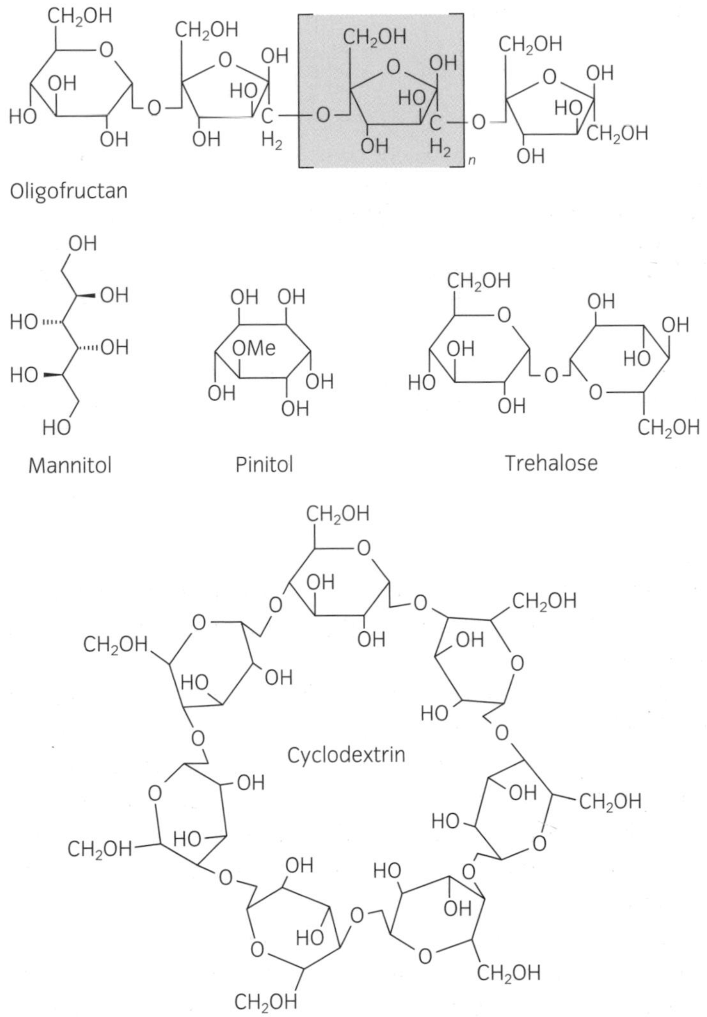

Figure 11.1 **Target carbohydrate products for molecular farming. The structure of some potential targets for molecular farming are shown. Oligofructans are polymers produced by the sequential addition of fructose on to an initial sucrose molecule. The sugar alcohols mannitol and pinitol were discussed in Chapter 9 with reference to their roles as osmolytes, and the potential for producing them in crops to enhance stress tolerance. Trehalose is a disaccharide with useful properties as a food preservative. Cyclodextrins are rings of six, seven, or eight glucose residues. The seven-sided ring shown here has a number of pharmaceutical applications for the solubilization of small hydrophobic molecules.**

involves two major classes of enzymes. Starch synthase catalyses the addition of glucose residues from ADP–glucose to the non-reducing end of the growing chain, forming $\alpha(1\rightarrow4)$ links. One form of starch molecule, amylose, is composed entirely of unbranched chains of glucose of about 1000 residues long. There are several starch synthase isoenzymes, some of which are soluble in the plastid stroma, and others that are bound to the insoluble starch granules (granule-bound starch synthase, or GBSS).

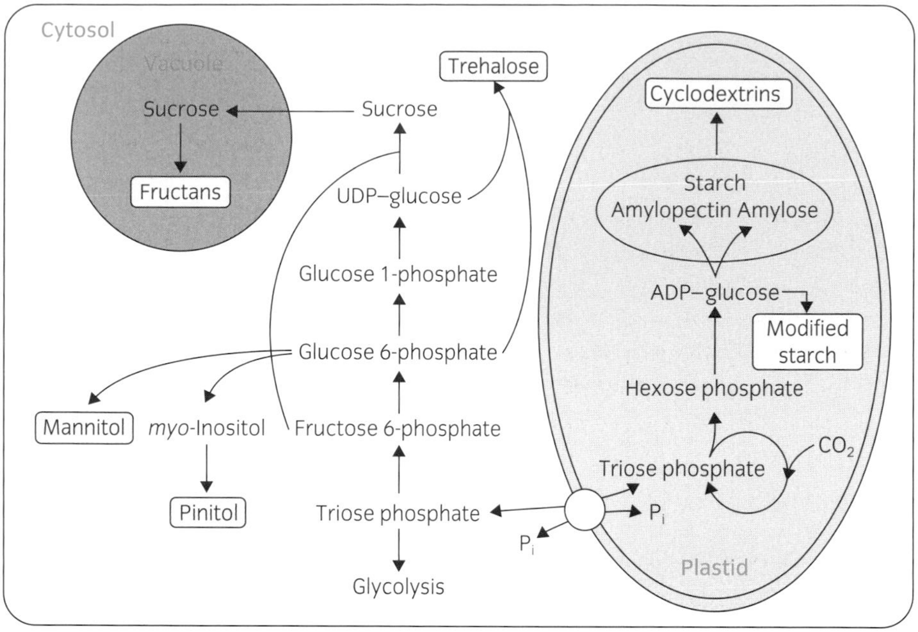

Figure 11.2 Metabolic pathways for the biosynthesis of carbohydrate products for molecular farming. The biosynthetic pathways for the production of the compounds shown in Figure 11.1 are shown schematically in this stylized cell. Starting from the pool of hexose phosphate sugars in the cytoplasm, sucrose is synthesized via UDP–glucose and transported into the vacuole, where it may be stored as fructans. UDP–glucose and glucose 6-phosphate are used to form trehalose in the cytoplasm, and glucose 6-phosphate is also the precursor of the sugar alcohols. In leaf cells, triose phosphates are transported into and out of the chloroplast, where they can be converted to hexose phosphate and then ADP–glucose for starch synthesis. Cyclodextrins are not found in plants, but starch is a substrate for some bacterial biosynthetic enzymes. (Redrawn with permission from Goddijn and Pen (1995).)

- Starch exists in two forms: amylose and amylopectin.

The other main class of starch biosynthetic enzymes is the starch-branching enzymes (SBEs), which create α(1→6) branches in the starch molecule. These branched starch molecules are called amylopectin and typically contain 10^4–10^5 glucose residues. The branches are actually formed by cleaving an α(1→4) linkage in the chain and joining the reducing end of the cut fragment to the C-6 position of a glucose residue some 20 residues downstream. This mode of action ensures that the branches in amylopectin are not random, but occur roughly every 20 glucose residues.

- Starch can be modified in three main ways: increased amount, high amylose, and no amylose.

The simple diagram of starch biosynthesis shown in Figure 11.2 illustrates some important principles in the genetic engineering of metabolic pathways. There is a single step from a pool of common primary metabolites (hexose phosphate sugars) to the immediate precursor of starch, ADP–glucose, catalysed by the enzyme ADP–glucose pyrophosphorylase. This precursor molecule is specific to the starch pathway, so this step commits hexose sugars to starch biosynthesis. Starch comprises amylose and amylopectin, so effectively there is a branch in the biosynthetic pathway leading to one or the other form of starch. Thus, this pathway could be manipulated at three points, in a number of different ways. The production of ADP–glucose could be altered, which would affect the overall level of starch biosynthesis. Thus, increasing the amount or activity of the enzyme ADP–glucose pyrophosphorylase should increase the amount of starch produced. On the other hand, manipulation of one of the branches in the pathway would affect the ratio of amylose to amylopectin.

The proportion of amylose:amylopectin is normally about 20–30% amylose to 70–80% amylopectin, and it is this ratio that has the greatest influence on the physicochemical properties of the starch. For some applications, it would be advantageous to increase the proportion of amylopectin. For example, many of the uses of starch in food production involve the formation of a gel after heating the starch in water and cooling. Amylose molecules tend to aggregate and crystallize on cooling, whereas amylopectin gels are more stable and generably more desirable for food processing. The *waxy* mutants of maize and other cereals have been studied for many years. The waxy phenotype is characterized by a starch with no amylose that gelatinizes easily, giving clear pastes that are ideal for thickening and stabilizing food products. It is now known that the *waxy* gene encodes the GBSS protein, and waxy starches have been produced in transgenic potatoes by antisense inhibition of GBSS1, and in sweet potato by sense suppression of the same gene.

A starch with high freeze–thaw stability has been produced in potato by antisense inhibition not only of GBSS, but also of two of the three soluble starch synthase isoforms, SSII and SSIII. With these three genes down-regulated, a waxy starch with short-chain amylopectin is produced with enhanced freeze–thaw stability suitable for frozen food products. Highly branched, short-chain amylopectins have also been produced in rice by overexpression of SBEIIb in an SBEIIb-defective mutant.

In contrast, high-amylose starches with limited branching have high gelling properties, which can be used for confectionary, and would also be a valuable feedstock for industrial purposes. In potatoes, there are two isoforms of SBE (SBE A and B). The engineering of a high-amylose starch required the antisense inhibition of both SBE A and SBE B to achieve greater than 60% amylose levels. An alternative strategy by expressing single-domain antibodies against SBE A along with antisense inhibition of SBE B has been used to achieve the same end. The resulting high-amylose starch also had much higher phosphate levels and could prove to be useful for food and industrial applications (e.g. paper manufacture).

Cyclodextrins from starch

Whereas starch itself is a bulk feedstock for industry, the potential exists for carrying out further biotransformations on the starch in the plant, rather than by chemical or fermentation processes after extraction. One type of high-value product that could be made from starch is the cyclodextrins. These compounds are typically six-, seven-, or eight-membered rings comprising glucopyranose subunits attached in $\alpha(1\rightarrow4)$ linkages (Figure 11.1). These compounds are normally produced by bacterial fermentation of maize starch, particularly for pharmaceutical applications. The structure of the cyclodextrins is such that they form a cone-shaped ring, with the hydrophilic residues on the exterior and a hydrophobic pocket in the centre of the ring. The seven-membered ring has the ideal dimensions to form a pocket for small hydrophobic compounds. Thus, in a concentrated suspension, the cyclodextrins will effectively solubilize hydrophobic pharmaceuticals such as steroids. At lower concentrations, for example after injection into the bloodstream, the therapeutic agent is released.

● Cyclodextrins are useful for solubilization of hydrophobic pharmaceuticals.

Having been identified as a suitable target for molecular farming, it is striking that only one report, in 1991, of an attempt to produce cyclodextrins has been published. A bacterial cyclodextrin glycosyltransferase gene from *Klebsiella pneumoniae* was fused to a plastid-targeting sequence and placed under the control of the promoter from the patatin gene. Patatin is a protein that accumulates in potato tubers, and the promoter directs high levels of expression in the tubers of transgenic potatoes. However, transformation of potatoes with this construct resulted in very little conversion (0.001–0.01%) of starch to cyclodextrins. It was concluded that the insoluble starch granules may have been inaccessible to the bacterial enzyme, or that the enzyme became trapped in the growing granule. Whatever the reason, no subsequent attempts to produce this commodity in plants have appeared in the literature.

CASE STUDY 11.2 **Polyfructans**

● Polyfructans are fructose polymers.

Another example of a carbohydrate targeted for production in transgenic plants is the polyfructans. These compounds are soluble polymers of fructose that are synthesized and stored in the vacuole. They have a typical structure of glucose–fructose–(fructose)$_n$ (G–F–F$_n$) as shown in Figure 11.3. The use of fructans as a carbohydrate reserve is widespread throughout the plant kingdom. As with glucose polymers, there are different glycosidic linkages possible between the fructose residues, giving different straight and branched polymers. The inulins are the major storage carbohydrate found in bulbs such as onion, and storage roots such as chicory and Jerusalem artichoke, and are formed by (1→2β) linkages. Levans are widespread in the leaves and stems of grasses, including major cereal crops such as wheat, and comprise (6→2β) linkages. Graminae-type fructans found, for example, in grasses, are a mixed type and have both (1→2β) and (6→2β) linkages.

The biosynthetic pathway of fructans in plants is a two-stage process. The first step involves the transfer of fructose from a donor sucrose molecule to an acceptor sucrose molecule to form kestose by the enzyme sucrose–sucrose fructosyltransferase (SST):

$$\text{G–F} + \text{G–F} \rightarrow \text{G–F–F} + \text{G}$$

Here, G is glucose and F is fructose. In the second step, the kestose (GFF or GF$_2$) acts as the fructose donor to the growing fructan chain, via fructan–fructan fructosyltransferase (FFT) activity, and a sucrose molecule is recycled.

$$\text{G–F–F}_n + \text{G–F–F} \rightarrow \text{G–F–F}_{n+1} + \text{G–F}$$

In certain bacteria, such as *Bacillus subtilis*, very high-molecular-mass levans are produced by a single reaction in which sucrose acts directly as the fructose donor to the growing chain (sucrose–fructan fructosyltransferase, SFT). Note that in all cases sucrose is the initial acceptor molecule of the chain, so the first sugar in the fructan chain is always glucose. However, for each

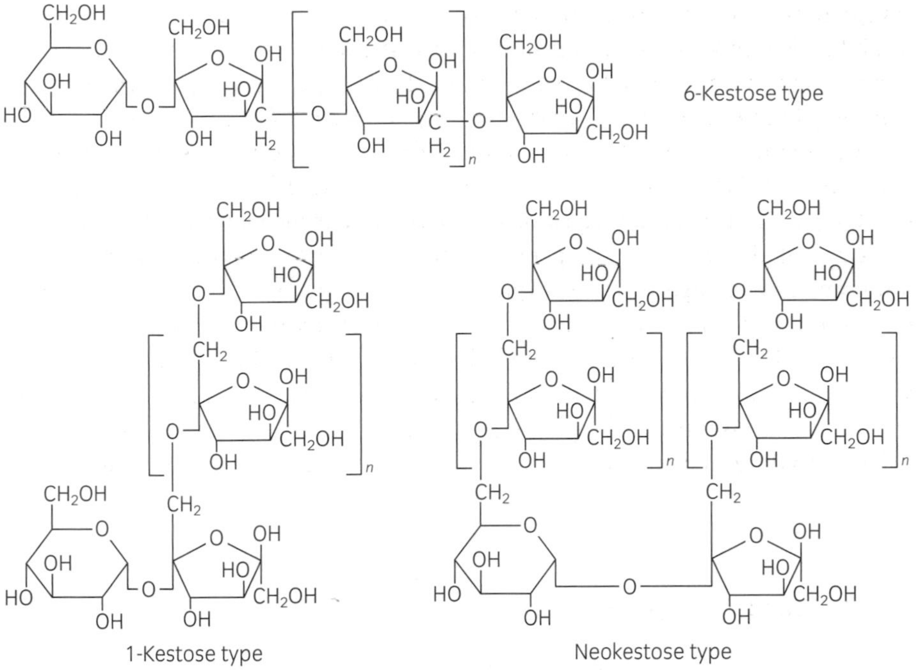

Figure 11.3 Structure and synthesis of polyfructans. The structures of three types of polyfructan are shown. The levan-type fructans of the 6-kestose class are formed by fructose-(6→2β)-fructose linkages. These fructans are typically found in grasses, with $n \approx 200$. (In bacteria, levans are much larger polymers than this). The 1-kestose type of fructans are also called inulins, and are formed by fructose-(1→2β)-fructose linkages. These are typically found in bulbs and the storage roots of plants such as Jerusalem artichoke and chicory, with $n<50$. Fructans of the neokestose type, with 1-kestose chains attached to both the glucose and fructose residues of the initial sucrose moiety, are the smallest fructans, with only between five and 10 fructose residues.

Figure 11.4 Construct for the synthesis of fructans in the vacuole of transgenic plants. A construct of the type used to transform tobacco in the first experiments to engineer fructan synthesis is shown. The bacterial *sacB* gene is flanked by an N-terminal vacuolar targeting sequence from yeast carboxypeptidase (*cpy*). Note the tandem-repeated cauliflower mosaic virus 35S promoter and alfalfa mosaic virus (AlMV) translational enhancer to obtain high levels of expression.

remaining fructose residue added to the chain a glucose residue is released, which in plants is transported back out of the vacuole into the cytosol.

A number of transgenic plants producing polyfructans have been developed. In earlier experiments, the *sacB* gene of *B. subtilis* or *Bacillus amyloliquifaciens*, which encodes a levansucrase catalysing a 6→2β linkage, was transformed into maize, tobacco, potato, and sugar beet. The construct used to express *sacB* is shown in Figure 11.4, demonstrating how the *sacB* gene was modified with a vacuolar targeting sequence from the yeast carboxypeptidase gene (*cpy*). The features of vacuolar signal sequences (see also Chapter 1) are described in Box 11.1.

● Vacuolar targeting sequences direct transgenic enzymes to the vacuole.

BOX 11.1 *Vacuolar targeting sequences*

There have been several examples of protein targeting to the chloroplast in previous chapters, but this is probably the first in which the final destination of the transgene product is the vacuole. A number of proteins are known to be targeted to the vacuole, but the targeting sequence appears to vary considerably among proteins. Examples of vacuolar targeting sequences are shown below.

NTPP (N-terminal) sequences

Sweet potato sporamin	HSRFNPIRLPTTHEPA
Barley aleurain	SSSSFADSNPIRPVTDRAASTLE

CTPP (C-terminal) sequences

Barley lectin	VFAEAIAANSTLVAE
Tobacco chitinase	NGLLVDTM
Tobacco β-1,3-glucanase	VSGGVWDSSVQTNATASLVSQM
Tobacco AP24	QAHPNFPLEMPGSDEVAK

Vacuolar targeting sequences, indicating the variation in location and amino acid sequence in different proteins. CTPP, C-terminal propeptide; NTPP, N-terminal propeptide; SP, signal peptide.

The targeting sequence is intended to direct the enzyme to the vacuole, where sucrose is stored and where polyfructan synthesis normally occurs in those plants that store this carbohydrate reserve.

In tobacco, expression of the *sacB* gene was responsible for the accumulation of significant amounts of fructans. The transgenic tobacco plants were found to be more tolerant to drought stress induced by growth in medium containing polyethylene glycol to reduce the water concentration. In Chapter 9, the production of osmoprotectants to provide stress tolerance against the effects of water deficit was described, and this provides another example of this principle in action. More recently, expression of the *sacB* gene in sugar beet has also been shown to improve tolerance to polyethylene glycol-mediated drought stress.

Polyfructans have also been produced in potatoes using the same vacuolar-targeted *sacB* gene, but under the control of the patatin gene promoter (see the earlier example of cyclodextrins). In potato, the formation of a new sink, by diversion of sucrose away from starch accumulation in the tuber and towards fructans in the vacuole, has provided a useful model system for studying the regulation of sucrose metabolism and the partitioning of assimilates. In subsequent work, the levansucrase gene from *Erwinia amylovora*, which was fused to different targeting sequences (apoplast, vacuole, and cytosol), was used to transform starch-deficient potatoes. The apoplast- and vacuole-targeted enzymes produced fructan to levels of 12 and 19% of tuber dry weight, respectively. However, the fructan did not behave as a new sink for increased allocation of carbohydrates in the starch-deficient tubers, and tuber yield was actually further decreased.

In maize, a *B. amyloliquifaciens sacB* gene was expressed in the endosperm under the control of a zein promoter (zein is a maize seed-storage protein). The enzyme was targeted to the vacuole of the endosperm cells using two different vacuolar targeting sequences. In one construct, the sweet potato sporamin signal peptide and vacuolar targeting sequences were fused to the N-terminal end of the enzyme (see Box 11.1). In the other, the barley lectin signal peptide was fused to the N-terminal end and the barley lectin vacuolar targeting sequence was fused to the C-terminal end of the enzyme. Seed from transgenic plants containing these constructs appeared no different from the controls. Fructan accumulated to 7–9 mg g^{-1} of seed in several different lines from either construct. This level could be increased by up to eight- to ninefold when the transgene was crossed into maize starch-mutant lines, which accumulate much higher levels of sucrose. On the other hand, expression of a *sacB* construct without vacuolar targeting produced kernels that were clearly distinguishable from the controls in terms of their small size and dry mass, and their darker colour. These factors are all indicative of very low starch accumulation, indicating a competition for sucrose in the cytoplasm between the fructosyl transferase and the starch biosynthetic pathway.

Oligofructans are nutraceuticals; long-chain fructans have industrial applications.

Plant genes encoding inulin biosynthetic enzymes have also been used for genetic manipulation. Short oligofructans (GF_2, GF_3, and GF_4) have been produced in sugar beet using a gene encoding the 1-SST (sucrose–sucrose fructosyltransferase) enzyme from Jerusalem artichoke. This enzyme catalyses the production of not only GF_2 (kestose) but also GF_3 and GF_4. It was found that the sugar stored in the storage root of sugar beet was almost totally converted into these oligofructans. This transgenic sugar beet has therefore been renamed the fructan beet. Interestingly, there is no obvious change to the phenotype or growth rate of the fructan beet storage root. The fructan beet has direct applications in the nutraceuticals market. Short-chain fructans are almost as sweet as sucrose, and can be used as substitute sugars in foods and drinks. Fructans are not digested in the gut, and can therefore be marketed as a low-calorie sweetener. The added health benefits are that they encourage the growth of beneficial gut flora (*Lactobacillus* and *Bifidus* species) and cause acidification of the digestive tract, which inhibits bacteria such as *Escherichia coli*. In the future, it is proposed that longer-chain fructans produced in sugar beet could be used as base materials for glues, textile coatings, and polymers.

Larger inulin molecules have been produced in potato tubers transformed with the 1-SST and 1-FFT (fructan–fructan fructosyltransferase) genes of globe artichoke. Globe artichoke produces larger inulins (degree of polymerization, DP, up to 200 subunits) than Jerusalem artichoke (DP≈60). The full size spectrum of molecule sizes naturally occurring in globe artichoke roots was obtained in the potato tubers. It has been possible to produce inulin of the neoseries (as well as the normal linear inulin) in chicory by transformation with the onion 6G-fructosyltransferase gene. Chicory inulin has also been modified by transformation with the barley 6-SFT (sucrose–fructan fructosyltransferase) gene to produce branched fructans of the graminan type. This gene encodes the enzyme sucrose:fructan 6-fructosyltransferase, which catalyses the formation and extension of the β-2,6-linked fructans typical of grasses. On the other hand, introduction of the barley gene into tobacco (which is unable to produce fructans) caused the accumulation of unbranched levan-type fructans.

The production of fructans in transgenic plants has been critically reviewed by Cairns (2003). His calculations indicate that although significant amounts of fructan may accumulate in transgenic plants over weeks or months, the rate of synthesis represents a miniscule fraction (0.00005–0.08%) of the primary carbohydrate flux. Far from indicating a dramatic modification of carbon partitioning in the plant, this represents a 'quantitatively minor leakage' into fructan synthesis. Cairns (2003) considers a number of possible explanations for this low fructan accumulation, including codon usage (of bacterial genes), low transgene mRNA concentration, low expression of enzyme protein, low enzyme activity, kinetic properties of transgenic enzymes inappropriate for host environment, product hydrolysis, and fructan toxicity. One conclusion is that correct vacuolar targeting is rarely checked, and that mis-localization of the transgene product to the cell perimeter or endomembrane system could explain the low conversion to fructan.

Trehalose

Trehalose is a valuable food-processing commodity.

Trehalose is produced in some plants and microorganisms (e.g. yeasts), often in response to osmotic stress (see Figure 11.1). It is therefore another potential target for the genetic manipulation of tolerance to abiotic stresses that create water deficit (see Chapter 9). However, it is also a valuable commodity for food processing, dehydration, and flavour retention. Genes for trehalose synthesis from yeast and *E. coli* have been introduced into tobacco, potato, and rice, but the primary purpose of these experiments was to manipulate drought tolerance (Chapter 9) rather than to manufacture bulk quantities of the sugar.

Metabolic engineering of lipids

The next class of compounds that we will explore as targets for molecular farming is the lipids and their derivatives. As with the carbohydrates, lipids are already produced in large quantities from major crops such as oilseed rape (canola), soybean, and maize, for industrial as well as food purposes. Molecular farming will therefore have a role in the same broad areas as shown for carbohydrates, that is:

- improvement of existing lipid products;
- engineering of novel lipid products.

Improvement of plant oils

In daily life, our closest association with plant oils comes by way of their rapid displacement of animal fats as cooking oils. This association alone should indicate that oils are already extracted from a number of different types of crop. Oilseed rape, or canola, and soybean are major sources of oil that are often unspecified, but sunflower oil, corn oil, and olive oil are often identified because of the improved characteristics of these oils for particular purposes. Plant oils are also used in a wide range of processed foods, and for animal feeds. They also have an increasing number of applications in industry for the manufacture of soaps and detergents, lubricants, and biofuel. However, the non-food applications currently comprise only about 10% of the total vegetable oils produced. One of the reasons for this relatively low level of industrial use is the complex and variable nature of plant oils, which generally makes them less suitable for oleochemical applications than cheaper alternatives from mineral oils. A major long-term goal of this sector is therefore the improvement of plant oils for industrial applications, since they must eventually replace non-renewable, petroleum-based products.

Storage oils that are used as carbon/energy reserves in plant seeds and some fruits (such as olives and avocados) are generally glycerol esters of fatty acids, called triacylglycerols (TAGs; Figure 11.5). The differences between oils from different sources are largely due to the particular proportions of different fatty acids comprising TAGs (Table 11.2). A few of the common fatty acids are also shown in Figure 11.5. These differ in their length, and the number and positions of double bonds. Saturated fats contain no double bonds: the more double bonds there are, the greater the degree of unsaturation.

- Plants store oils as triacylglycerols.

The site of *de novo* fatty acid biosynthesis is exclusively in the stroma of the plastid (see Box 5.1), whereas most modification of the fatty acids occurs in the cytoplasm and on the endoplasmic reticulum (ER). Assembly of TAGs takes place in the membrane of the ER. The TAGs are stored in oil bodies, which are essentially oil droplets surrounded by a lipid monolayer formed from one half of the ER membrane lipid bilayer. The oil bodies of seeds (and other tissues that undergo extreme desiccation) also contain proteins, called oleosins, in the lipid monolayer (oleosins are discussed later in this chapter).

- Oils differ in the length and degree of saturation of their fatty acids.

The first committed step in fatty acid biosynthesis in the plastid is the formation of malonyl-coenzyme A (CoA) from acetyl-CoA, in an ATP-dependent reaction catalysed by acetyl-CoA carboxylase (ACCase). The CoA residue is then exchanged for the acyl carrier protein (ACP). Acetyl-CoA is then condensed to the malonyl-ACP,

- Fatty acids are synthesized in the plastid but modified in the cytoplasm and ER.

Table 11.2 The fatty acid composition of major oil crops

Fatty acid	Composition (%)				
	Soya oil	Palm oil	Rape-seed oil (LEAR)	Sunflower oil	Peanut oil
16:0	11	42	4	5	10
18:0	3	5	1	1	3
18:1	22	41	60	15	50
18:2	55	10	20	79	30
18:3	8	0	9	0	0
20:1	0	0	2	0	3
Others	1	0	2	0	0

LEAR, low-erucic-acid rape.
Percentages may not add up to 100% due to rounding and non-inclusion of other constituents.

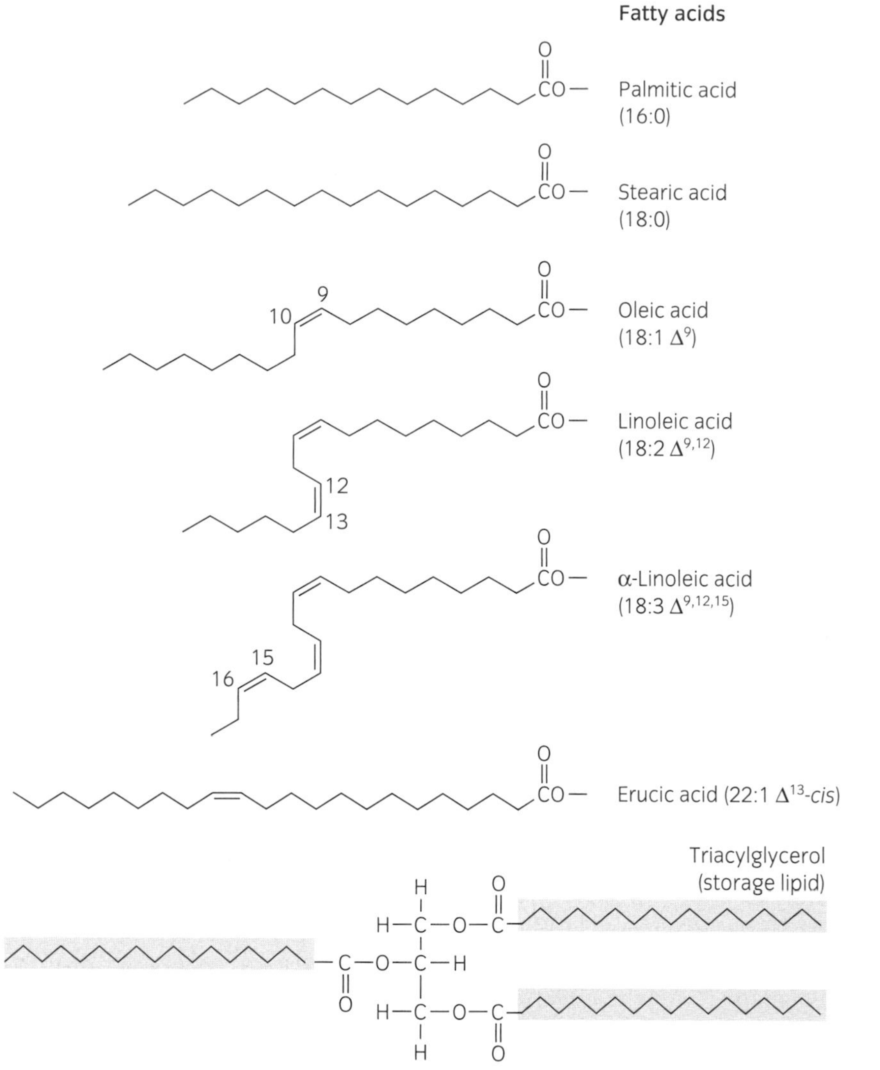

Figure 11.5 Fatty acid and triacylglycerol (TAG) structure. The structures of some of the common fatty acids found in plant oils are shown. The standard nomenclature in brackets denotes the number of carbon atoms, the number of double bonds and the position of the double bonds, counting from the carboxyl-, or Δ-end. Thus, linoleic acid (18:2 $\Delta^{9,12}$) has 18 carbons and two double bonds, one in position C_9–C_{10} and the other at C_{12}–C_{13}. The structure of a TAG storage lipid is also shown, indicating the attachment of three fatty acid chains to a glycerol molecule. Note that apart from the glycerol residue, these molecules are completely hydrophobic: compare this structure with the phospholipids of biological membranes, where one of the fatty acids is replaced by a polar head group.

forming acetoacetate-ACP, and liberating CO_2. Three subsequent steps (reduction of the carbonyl group, removal of water to form a double bond, and reduction of the double bond) produce an acyl-ACP, which is two carbons longer than the original. The entire sequence of elongation reactions from the initial binding to ACP is catalysed by a fatty acid synthase multienzyme complex. The elongated fatty acid chain is then transferred to another ACP protein and this is then condensed to a new malonyl-ACP. Thus, the formation of fatty acids occurs by the stepwise addition of two-carbon units at the carboxyl end, hence the even numbers of carbon atoms in the fatty acids shown in Figure 11.5. In many of the oil-producing crops, this process stops at the 16-carbon stage, and the palmitoyl-ACP (16:0) is elongated to stearoyl-ACP (18:0) by a specific synthase. (The standard nomenclature of fatty acids indicates the number of carbon atoms and the number of double bonds.) At this stage, desaturation by a soluble Δ^9C-stearoyl desaturase in the plastid stroma converts most of the stearoyl-ACP to oleoyl-ACP (18:1 Δ^9). (The Δ^9 indicates that the double bond starts at the ninth carbon counting from the carboxylic acid front end of the molecule.)

After termination of synthesis in the plastid, the fatty acids (mainly palmitic, stearic, and oleic acids) are released from the ACP and exported to the cytoplasm, where they are converted to acyl-CoA esters. TAGs are formed by stepwise acylation of glycerol 3-phosphate in the ER membrane. Further modifications to the fatty acids normally occur after they have been attached to various glycerophosphatides. Additional double bonds may be added (typically to Δ^{12}, Δ^{15}, and Δ^6 as well as Δ^9). Other modifications may include hydroxylation by the addition of water across a double bond.

Production of shorter-chain fatty acids

The oils produced by the major oil crops of the world consist mainly of palmitic, stearic, oleic, linoleic, and linolenic acids, which are all C_{16} or C_{18} in length (Table 11.2). Coconut and palm kernel oils are largely C_8–C_{14}, and lauric acid (12:0), in particular, is an important raw material for the production of soaps, cosmetics, and detergents (think of the widespread use of sodium lauryl sulphate, or SDS, in these products). The synthesis of fatty acids at a particular length is terminated by the hydrolysis of the acyl-ACP by a thioesterase. The California bay tree contains a very high proportion of lauric acid in its seeds, and an acyl-ACP thioesterase that specifically hydrolyses lauroyl-ACP has been cloned from this source. The introduction of this gene into oilseed rape causes fatty acid synthesis to terminate at the 12:0 stage, and a high proportion of lauric acid (58% of total fatty acids) to accumulate in the seed oil. Most importantly, field tests show that these plants grow normally and produce normal yields. Attempts to produce caprylic acid (8:0) in oilseed rape transformed with the acyl-ACP thioesterase gene from *Cuphea hookeriana* were less successful, with only 12% of fatty acids represented by caprylic acid. It is thought that this could be due to the low availability of short acyl-ACP pools, and further work indicated that crossing plants expressing condensing enzymes (3-ketoacyl-ACP synthase, KAS) from *Cuphea* (with unique specificity for 6:0- (caproic) and 8:0-acyl-ACPs) with plants expressing a *Cuphea* acyl-ACP thioesterase gene improved the proportion of short fatty acids.

- Most major crop oils are C_{16} or C_{18}.
- $C_{12:0}$ (lauric acid) is widely used in soap and detergents.

Production of longer-chain fatty acids

One of the important targets is to produce fatty acids longer than C_{18} for use as industrial oils. In oilseed rape and other *Brassica* species, there is a two-step elongation pathway from oleoyl-CoA (18:1 Δ^9) to erucoyl-CoA (22:1 Δ^{13}), such that erucic acid is

- Erucic acid is a valuable industrial oleochemical.

● Identity-preservation systems are required to keep HEAR and LEAR separate.

one of the constituents of brassica oils. However, although erucic acid is valuable as an industrial oleochemical, it is nutritionally unsuitable for human consumption. Conventional breeding has led to the development of two distinct oilseed rape crops: high-erucic-acid rape (HEAR) for industrial purposes, and low-erucic-acid rape (LEAR), with virtually no erucic acid, for food products (see Table 11.2). (NB: the existence of two distinct crops that need to be kept apart means that systems are in place in the seed production, agriculture, and processing industries to ensure minimal cross-contamination, a process called identity preservation. Thus, requirements to keep genetically modified (GM) and non-GM seeds and products separate do not create entirely new demands on this sector.) However, the highest erucic acid content of HEAR is about 50% of the total fatty acids, which makes the cost of separating out and disposing of the other fatty acids uncompetitive with mineral oil sources. Attempts to increase the erucic acid content are already showing promise. Transformation of oilseed rape with a combination of β-ketoacyl-CoA synthase and 22:1 acyl-CoA:lysophosphatidic acid acyltransferase produced 60% erucic acid.

Modification of the degree of saturation

● High-stearic-acid ($C_{18:0}$) oil is an important goal for industrial feedstocks.

As the erucic acid example shows, the ideal products for industrial purposes are as uniform as possible. This requires uniformity of desaturation as well as length. (The same does not necessarily apply to food constituents, where complex mixtures may produce the ideal product. In this case it is consistency, rather than uniformity, that is required.) One of the first attempts in this area was to increase the level of stearic acid (18:0) at the expense of oleic acid (18:1 Δ^9) by inserting an antisense construct of the Δ^9 desaturase gene into oilseed rape. The antisense gene was under the control of the napin gene promoter (napin is a seed-specific protein of brassicas). There was a marked decrease in the amount of the desaturase enzyme, resulting in a decreased formation of oleic acid, and a rise in the proportion of stearic acid from 1–2% up to 40% of total fatty acids. This high stearic acid oil has potential as a cocoa butter substitute.

● High-oleic-acid ($C_{18:1}$) oil is of value for food purposes.

On the other hand, there is considerable potential in the production of a very high (>90%) oleic acid oil for food purposes and as a uniform oleochemical feed stock. Conventional breeding has produced mutant oilseed rape lines with an oleic acid content close to 80%, but attempts to increase this level have produced plants with poor cold tolerance, presumably as a result of the lack of unsaturated fatty acids in the cellular membranes (see Chapter 9). Antisense repression and co-suppression of the Δ^{12} desaturase in the seeds of oilseed rape have, however, facilitated the raising of oleic acid levels to 87–88%, without affecting cold tolerance. Similar experiments in soybean increased the oleic acid level from 22 to 79%.

Production of rare fatty acids

● Petroselenic acid ($C_{18:1}$; coriander oil) is a high-value oil.

There are 210 known types of fatty acid produced in plants, but most of these are not found in the major crop plants and would be difficult to produce commercially in their host plants. However, it is possible to transfer genes from these plants to manipulate the profile of fatty acids produced in the major oil crop plants. One such fatty acid is petroselenic acid (18:1 Δ^6), which is found in certain Umbelliferae, such as coriander, and has potential as a raw material for industry. Oxidation of petroselenic acid by ozone produces lauric acid (12:0), for soaps and detergents, and adipic acid (6:0), which can be used for nylon production. Transformation of tobacco with a coriander acyl-ACP desaturase cDNA led to the production of petroselenic acid in calli to a level of 5% of total fatty acids (see below, in Box 11.3, for a discussion of the economics of the process).

Certain polyunsaturated fatty acids have pharmaceutical or nutraceutical value. These include γ-linolenic acid (18:3 $\Delta^{6,9,12}$) and arachidonic acid (20:4 $\Delta^{5,8,11,14}$), which are essential fatty acids for humans and precursors of eicosanoids (including prostaglandins, leukotrienes, and thromboxanes). Recently, it has become clear that the synthesis of these compounds involves a specific subclass of microsomal desaturases called front-end desaturases. These enzymes insert additional double bonds between existing bonds and the carboxyl front-end of the fatty acid, whereas most desaturases in plants add sequentially towards the methyl end. The transformation of a front-end desaturase from borage (which is one of a few plant species to produce γ-linolenic acid) into tobacco resulted in the accumulation of high levels of Δ^6 unsaturated fatty acids. More recently, oilseed rape transformed with 18:1 Δ^{12}-desaturase produced up to 46% linoleic acid in the seeds, whereas transformation with both 18:1 Δ^{12}-desaturase and 18:2 Δ^6-desaturase produced up to 43% γ-linolenic acid.

● Polyunsaturated fatty acids have pharmaceutical/nutraceutical value.

Other polyunsaturated fatty acids have been produced in transgenic plants. A Δ^5-desaturase was cloned from the filamentous fungus *Mortierella alpina* (which accumulates arachidonic acid) and expression in oilseed rape produced significant levels of polyunsaturated fatty acids. Ricolenic acid (Δ^{12}-hydroxyoleic acid) is produced in castor beans to a level of 90% of the total fatty acids. However, the castor oil crop has a number of problems, including the presence of toxic compounds such as ricin in the residual meal. The synthesis of ricolenic acid involves the direct hydroxylation of oleic acid bound to phosphatidylcholine on the ER membrane. The cDNA for the Δ^{12}-hydroxylase has been cloned from castor bean and could be used to produce castor oil in major oil crops.

Epoxygenases are another useful group of fatty acid-modifying enzymes that present a valuable way of inserting chemically reactive epoxy sites into oils. Epoxygenated fatty acids would be useful for a range of industrial applications as glues, resins, plastics, and other polymers. One class of Δ^{12}-epoxygenases has been cloned from *Crepis* species and shown to be a non-haem, di-iron protein, similar in sequence to a Δ^{12}-acetylenase from the same plant, and with a similar structure to fatty acid Δ^{12}-desaturases. Transformation of *Arabidopsis* with the Δ^{12}-epoxygenase of *Crepis palaestina* under the control of a seed-specific promoter resulted in only approximately 6% epoxygenated lipids (12-epoxy-18:1 Δ^9 and 12-epoxy-18:2 $\Delta^{9,15}$) but this could be improved by co-transformation with a Δ^{12}-desaturase gene. The *Euphorbia lagascae* epoxygenase is a cytochrome P450 enzyme, and expression in tobacco callus and soybean somatic embryos resulted in the accumulation of epoxygenated fatty acids.

CASE STUDY 11.3 **Bioplastics**

One of the most imaginative examples of molecular farming has been the attempt to produce biodegradable plastics (bioplastics) in plants. These compounds are currently produced by microbial fermentation, but a number of experimental studies have been carried out to determine the feasibility of producing them in bulk in plants. The structure of the polyhydroxyalkanoates (PHAs) that form bioplastics is shown in Figure 11.6. The length of the R side chain alters the properties of the plastics, and can vary from no carbons (3-hydroxypropionate), one carbon (3-hydroxybutyrate), or two carbons (3-hydroxyvalerate) up to long carbon chains. Polyhydroxybutyrate (PHB) is the best-characterized PHA and is found as intracellular inclusions in a wide variety of bacteria. In *Alcaligenes eutrophus*, PHB accumulates as a high-molecular-mass polymer up to 80% of the bacterial dry weight.

● PHB biosynthesis involves three steps from acetyl-CoA.

The pathway for PHB synthesis is shown in Figure 11.7. It can be seen that this is a relatively simple three-stage pathway starting from acetyl-CoA. The genes for the three enzymes involved in the pathway (successively the *phaA*, *phaB*, and *phaC* genes) have been cloned from *A. eutrophus*. In the initial experiments, it was recognized that the first step of the pathway, production of acetoacetyl-CoA, occurs in the cytoplasm of plants, at the start of the isoprenoid pathway (see Chapter 10). Thus, only *phaB* and *phaC*, encoding acetocetyl-CoA reductase and PHB synthase, respectively, were transformed into *Arabidopsis*, without targeting sequences (Figure 11.8a). Microscopic observation of the *Arabidopsis* leaves indicated the formation of microbodies of bioplastics accumulating in the cytosol, nucleus, and vacuole. However, the amount of bioplastic was relatively low (20–100 μg g^{-1} fresh weight) and the plants were severely stunted in growth.

Subsequently, all three genes were transformed into *Arabidopsis*, and targeted to the chloroplast (Figure 11.8b). In the first generation of experiments, each gene was separately fused to a sequence encoding the transit peptide plus N-terminal fragment of the ribulose-1,5-bisphosphate carboxylase/oxygenase (Rubisco) small subunit (rbcS) protein, and expression of each construct was directed by the cauliflower mosaic virus 35S promoter. Each construct was transformed into separate *Arabidopsis* plants, and brought together by a series of sexual

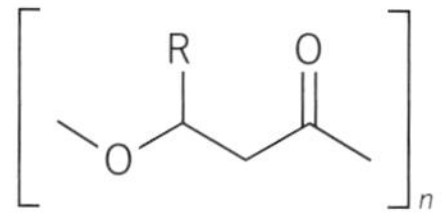

Generic polyhydroxyalkanoate

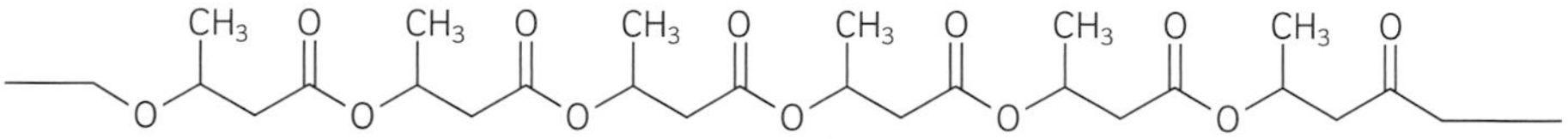

Polyhydroxybutyrate

Figure 11.6 **Chemical structure of the major polyhydroxyalkanoates (PHAs). The repeating subunit of PHAs is shown, along with a short section of a polyhydroxybutyrate (PHB) chain (where R = methyl group). The R group can be anything from no carbons (i.e. H) in hydroxypropionate, to >10. The length of the R-group side chain affects the physical properties of the plastic formed by these polymers.**

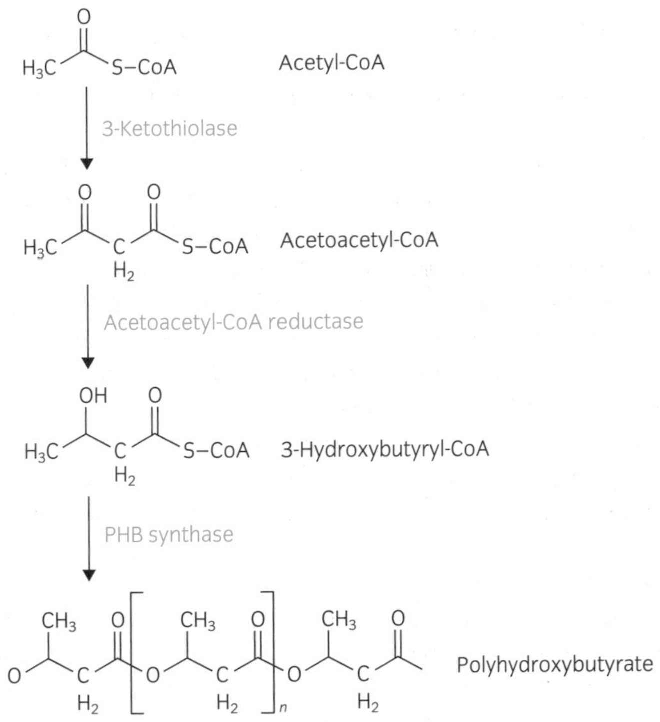

Figure 11.7 **The PHB biosynthetic pathway of bacteria. The biosynthesis of PHB requires only three steps from acetyl-CoA. In the first step, two acetyl-CoA residues are condensed by 3-ketothiolase to form acetoacetyl-CoA. This is reduced by acetoacetyl-CoA reductase to give 3-hydroxybutyryl-CoA. In the final stage, this compound is polymerized by PHB synthase.**

crosses between the individual transformants. In this case, the bioplastics accumulated as 0.2–0.7-μm granules in the plastids, to levels of up to 14% of plant dry weight, and there was no observable effect on growth or fertility.

Increased PHB production in *Arabidopsis* has been achieved using a triple construct, so that all three genes are transferred into the plant in a single transformation event. A rapid gas chromatography–mass spectrometry (GC–MS) procedure was used to screen a large number of transgenic plants, permitting the selection of plants accumulating PHB in the leaf chloroplasts to more than 4% of their fresh weight (40% dry weight). However, the high-producing lines showed stunted growth and loss of fertility. Interestingly, the production of PHB did not affect fatty acid accumulation or composition, but there was a significant impact on the levels of various organic acids, amino acids, sugars, and sugar alcohols.

Although *Arabidopsis* is a valuable model species for these analyses, some research has been focused on the potential for the commercial production of plastics in oil crops, effectively diverting the pool of acetyl-CoA away from fatty acid biosynthesis (for oil-body production, see above) towards bioplastic production. A team from Monsanto used the three genes for the PHB biosynthetic pathway from the bacterium *Ralstonia eutropha* and fused each one to a seed-specific promoter (from *Lesquerella fendleri* oleate Δ^{12}-hydroxylase). These were transferred into a single, multigene vector, which was used to transform oilseed rape. PHB was found

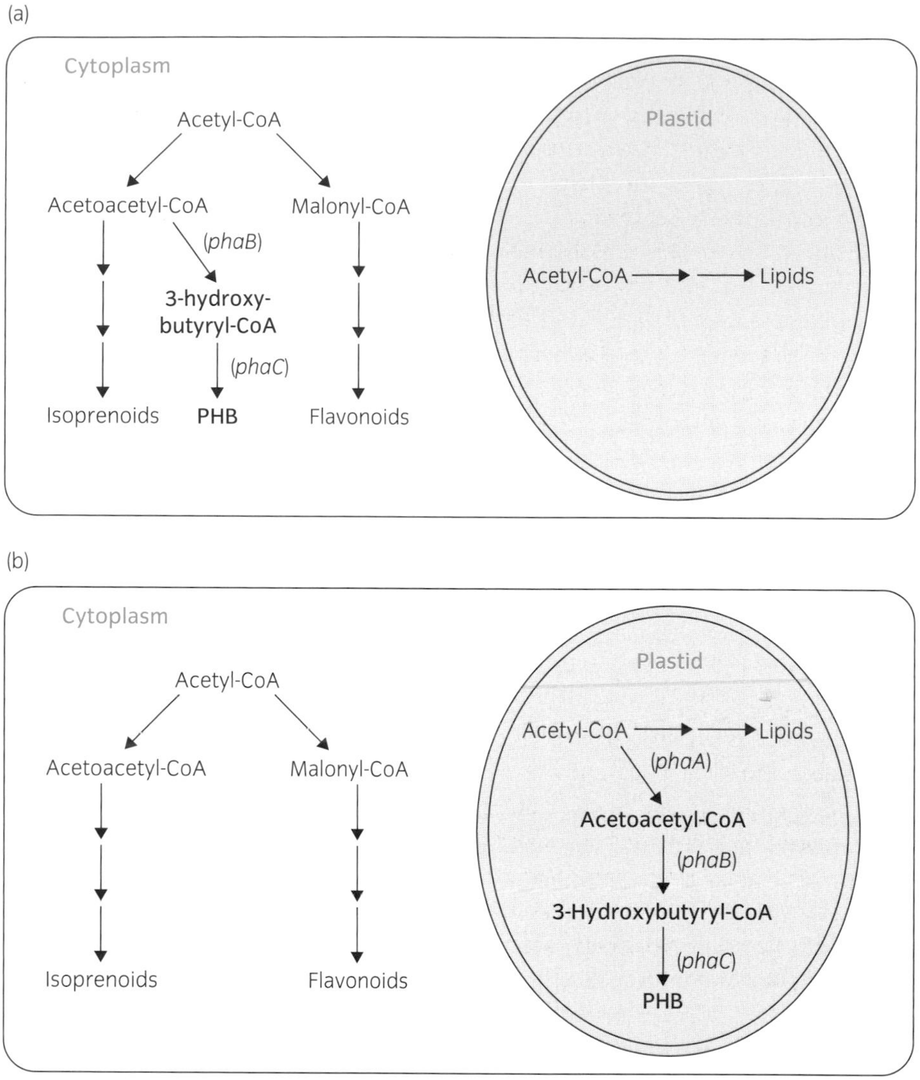

Figure 11.8 Expression of the PHB biosynthesis pathway in transgenic *Arabidopsis*. In scheme (a), the presence of 3-ketothiolase activity in the plant cytoplasm removes the necessity to transfer three genes to the plant. The *phaB* and *phaC* genes from *Alcaligenes eutrophus* were transformed into *Arabidoposis* without protein-targeting sequences. PHB granules were produced in the cytoplasm, but they also accumulated in the nucleus and vacuole. In scheme (b), all three genes of the pathway are targeted to the chloroplast using Rubisco small-subunit transit peptide sequences. In this case, PHB accumulates in the plastids. (Redrawn with permission from Poirier *et al.* (1995).)

to accumulate in mature oilseed leucoplasts to levels up to 7.7% of fresh seed weight. However, apart from in *Arabidopsis*, it has proved difficult to obtain very high levels of PHB in transgenic plants. Recent work indicates that constitutive expression of the β-ketothiolase gene is detrimental to the efficient production of PHB in some plant species, and the use of inducible or developmentally regulated promoters to drive the *phbA* gene permitted some production of PHBs in tobacco and potato.

Plastic cotton fibres have enhanced insulation properties.

Another interesting direction taken by this research has been the production of bioplastics in cotton and flax fibres. It was recognized that cotton fibres contain 3-ketothiolase activity (the first enzyme in the PHB pathway, see Figure 11.7). Therefore, the *A. eutrophus phaB* and *phaC* genes were transformed into cotton by particle bombardment (see Chapter 3) of seed-axis meristems. The *phaB* gene was driven by a cotton fibre-specific gene promoter, whereas *phaC* was fused to the 35S promoter. Clusters of small PHB granules were found in the cytoplasm of the fibre cells (i.e. in the fibre lumen). The thermal properties of these fibres were found to be altered, indicative of enhanced insulation characteristics. In the case of flax, a stem-specific promoter was used to direct the expression of the *phaA* gene, whereas *phaB* and *phaC* were under the control of the 35S promoter. All three enzymes were targeted to the plastids and PHB accumulation in the stem was observed. Modification of the mechanical properties of the stems was observed, although with considerable variation among transgenic lines.

One of the limitations of this approach to bioplastic production is that PHB is a highly crystalline polymer, which produces rather stiff and brittle plastics. PHA co-polymers made from longer monomers have better physical properties in terms of being less crystalline and more flexible. Pseudomonads accumulate medium-chain-length PHAs when grown on fatty acid substrates. These PHAs are synthesized from 3-hydroxy-acyl-CoA intermediates generated by the β-oxidation of the fatty acids; hence the PHA monomer size is related to the length of the fatty acid substrate. Medium-chain-length PHAs have been produced in *Arabidopsis* using the *phaC1* gene from *Pseudomonas aeruginosa* with a peroxisome-targeting sequence from an oilseed rape isocitrate lyase. The enzyme was targeted to leaf-type peroxisomes in light-grown plants and glyoxisomes in dark-grown plants. These plants accumulated PHAs in the glyoxisomes and peroxisomes, and in the vacuole, to a level of 4 mg g^{-1} dry weight. The PHAs contained saturated and unsaturated 3-hydroxyalkanoic acids ranging from six to 16 carbons, with 41% being a mixture of the eight-carbon saturated and monounsaturated monomer.

PHAs are not the only type of bioplastic that could be produced in plants. Amino acid polymers are found in certain bacteria and are already exploited for various commercial purposes. Box 11.2 describes the development of strategies to produce one such polymer, polyaspartate, in transgenic plants.

Molecular farming of proteins

Throughout this book we have presented to you examples where transgene expression is used as a tool to, for example, modify a biosynthetic pathway, manipulate growth and development, or alter some other property of the plant such as herbicide resistance. However, in some cases it is the transgenic protein itself that is important. In Chapter 10, we saw how attempts to improve the amino acid content of grain are being addressed by expression of proteins with a high content of lysine. In this section, we will consider the production of bulk enzymes and potentially high-value proteins such as antibodies and vaccines. The production of functional proteins in plants on an industrial scale does not, at first sight, suggest too many difficulties. However, the production of a complex protein that is folded, processed, easily purified, and shown to be safe for pharmaceutical use, all at low cost, is a real challenge.

BOX 11.2 *Polyaspartate polymers*

Plastic-like materials can be produced by polymerizing certain amino acids. Polymers such as poly-γ-glutamate and poly-ε-lysine are synthesized in various bacteria and are used for thickening and the production of hydrogels. Polyaspartate is a soluble, non-toxic, biodegradable polymer with similar applications, but has not been detected in any known organism. However, it can be produced by mild hydrolysis of cyanophycin, a polymer of repeating β-Asp-Arg subunits that accumulates in cyanobacteria. Cyanophycin synthesis requires only one enzyme, cyanophycin synthetase, which catalyses both the formation of the isopeptide bonds that connect arginine to the β-carboxy groups of the aspartate residues, and the peptide bonds that link the aspartate residues together to form the backbone of the polymer.

```
         Arg   Arg   Arg         Arg   Arg   Arg
          |     |     |           |     |     |
+H3N—Asp—Asp—Asp—. . .—Asp—Asp—Asp—COO-
```

Tobacco and potato plants have been transformed with a cyanophycin synthetase gene from *Thermosynechococcus elongatus* controlled by the constitutive cauliflower mosaic virus 35S promoter, without protein targeting sequences. A cyanophycin-like polymer was detected in the cytoplasm and nucleus and accumulated to levels up to 1.1% dry weight.

● The size of the market for pharmaceutical proteins is a major attraction for plant scientists.

The economics of protein production in plants is complicated. The actual cost will depend on many factors, among them being the cost of growing the plant, transport costs, and processing and protein-purification costs. The costs of proteins produced in plants may undercut the costs of producing proteins by existing methods, but if too much (in economic terms) is produced, then the market price will fall. This has already proved to be the case for some vaccines produced by existing methods. However, the biopharmaceutical market is worth some US$40 billion so there is significant attraction for the plant biotechnologist towards this market.

In this remaining section, the protein products that will be considered fall into two main types. First, enzymes for industrial and agricultural uses will be looked at. Second, we will consider medically related proteins, such as:

- antibodies: full immunoglobulins and engineered types such as scFvs (single-chain variable fragment antibodies);
- subunit vaccines: vaccines based upon short peptide sequences that act as antigens rather than whole organism-based vaccines; and
- protein antibiotics.

Production systems

The interest in producing such proteins in plants comes in part from the problems associated with existing fermentation or bioreactor systems. Mammalian cell systems are expensive and cannot easily be scaled up (input costs can approach US$1 million kg^{-1} of crude product). Bacterial and yeast systems can be scaled up to some extent, but in the case of the prokaryote, the recombinant proteins are often not processed properly (e.g. they are not properly folded and disulphide bridges are not formed), so intracellular precipitation of non-functional proteins can occur. Plant systems can be scaled up so that large amounts of material can be harvested and processed, allowing industrial-scale amounts of protein to be purified. In some cases, it may be possible to omit purification, as is the case with vaccines from plant sources, and where plant material containing recombinant enzymes can be added directly to animal feed or industrial processes. There are some differences between plant and animal post-translational modification systems, and the difference in glycosylation patterns between plants and animals has caused some concerns, particularly when expressing antibodies in plants. However, plants are able to fold, cross-link, and post-translationally modify (by, for example, glycosylation) non-plant proteins sufficiently well to ensure that, in most cases, functional proteins are obtained. The production of recombinant proteins in plants also benefits from the ability to target the proteins to cellular compartments where processing occurs, or where the proteins may be more stable. Another advantage of using plant systems is that the potential for contamination of the protein products with toxins or pathogens of animals and humans, and now prions, is very much reduced.

The plants initially chosen for expression studies related to the ability of scientists to transform the plant material. A lot of work was first done with tobacco, but this was not a suitable plant for feeding to animals because of the alkaloids the leaves contain. For work in which edible products are being developed, plants such as potatoes, tomato, maize, and lettuce have been used. For bulk production, the leaves of plants such as tobacco and alfalfa remain a favourite, because they can be harvested several times a year. However, plants such as rice, wheat, maize, and soybean have also been used, although their biomass yields are lower (Table 11.3).

Strategies for protein production

Several major strategies have been developed for the production of proteins in plants; the extremes are the stable integration approach, which has been used for many systems, and the use of plant viruses as transient vectors. Now, as we will discuss below, the edges between these extremes have been blurred as systems in which virus

● Stable integration and transient expression systems have now been combined to optimize expression.

Table 11.3 Potential biomass yields from various plants considered as potential vehicles for protein production

Crop	Potential annual biomass yield (t ha^{-1})
Tobacco	>100
Alfalfa	25
Maize	12
Rice	6
Wheat	≈3

sequences are stably introduced into plants are now available. To these we must add the use of the chloroplast.

Stable expression of the transgene in the nucleus

The major advantage of this approach is the inheritability of the transgene. Once a line has been selected that has optimal expression levels, it can be used to produce the protein over a number of growing seasons. The disadvantages include the potentially long time it takes to produce the lines, gene-silencing effects, and less-than-good expression levels. The stable transgene expression approach, in which the transgene is regulated by a strong, constitutive promoter (such as the 35S promoter), is perhaps the most suitable for the bulk production of soluble proteins in leaves, although yields can be low using this approach. A more sophisticated approach has been to target gene expression and protein production to specific tissues (see Figure 11.9). Restricting gene expression to particular organs or tissues often leads to higher yields

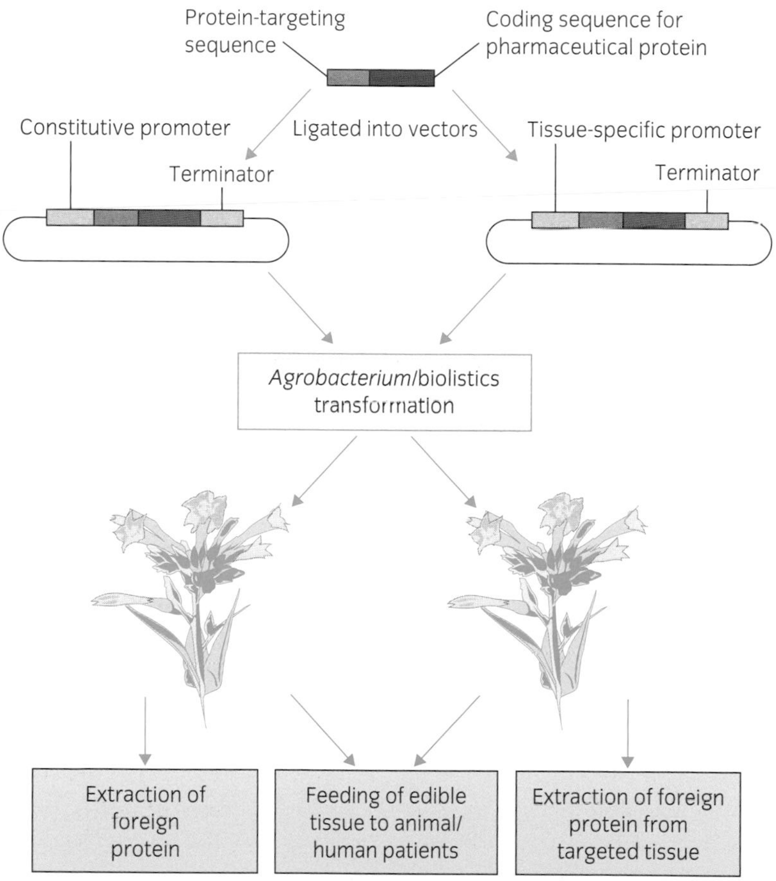

Figure 11.9 Different routes of protein manufacture in stably transformed plants. Constitutive (from, for instance, the 35S promoter) or targeted (to particular organs or tissues) expression and accumulation results in recombinant proteins that can be used either after purification or directly as plant products.

of recombinant proteins. Targeting expression has a number of advantages, some of which are related to the increased yield. The resources of the plant are effectively wasted if proteins are produced in parts of the plant that are not harvested. So, for vaccines that are to be ingested, after the minimum amount of processing, it makes sense to target the protein only to the parts of the plant that are normally eaten. Another important point is that storage organs, such as tubers and seeds, are designed to maintain their biological integrity over long periods. It has been found that in some instances using seed-specific promoters increases protein production 36-fold (over systems using the 35S promoter) and that targeting proteins to these structures has been shown to increase the stability of recombinant proteins from weeks to years. Seeds are also of great interest because they can offer simple strategies for protein purification. In tobacco, seeds contain a simpler mixture of proteins and lipids, with fewer phenolic compounds, than do green leaves.

As well as targeting expression to particular organs or tissues, intracellular targeting has been used to optimize the production of active proteins. Post-translational modification has been shown to be associated with the ER and/or apoplastic export, and the inclusion of transit peptides has been shown to improve both yield and activity. Tissue and intracellular targeting are brought together in the use of oleosins as vehicles for protein purification. These are discussed in Case study 11.4. Much effort is now going into optimizing the various approaches for protein production and several other examples are given in the text. These include the use of fusion proteins and engineering the coding sequences of the genes to make them more plant-optimized.

CASE STUDY 11.4 The oleosin system: hirudin and insulin production

In this case study, we will look at a purification system that was developed to improve the efficiency and economic viability of pharmaceutical proteins produced in plants. We will look at the application of the technology to the production of hirudin, an anticoagulant peptide, and the hormone insulin. Although pharmaceutical protein production is discussed in more detail later in this chapter, the case study here illustrates how novel purification procedures may help to make pharmaceutical proteins produced in plants a widely adopted method of production.

Background

Oil bodies, the subcellular organelles that store TAGs, have already been met in this chapter (see the section on Improvement of plant oils). They are especially abundant in oilseeds, like the seeds of oilseed rape, where they are the primary source of energy for seedling germination and the early stages of growth. Oil bodies from seeds are composed of TAGs surrounded by a phospholipid monolayer. Proteins called oleosins are associated with the surface of the oil bodies. Studies have shown that if the *gusA* reporter gene (see Chapter 4) was fused to the C terminus of either a native or C-terminal-deletion-mutant oleosin, then β-glucuronidase (GUS) activity is partitioned with the oil body fraction, indicating that it is targeted to the oil body by the oleosin. The N-terminal and central domains of the oleosin are required for correct targeting of GUS to the oil body.

Purifying oil bodies

Oil bodies can be purified extremely simply by a series of centrifugations. Seeds are ground in an aqueous buffer and centrifuged. After centrifugation, three fractions are evident. A pellet of insoluble material is formed at the bottom of the tube. An aqueous phase, consisting of the soluble components of the seed, is present; and floating on top of this is a layer composed almost exclusively of oil bodies with their associated oleosins. This oil body layer can be removed, resuspended in buffer and re-centrifuged to increase the purity of the oil bodies. This process is illustrated in Figure 11.10.

Hirudin

Hirudin is a small (≈7-kDa) anticoagulant peptide found in the salivary glands of the medicinal leech *Hirudo medicinalis*. Hirudin is the most potent inhibitor of thrombin (a serine protease that catalyses the initial steps in blood clotting) so far identified, and has high therapeutic value in the treatment of thrombosis. Thrombosis (in heart attacks and strokes) is the biggest killer in the Western world. Hirudin has several advantages over other, more commonly used, anticoagulants such as heparin. It has a low immunogenicity, does not require endogenous cofactors for activity, and does not interact with other proteins in the blood. Its use is limited, however, due to difficulties in obtaining large quantities. Recombinant hirudin has been produced in both *E. coli* and yeast. Production in *E. coli* resulted in low yields of the protein and reduced activity. Some yeast production systems have proved quite efficient, producing large amounts of functional protein.

Production and purification in plants

A synthetic gene encoding hirudin, with codon usage optimized for expression in *Brassica napus*, was fused to the 3′ end of an *Arabidopsis* oleosin gene with its own promoter. An endoprotease Factor Xa cleavage site was engineered between the C-terminal end of the oleosin and the N-terminal end of the hirudin. This construct was used to transform *B. napus*. Hirudin was purified using a modification of the method for oil-body purification outlined earlier in this case study. After the oil bodies have been centrifuged, they are treated with the endoprotease Factor Xa, which cleaves the hirudin from the oleosin, as shown in Figure 11.10. The oil bodies can then be re-centrifuged, after which the hirudin will be present, highly purified, in the aqueous phase (Figure 11.10).

Prospects

There appears to be a good prospect that purification systems based on oil-body/oleosin technology could make pharmaceutical protein production in plants an economic reality, especially if cheaper alternatives to Factor Xa, for example, are found. A company called SemBioSys has been established in Canada that hopes to successfully market pharmaceutical proteins, like hirudin, produced using this technology.

● Complex proteins can be expressed with this system.

One initial drawback of the technology is that it cannot be used to produce proteins that require extensive processing and modification in the ER, which unfortunately applies to many clinically important proteins. To develop this technology further and show it is not restricted to single peptides, it has recently been used to produce insulin. *Arabidopsis* plants have been transformed with an oleosin fusion construct containing the oleosin protein linked via a Klip (trypsin-cleavable) sequence to a mini-insulin molecule. This is also a fusion molecule of the

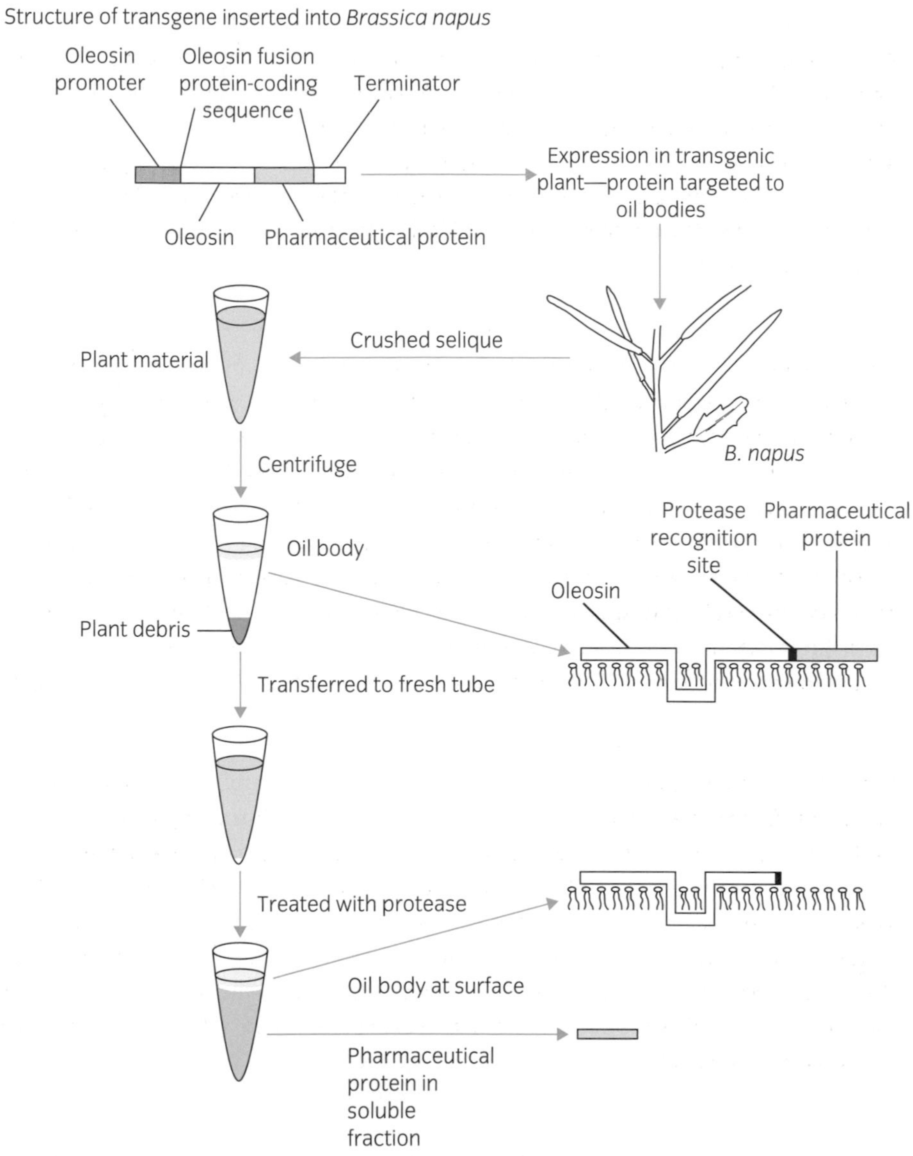

Figure 11.10 A scheme for pharmaceutical protein purification by production in oil bodies. The pharmaceutical protein is produced as a C-terminal fusion to oleosin (an oil-body-associated protein). The pharmaceutical protein and the oleosin are separated by a protease cleavage site. Seeds are crushed in an aqueous buffer and centrifuged, resulting in three fractions; the oil bodies, with the oleosins and the pharmaceutical protein, are the topmost layer (which floats on the aqueous layer). This oil-body layer can be removed, re-centrifuged, and then treated with a protease, which cleaves the pharmaceutical protein from the oleosin. Re-centrifugation results in highly purified protein being present in the aqueous phase. Full details can be found in the text. (Redrawn from Goddijn and Pen (1995).)

A-chain and a DesB30 chain linked with a Mini-C propeptide. Mature seeds have been harvested from these plants and partial purified to remove most of the proteins using the method described above. The oil body was treated with trypsin to release the insulin protein and to remove the Mini-C fragment. Some further downstream processing was required to remove further contamination but an apparently crosslinked (two disulphide bridges between the two insulin chains, and one within the A-chain) and functional insulin molecule was obtained. The company has introduced high-level production constructs into its commercial platform crop safflower (*Carthamus tinctorius*) to provide a high-capacity economical source of insulin. With this in place, they are now looking at reducing the production costs further.

Virus expression systems: transient and stable systems

Plant viruses have been used as vector systems for the production of recombinant proteins. The main ones used have included tobacco mosaic virus (TMV), potato virus X (PVX), and alfalfa mosaic virus (AlMV). There are several advantages to using viruses. These include short production time scale, scale-up capacity, and very high production levels. Initial work in this area did not involve producing transgenic plants. Rather, plant virus capsids were used as carriers of recombinant proteins, particularly vaccines. In one approach, coding sequences for epitopes (20 amino acids) or proteins have been introduced into the coat protein (CP) gene of the virus genome so that it is presented externally when virus particles are made. This approach is illustrated in Figure 11.11.

Another approach that has been used is to construct viral vectors to produce recombinant proteins that are targeted to the ER for processing. The virus can be replicated in the host plant, and through serial passage enough protein can be generated for clinical use.

The limiting factor on time for the process is identifying the epitope and constructing the vector. The virus replicates and moves through the plant as in a normal infection, which for some viruses takes 2–3 weeks. The yield reflects the normal levels of CP production and in some systems can be as much as 10% of the total soluble protein. This level has been achieved by using recombinant (sequences have been removed or replaced using parts of other viruses) and molecular evolution techniques to optimize viral replication and movement through the plant. Various strategies have been used to overcome some of the other disadvantages of using complete virus sequences. These problems include limited host range, the problem of gene silencing, recombination, virus sequences being read as introns and therefore being processed, and long sequences not expressing properly. The gene-silencing problem can be addressed; that is, it can be switched off by introducing an inhibitor of the process. Alternative strategies have involved introducing movement proteins into the plant. In this area, one of the goals is also to create a better host, one that allows virus replication to occur optimally. This is the opposite of engineering for defence. However, a lot can be learnt from both sides of these approaches that will help the other. Further examples of optimizing expression will be given below.

● *Agrobacterium* is used to carry engineered viruses into plants.

Other interesting strategies are being developed to improve protein production systems using viruses. One is the deconstructed virus approach, another is to use

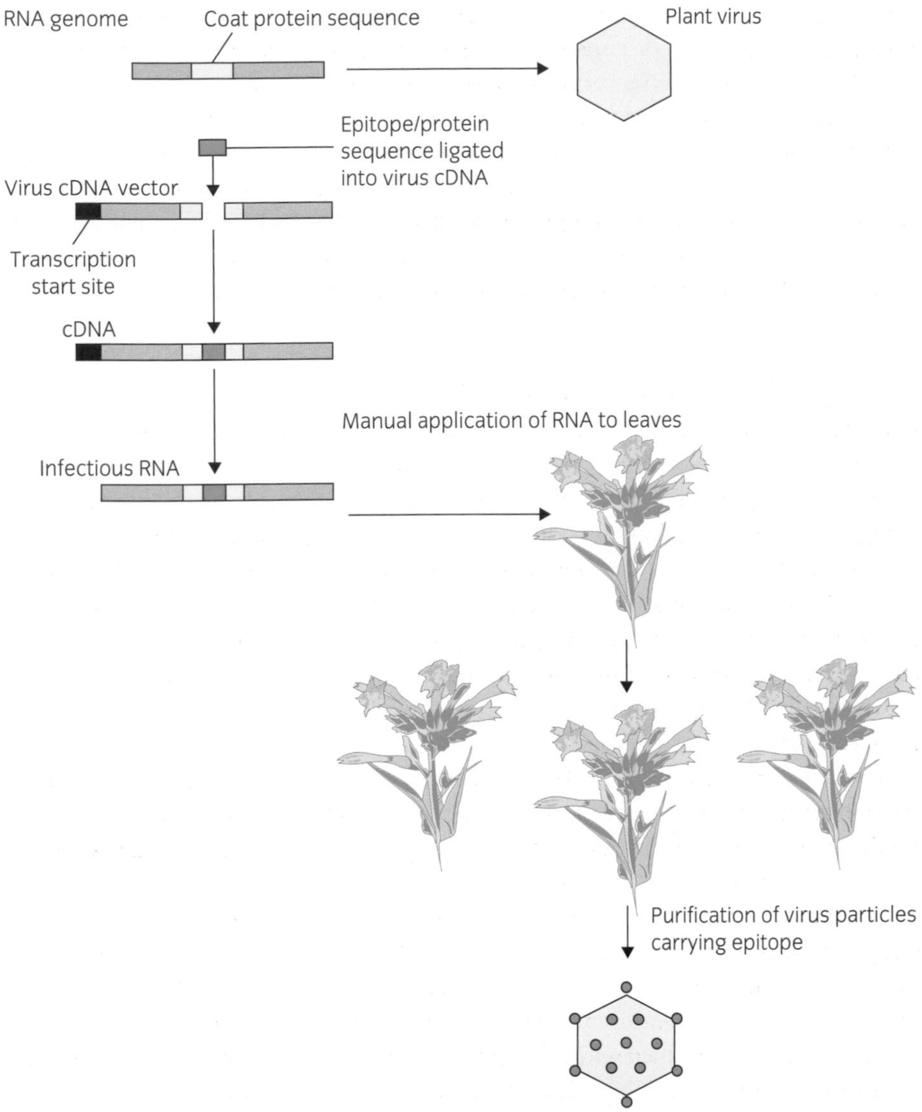

Figure 11.11 Construction of a recombinant plant virus carrying a protein epitope. The gene for the epitope sequence was introduced into the coat protein (CP) sequence within the vector. Infectious RNA was made and then introduced into plants (transfection) where virus capsids, displaying the expressed protein, were produced. This is known as the **overcoat system** and can be used for a range of proteins.

stable integration of the virus sequence (complete or deconstructed) in plants and induce its expression in a controlled fashion. This could either be the synthesis of the virus RNA or switching on the expression of the gene of interest. A system has been developed for the bipartite RNA virus cowpea mosaic virus (CPMV). In this, the RNA2 sequence (complete or partial), which has the CP coding sequence and the protein of interest, is integrated into the host genome and is activated either by infecting the plant with RNA2 or by crossing with a plant that has this sequence. The deconstructed virus can take the engineering to the point where all non-essential sequences are removed or engineered into other genomes (possibly the host plant). For instance, a short virus vector may be introduced into the plant with *Agrobacterium* (termed Agroinfection), the coding sequence of the protein of interest may be linked to a plant promoter, and movement functions may be provided by the host plant. These approaches also reduce potential containment problems and interactions with other virus sequences. Systems that can be used to produce small amounts of protein have been given the name 'desktop systems'.

There are a lot of different virus expression systems available; one is described in Case study 11.5 (see below). Another one that has attracted a great deal of interest is based on engineered TMV virus. The magnICON system (Icon Genetics) uses TMV sequences that can be recombined together to give functional viral replicons. The gene for the protein of interest is carried on one module and the other viral genes are carried on another and they are introduced separately into the plant nucleus by Agroinfection. In the nucleus, they are assembled if they are co-expressed with a specific recombinase. Using versions of this system, a number of proteins have been made. Authentic human growth hormone has been expressed in transfected *Nicotiana benthamiana* and targeted to various compartments of the plant cell: the apoplast, the chloroplast, and the cytosol. The apoplast gave the best expression levels of 10% total soluble protein (1 mg g^{-1}). Also using this system, F1 and V antigens from *Yersinia pestis* have been expressed. The purified antigens were administered to guinea pigs, subcutaneously, generating a systemic immune response that provided protection against an aerosol challenge of virulent *Y. pestis*. A similar approach has been used where two non-competing viruses have been used to carry separate proteins that can be assembled into a complex protein. In this case, TMV and PVX have been used to produce a fully functional IgG. To do this, six separate vector constructs were introduced into the plant host. The viruses were engineered to reduce the protein load on the cell, the PVX CP was removed, and it was expected that the TMV movement protein would complement missing functions. However, this did not work completely so the *CP* gene was infiltrated on a separate vector. Yields of 0.5 g kg^{-1} fresh-leaf biomass were obtained with this combination. Work continues to overcome the problems associated with the virus approach and to scale up the process.

The production of enzymes for industrial uses

The global industrial enzyme market has been estimated to be worth something approaching US$2 billion per annum. Proof-of-concept incursions by plant biotechnology into this market have been on a small scale. The first commercialized industrial proteins produced from transgenic plants were two proteins familiar to the molecular biologist—avidin (from chicken) and β-glucuronidase (or GUS, from *E. coli*)—both of which were produced in maize (Table 11.4 summarizes recombinant-enzyme production in plants). Following on from these successes, ProdiGene, the company responsible for their development, is leading the race towards large-scale production of plant-derived transgenic enzymes for commercial use. They announced in spring 2002 that they would be going into large-scale production of trypsin, a proteolytic enzyme, which by its very nature is difficult to produce in conventional

Table 11.4 Examples of enzymes produced in plants

Enzyme	Use
Avidin	Diagnostic kits
β-Glucuronidase	Diagnostic kits
Trypsin	Pharmaceuticals, wound care
Cellulase	Ethanol production from cellulose waste
Thermostable xylanase	Biomass processing
Phytase	Phytate breakdown, improved phosphate utilization
α-Amylase	Food processing
1,3–1,4–β-Glucanase	Brewing
Lignin peroxidase	Paper manufacture
Gastric lipase	Pharmaceuticals

(Data taken with permission from Hemming (1995), Kusnadi *et al.* (1997), Giddings *et al.* (2000), and Daniell *et al.* (2001).)

recombinant systems. This protein is currently harvested from bovine and porcine pancreases, and is used for a number of applications, including the production of pharmaceuticals such as insulin, vaccine production, and wound care. The company has predicted that the worldwide demand for the enzyme will increase fivefold in the next 5 years. Existing sources may be unable to supply this increase, and there is the added worry of the spread of pathogens (like bovine spongiform encephalopathy) with products derived from animals. ProdiGene aims to meet this demand by producing kilogram quantities of the protein from plant sources.

Some of the other important enzymes that may be produced in plants are, interestingly, normally targeted against components of the plant cell. Cellulases and xylanases, which are normally produced by the microorganisms associated with digestion in ruminants, have been produced in a number of different plants. There is a large market for these enzymes in the bioethanol, textile, pulp, and paper industries and for the production of animal feed. It may, at first sight, seem a strange concept to express such enzymes in plants, due to the risk of so-called autodigestion, but engineered, thermostable forms of the enzymes with high temperature optima were used. Therefore, no deleterious effects were detected at the temperatures at which a plant normally grows, but high levels of enzyme activity were detected when plant extracts were heated to the temperature optima of the enzymes. These were only small-scale initial trials, but the process could easily be scaled up.

ProdiGene has recently expressed manganese peroxidase (from white-rot fungus) in maize. This enzyme is a lignin-degrading enzyme that belongs to a group of enzymes known as the oxidoreductases. It is not an easy enzyme to produce by biotechnological means. Attempts with fungal fermentations have been hampered by the need to add vast quantities of haem cofactor. In plants, previous attempts have given unhealthy specimens. The company were successful because they expressed the enzyme in seed. They obtained high levels of expression and the plants were healthy. Again the market for this protein is the pulp industry where its use would help to reduce the vast amounts of bleach used.

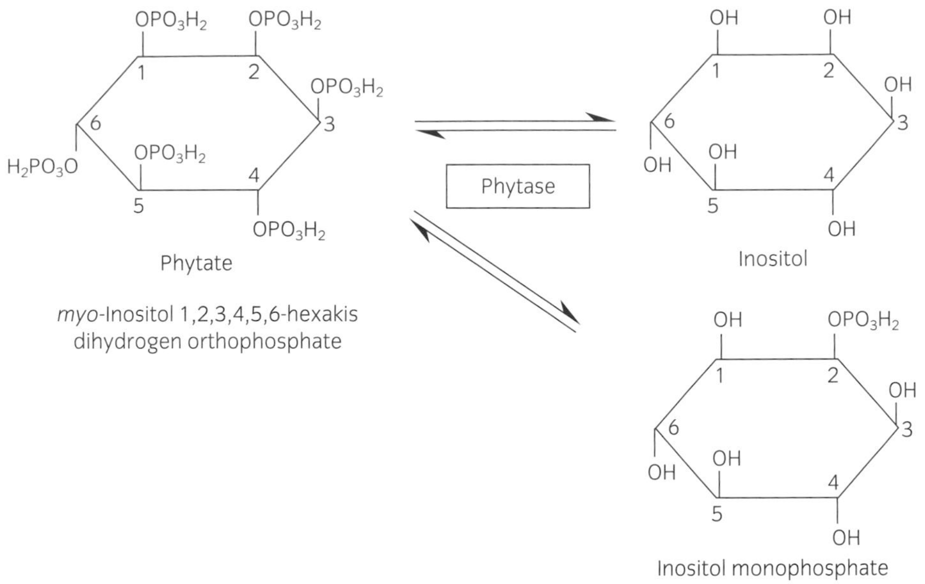

Figure 11.12 The action of phytase. Phytate is broken down either to inositol monophosphate (which is believed to be the main product of phytate breakdown) or to inositol.

An interesting point to note is that these enzymes would require only the minimum of purification (an important consideration when determining applicability and economic viability). This is also the case for another important hydrolytic enzyme, phytase. This enzyme catalyses the hydrolysis of inositol hexaphosphate (phytate) to its constituents, the sugar-alcohol inositol and inorganic phosphate (see Figure 11.12). Phytase is a very useful enzyme as it releases phosphate from a substrate that is normally indigestible to monogastric animals. As phytate is found in significant quantities in seeds used to formulate feed meal for pigs and poultry, these animals do not benefit from this source of phosphate, which is required for proper skeletal growth. Problems with phytate are exacerbated by the fact that monogastric animals excrete the indigestible phytate in their faeces, which contributes to an excessive phosphate build-up in ground water, leading to eutrophication. Transgenic plants have been produced expressing phytase in, for example, seeds. When these transgenic seeds are added to the diet of pigs and poultry, phosphate utilization is increased (negating the need for supplements) and phosphate excretion is decreased. Thus, the addition of the enzyme in a simple plant-derived formulation solved both a nutritional and an environmental problem.

Medically related proteins

Medicine is an important, expanding area for plant biotechnology, in which there is great hope that recombinant proteins made *in planta* will be accepted and widely utilized. This section is divided into three parts:

1 antibodies;

2 vaccines; and

3 other medically related proteins.

Antibodies or 'plantibodies'

In mammals, there are a number of different types of antibodies, or immunoglobulins (Igs). The immunoglobulins have been divided into five major classes on the basis of their physical, chemical, and immunological properties: IgA, IgD, IgE, IgG, and IgM. Figure 11.13 shows the structure of an IgG (discussed later in this chapter), which can be used as an example. Immunoglobulins are composed of two heavy chains and two light chains. Each class of immunoglobulin has its own class of heavy (H) chain, termed α, δ, ε, γ, and μ, respectively (IgA molecules have two α-class heavy chains, for example). Each immunoglobulin also has two light (L) chains, which are either κ or λ, but not a mix of the two. IgA, the main antibody found in secretions, can have a more complicated structure. In secretions, IgA exists as a dimer of two IgA molecules joined by a joining (J) chain and associated with a secretory component. Each antibody has two antigen-binding sites, formed from the variable (V) regions of the light and heavy chains. Also shown in Figure 11.13 is the structure of a single-chain variable fragment antibody or scFv. As the variable regions of the light and heavy

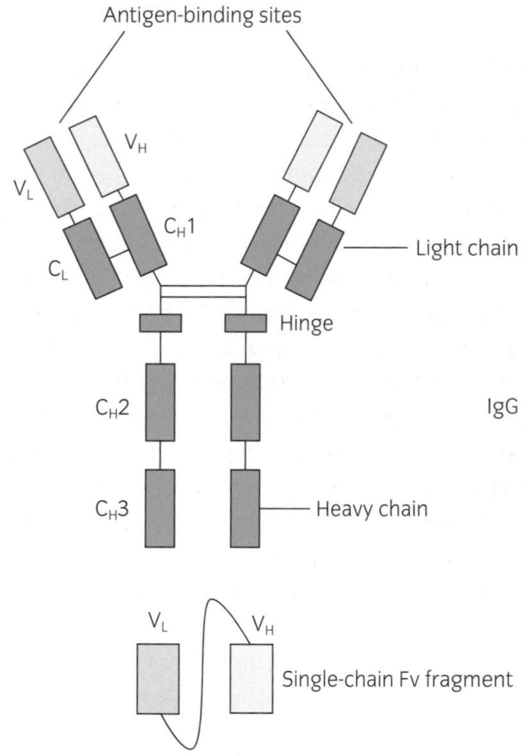

Figure 11.13 Structures of IgG and an engineered single-chain variable fragment antibody (scFv). A typical antibody is composed of two identical heavy (H) chains and two identical light (L) chains. Antigen-binding sites are found at the N-terminal regions of both the heavy and light chains. The hinge and tail regions are formed only from the heavy chains. The light and heavy chains are composed of constant (C) and variable (V) domains. The variable domains of the heavy (V_H) and light (V_L) chains make up the antigen-binding sites. scFvs are engineered by joining together light- and heavy-chain variable regions. Antigen binding is determined solely by these variable regions.

chains are solely responsible for antigen binding, simpler, smaller molecules that still bind antigens can be produced. scFvs are produced from synthetic genes made by fusing the sequences for light- and heavy-chain variable regions. Although not the only example of engineered antibodies, scFv antibodies are the most commonly used and most successful.

The first example of a functional antibody (a mouse immunoglobulin IgG1) being produced in plants was reported in 1989. The construction of this antibody was a two-step process. In the first stage, two separate transgenic tobacco lines were produced: one contained the gene for the γ heavy chain and the other contained the gene for the κ light chain. The second stage of the process was to cross these plants to generate progeny that expressed both genes. It was found that in the F_1 progeny expressing both chains, significant amounts (up to 1.3% of total leaf protein) of assembled antibody were produced. These transgenes contained native immunoglobulin signal sequences, which targeted the nascent light and heavy chains to the ER and were found to be necessary for efficient assembly of the antibody in plants.

Several important observations, which underpinned much of the future work in this area, were made in this original research. Targeting of the antibody chains to the lumen of the ER was found to be necessary for efficient chain assembly and stability. Conditions in the lumen of the ER favour the correct processing of the antibody chains, especially disulphide-bridge formation, and molecular chaperones also help to direct antibody chain folding and assembly. Another important observation was that the formation of the complete antibody in the F_1 plants increased the protein yield over that of either of the parents that contained single chains, reflecting the stable nature of the complete antibody in the plant system. Although a pioneering study demonstrating that functional antibodies could be assembled in plants, the antibody in question had no practical applications.

● Plants are capable of synthesizing complex transgenic protein molecules such as the secretory antibody sIgA.

Antibodies produced in plants are thought to be particularly suitable for topical immunotherapy. In an extension of the strategy of producing the various antibody subunits in different plant lines that are subsequently crossed to produce a functional antibody, a secretory IgA (sIgA) that protects against dental caries produced by *Streptococcus mutans* has been expressed in plants. The antibody does this by recognizing the native streptococcal antigen (SA) I/II cell-surface adhesion molecule, thereby preventing colonization.

This multiple-component sIgA antibody required four parental transgenic tobacco lines to be produced, in which cDNAs for the heavy, light, or joining chains (the heavy and light chains were derived from a murine antibody termed Guy's 13 that recognizes the SA I/II cell-surface adhesion molecule), or the secretory component, were expressed under control of the 35S promoter, and then crossed. The strategy used is illustrated in Figure 11.14. The success of the approach has been quite an impressive technical achievement. In mammalian systems, both plasma cells and epithelial cells are involved in the synthesis and assembly of secretory antibodies, yet single plant cells containing all four transgenes are able to produce them efficiently. Secretory antibodies have many theoretical advantages in situations like the one described here. Secretory IgAs are the predominant antibody type that protects against microbial

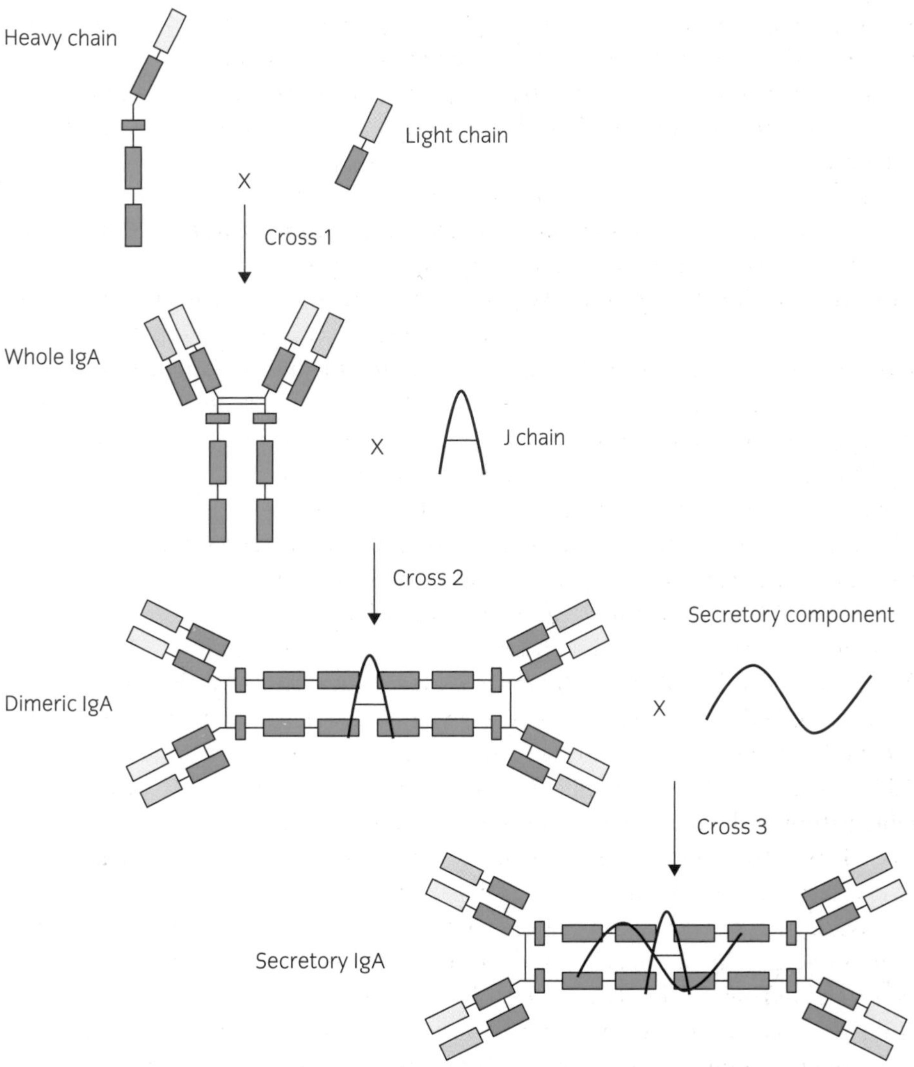

Figure 11.14 Synthesis of a secretory IgA (sIgA) molecule in transgenic plants. Four separate parental plant lines were constructed. Each expressed a separate piece of the antibody (the heavy, light, or joining chain, or the secretory component). The lines expressing the heavy and light chains were crossed, giving F_1 plants that accumulated IgA. These plants were further crossed with plants expressing the joining (J) chain, which, in progeny, resulted in the accumulation of dimeric IgA. These plants were finally crossed with plants expressing the secretory component. Progeny of this cross accumulated functional sIgA.

infections at mucosal sites. They are also more resistant to proteolysis (a quality that contributes to high yield), and they bind antigens with a higher efficiency than do the original antibodies produced in plants.

These secretory antibodies have now been tested on humans. They have been topically applied to teeth and found to be effective at preventing recolonization by *S. mutans* for up to 4 months, which is as effective as an IgG produced in a murine

Table 11.5 Examples of antibodies that have been produced in plants

Application	Plant	Antibody	Signal sequence
Immunoglobulins			
S. mutans SA I/II (dental caries)	Tobacco	sIgA (hybrid)	Murine IgG
S. mutans SA I/II (dental carries)	Tobacco	IgG (Guy's 13)	Murine IgG
Surface antigen (colon cancer)	Tobacco	IgG Co17–1A	Murine IgG/KDEL[a]
Herpes simplex virus	Soybean	IgG (anti HSV-2)	Tobacco extensin
Single-chain Fv			
Lymphoma	Tobacco	scFv (38C13)	Rice α-amylase
Carcinoembryonic antigen (cancer)	Cereals	scFvT84.66	Murine IgG/KDEL[a]

[a]ER targeting sequence (see Chapter 1).
(Data taken with permission from Hemming (1995), Kusnadi *et al.* (1997), Giddings *et al.* (2000), and Daniell *et al.* (2001).)

hybridoma. Despite the differences in structure, there appear to be no differences in binding properties between the proteins.

Plants have been used to produce a variety of antibodies, including whole antibodies, antigen-binding fragments, and scFvs. Despite the now well-characterized differences in the glycosylation pattern seen between antibodies produced in plant and mammalian expression systems, antibodies produced in plants generally seem to exhibit similar properties to antibodies produced in other, mammalian, systems. Some of the most important examples are listed in Table 11.5. Although there have been a number of successes, studies to optimize protein production still continue. One particular focus has been on glycosylation, and many approaches have also been tried to increase the level of protein accumulation. As discussed above, and in the section on vaccines (see below), protein fusions have been used in many studies. scFvs have recently been made in tobacco as a C-terminal fusion to an elastin-like polypeptide that is expressed in seed. This approach was developed to take advantage of high levels of protein expression and the seed's ability to keep proteins functional during extended storage periods at ambient temperature. The strategy lead to a 40-fold increase in scFv accumulation with levels approaching 25% of the total soluble protein.

CASE STUDY 11.5 Custom-made antibodies

● These custom-built antibodies are undergoing clinical trials.

The strategy used to generate most of the antibodies listed in Table 11.5 is to generate stably transformed plants that express the protein. This has not been the approach used by Large Scale Biology Corp. for the development of an idiotype vaccine. A TMV-based vector has been developed for the production of a secreted scFv protein during virus infection of non-transgenic tobacco plants. In this vector, a virus promoter regulates the expression of the *scFv* gene and the antibody coding sequence is inserted downstream from a rice α-amylase leader sequence, which targets the protein to the extracellular compartment of the plant (see Figure 11.15). The advantage of this transient expression system is that it is very rapid and can be used to produce small batches of customized antibody. The company has targeted its use towards the treatment of B-cell lymphoma: in this disease, these cells produce a particular antibody that is unique to each patient. This idiotype vaccine system is one where the patient is treated with a copy of the lymphoma antibody to generate immunity against the original antibody-producing lymphoma

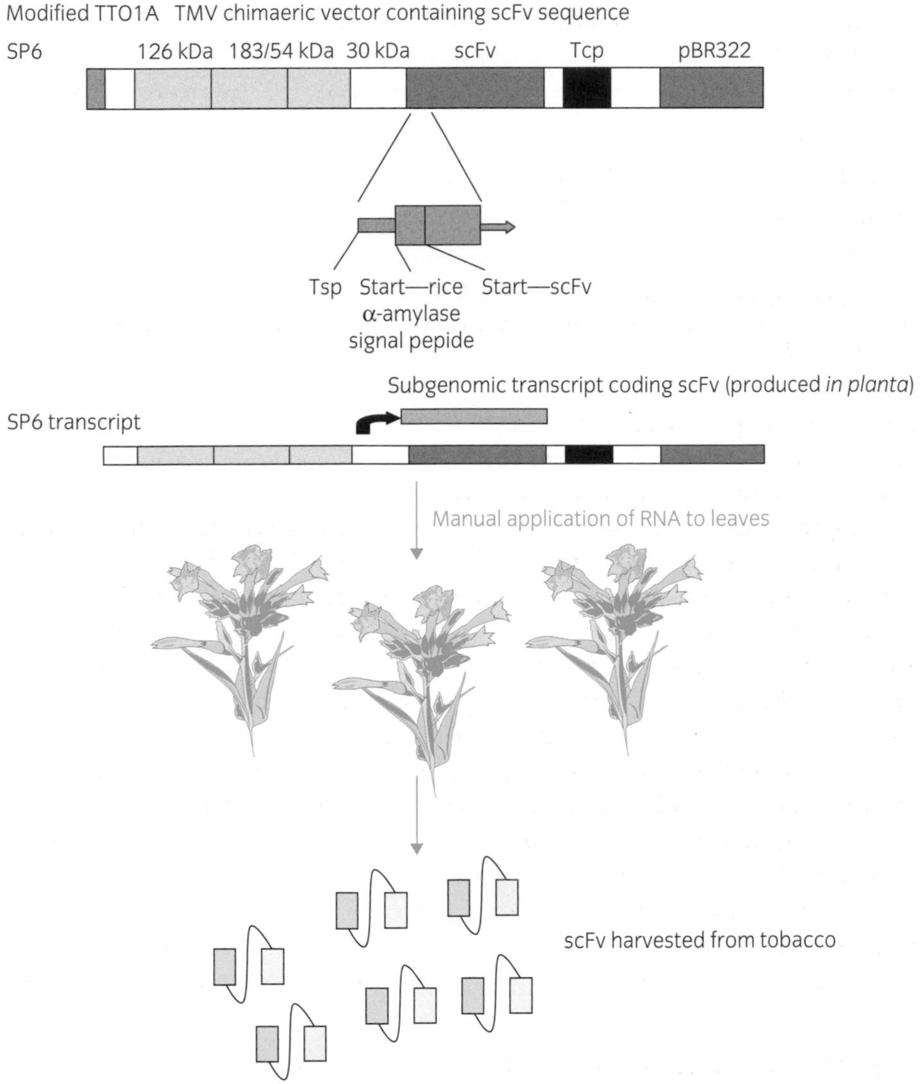

Figure 11.15 The use of engineered TMV vectors for the expression of pharmaceutical proteins. An scFv antibody sequence has been located adjacent to a translation start sequence and a rice α-amylase signal peptide. Infectious RNA is produced from an SP6 promoter and used to infect plants (transfection). Subgenomic RNA (see Chapter 8) is produced, and this acts as a template for expression of the scFv. scFv proteins can be isolated. pBR322, bacterial replication sequence; Tcp, tomato CP; Tsp, translation start position.

cells. This works because each individual antibody has a variable region that has a unique antigenic constitution, the idiotype. Thus, the idiotype vaccine will only raise a specific response against the lymphoma cells and not against any of the patient's other cells. It is now possible, using the TMV vector system, to generate the idiotype region of a tumour-specific immunoglobulin from a patient, as a scFv, in a matter of 6–8 weeks. The system was developed using mice as a model, but it has now entered clinical trials with 16 patients. This phase 1 testing is the first and smallest required by the US Food and Drug Administration (FDA) for any drug. It is interesting that each of the antibodies in the clinical trials is unique, and yet can be considered as a single drug. One important advantage of this system is that it is possible to harvest enough material for clinical use from greenhouse-grown plants, thus ensuring high levels of containment.

Plant-derived vaccines

Both stable and transient expression systems have been used to produce vaccines in plants. One of the first successful attempts in this area used a transient expression system. Recombinant CPMV was used as a vector for a linear epitope (antigenic determinant) from the VP2 capsid protein of mink enteritis virus. The recombinant CPMV was replicated in black-eyed beans (*Vigna unguiculata*), from which chimaeric virus particles (CPMV–VP2), displaying the VP2 epitope, were isolated and used for the subcutaneous injection into mink. The vaccine was able to induce resistance to clinical infection by mink enteritis virus. A vaccine to classical swine fever has been produced *in planta* using a fusion of the E2 glycoprotein fused to the CP of PVX. Many other vaccines have now been produced in plants and Table 11.6 lists some examples.

Table 11.6 Examples of vaccines produced in plants

Origin	Recombinant protein	Plant	Production level[a]
Human			
Escherichia coli	Heat-labile enterotoxin B	Tobacco	0.001% SLP
Vibrio cholerae	Cholera CtoxA and CtoxB subunits	Potato	0.3% TSP
Hepatitis B	Envelope surface protein	Tobacco/potato	<0.1% FW
Norwalk virus	Capsid protein	Tobacco/potato	0.23/0.37% TSP
Rabies virus	Rabies virus glycoprotein	Tomato	1% TSP
Animal			
Foot-and-mouth disease virus	Virus epitope VP1	Alfalfa/*Arabidopsis*	N/A
Porcine coronavirus	Viral glycoprotein	Tobacco/maize	0.2% TSP/0.01%
Mink enteritis virus	Viral epitope VP2	Black-eyed bean	N/A (CPMV)[b]
Canine parvovirus	Peptide from VP2 capsid protein	*Arabidopsis*	3% SLP

[a]FW, fresh weight; N/A, not available; SLP, soluble leaf protein; TSP, total soluble protein.
[b]Produced using a CPMV expression system.
(Data taken with permission from Hemming (1995), Kusnadi *et al.* (1997), Giddings *et al.* (2000), and Daniell *et al.* (2001).)

The main aim of many of these studies is the generation of edible vaccines. The driving force behind the use of edible vaccines has been the need for cheap, effective treatments for enteric disease in the developing world. Plant-based systems have the potential to provide low-cost, easily administered, locally produced edible vaccines that could provide mucosal immunity against the infectious agents that are responsible for the deaths of millions of people, particularly children, annually. To bring this about, it is necessary to present, in a practical, edible form, a quantity of vaccine that will stimulate the gut-associated lymphoid tissue (GALT) system in the gut to generate secretory antibodies.

The questions that have had to be considered relate to delivery of the vaccine to the gut through the stomach. As proteins, these vaccines are likely to be degraded in the stomach. The experiments carried out so far indicate that orally administered plant material expressing a vaccine can induce an immune response, with particulate material the most effective. Interestingly, when virus CPs are expressed in plant leaves, they can form virus particles that may also improve the stability and effectiveness of the vaccine by presenting the protein in its native form.

The main problem with the examples given is that expression levels in many of the original studies were poor. Much work is now focused towards that problem. One area of interest is the use of protein fusions. We have already considered virus fusions; here the examples are for the production of vaccines by stable integration. A virus against a virulent canine parvovirus was produced by linking the gene for a 21-mer peptide to the gene for the GUS protein. This construct was expressed in *Arabidopsis* where the fusion protein was made. Plant material was fed to mice, which became immunized. Similar approaches have been used but with mixed success.

● Protein fusions may stabilize the transgenic proteins and increase accumulation.

Recently, the human immunodeficiency virus type 1 (HIV-1) p24 capsid protein fused with an IgA (two constant region sequences of the α chain) has been expressed in tobacco where it was targeted to the endomembrane system. These developments increased the production level 13-fold; much of the increase was put down to dimerization of the protein, increasing stability.

Another interesting study was done to make a vaccine for rabbit hemorrhagic disease virus. The virus VP60 protein was fused to either ubiquitin or rbcS. Both of these proteins can have an effect on protein stability and have been shown to have beneficial effects in some systems. The first is an extremely stable, proteolysis-resistant protein that folds rapidly, the second targets proteins to the chloroplasts. Non-fused and fused proteins were introduced into plants and proteins were expressed. However, as one might expect, only the non-fused protein constructs produced the virus-like particles (VLPs) in the transgenic plants that are necessary for immunogenic effects. The translational fusion upset the formation of the correct protein structure that produces VLPs.

Much of the initial work on producing vaccines in plants was carried out in inedible plants, such as tobacco. There has been a move by some scientists towards producing vaccines in plants that could be eaten raw, such as tomato and banana. Bananas are considered to be the ideal production system by some scientists as they are grown widely in countries of the developing world and can be eaten raw. For use in animals, it is possible to consider the use of fodder crops or other food crops. In Case study 11.6, a recent study on using tomatoes as the food plant is presented.

CASE STUDY 11.6 Edible vaccines

Norwalk virus is a major cause of acute gastroenteritis in humans. Estimates suggest it is responsible for greater than 96% of the outbreaks of non-bacterial forms of the disease in developed and developing countries. Norwalk virus is a positive-sense, single-stranded, non-enveloped RNA virus with its second open reading frame coding for a single capsid protein (NVCP). When the coding sequence for the CP (rNV) is expressed either in plants, such as potato, or in baculovirus/insect cell-suspension systems the protein self-assembles into empty VLPs lacking viral RNA (see Chapter 8; arabis mosaic virus). These particles contain 90 dimers of the NVCP, and are stable at low pH when lyophilized and stored at 4°C. The insect-derived material is also immunogenic in experimental animals and human volunteers. This indicates that the VLPs are potentially useful for vaccines. However, the aim is reduce the cost of production, to get away from needles, and to provide raw material that can be stored without using refrigeration. This precludes the use of the insect cell system and to some extent the use of potato as a source of vaccine. Potato fails on several counts: in previous studies the expression levels in the potato tuber were low and the volunteer had to consume raw potato. In a recent study, several of these problems were addressed by using a plant-optimized gene transformed into tomato plants. The gene (*sNVCP*) was engineered to contain dicotyledonous-plant-favoured codons, and 28 CG dinucleotides were removed along with several methylation sites and potential polyadenylation motifs. The engineered gene was introduced into tomatoes and potatoes where an increase in protein levels over previous amounts was found. In tomato fruit, the protein levels were up to 8% of the total soluble protein. However, in potato tubers, it only reached 0.4%. The protein was shown to assemble into VLPs and was subsequently used with a mouse model to establish whether they would bring about an immune response. It was found that four doses of 0.4 g freeze-dried tomato material containing 64 μg rNV (equivalent to 40 μg VLPs) induced Norwalk virus-specific serum IgG and mucosal IgA in up to 80% of the mice tested. Doses of 0.8 g elicited a response in all mice. For the potato, 1 g freeze-dried potato tuber containing 120 μg rNV (90 μg VLPs) was required. Although there is not a big difference between the values, it was noted that on rehydration the freeze-dried potatoes displayed oxidation of the phenolic compounds that reduced the stability of the VLPs, reducing immunogenicity. Freeze drying also may have damaged the tomato fruit membranes and cell matrix, so air-dried material was also tested. Air-dried fruit gave a much stronger immune response for the same amount of material compared with freeze drying.

These observations have certain consequences when taken out of the laboratory into the real world. Tomato seems to be a better production system than potato tubers, giving higher levels of expression and immune response. It may have drawbacks with growing in that it may have to be grown in a greenhouse, but from a regulatory or safety point of view, containment seems the best option at present. Also, air drying represents a better low-tech option for use in a developing country. Perhaps the most attractive point of this system is that in some people's view eating tomato fruit is more acceptable to children than eating raw potato tubers.

These so-called edible vaccines do work in the laboratory, but would they really work as originally envisaged? The major flaw in the process is that one needs to control very carefully the dose being given to patients if effective and safe protection against disease is to be ensured. One of the drawbacks of edible vaccines is that it may prove very difficult to control the dose administered if consumption as part of a foodstuff is the means of delivery. Charles Arntzen, one of the main lights in this field, has recently expressed the view that the phrase 'edible vaccines' is not acceptable as it implies plant products that can be eaten directly after harvest. Bananas containing vaccines could be picked and eaten without any processing. His group produces vaccines in tomato, but they are partially processed so that food-based tablets containing a known dose of vaccine can be produced. Not only is this better science but it will be easier to license the use of these vaccines from plant sources. Such vaccines should be considered as vaccines that come from a plant source and the term edible vaccines should not be used. Bananas may, of course, still prove to be a useful system that allows locally based production facilities to be established in developing countries.

Other proteins: biopharmaceuticals

A wide range of proteins of pharmaceutical interest has been expressed in transgenic plants in the hope of producing an economically viable system for large-scale production (Table 11.7). Although active recombinant proteins have been produced, one problem associated with production in plant systems is that these often give a

Table 11.7 Examples of biopharmaceuticals produced in plants

Recombinant protein	Origin	Plant	Application
Protein C	Human	Tobacco	Anticoagulant
Hirudin	*Hirudo medicinalis*	Canola	Anticoagulant
Somatotrophin	Human	Tobacco	Growth hormone
β-Interferon	Human	Rice/turnip/tobacco	Treatment for hepatitis B + C
Serum albumin	Human	Tobacco	Burns/fluid replacement, etc.
Haemoglobin-α and -β	Human	Tobacco	Blood substitute
Homotrimeric collagen	Human	Tobacco	Collagen
α_1-Antitrypsin	Human	Rice	Cystic fibrosis, haemorrhages
Aprotinin (trypsin inhibitor)	Human	Maize	Transplant surgery
Lactoferrin	Human	Potato	Antimicrobial
ACE	Human	Tobacco/tomato	Hypertension
Enkephalin	Human	*Arabidopsis*/canola	Opiate
Trichosanthin-α	*Trichosanthes kirilowii*	Tobacco	HIV therapy, cancer

ACE, angiotensin-converting enzyme; HIV, human immunodeficiency virus.
(Data taken with permission from Hemming (1995), Kusnadi *et al.* (1997), Giddings *et al.* (2000), and Daniell *et al.* (2001).)

relatively low product yield and recovery. Various strategies to overcome this problem are being developed, among them the use of novel purification systems (see the section on oleosins for an example) and chloroplast transformation (which is discussed in Chapters 3 and 12). Despite some difficulties, plants hold out great promise as production systems for pharmaceutical proteins, and we will look at some examples in a little more detail.

Trichosanthin

Trichosanthin is a component of the tuber of the plant *Trichosanthes kirilowii*, which is used in Chinese medicine. A ribosome-inactivating protein, trichosanthin inhibits tumour growth and the immune response, and may be useful as a treatment for HIV/acquired immunodeficiency syndrome (AIDS). Relatively high levels of trichosanthin accumulation (2% of total soluble protein 2 weeks after inoculation) have been achieved in *N. benthamiana* using a viral RNA-based transfection system. This system offers the possibility of rapid, large-scale production of trichosanthin. The rapid production also offers the opportunity to rapidly produce mutated forms of trichosanthin for screening in trials aimed at identifying forms with increased efficacy or reduced side effects.

Glucocerebrosidase

Gaucher's disease is an inherited disorder in which glucocerebroside (a component of red blood cells that is normally broken down into glucose and ceramide as old and damaged red blood cells are removed by the body) accumulates in lysosomes, due to a deficiency in the enzyme glucocerebrosidase. Gaucher's disease, which results in swelling of the spleen and liver and severe bone damage, can be extremely debilitating and painful. Currently, treatment is based on managing the symptoms or by treatment with a drug developed from glucocerebrosidase purified from human placentas. However, a huge amount is required (9–11 t of placentas a year for one patient), making this one of the world's most expensive drugs. Recombinant approaches to production in mammalian cell cultures have reduced the cost to some extent, although it remains an extremely expensive drug.

A process to produce glucocerebrosidase in tobacco has recently been patented, which will hopefully result in cheaper glucocerebrosidase being available for patients with Gaucher's disease.

Human serum albumin

Human serum albumin (HSA) has many potential medical applications (the treatment of burns and in liver cirrhosis, for example). HSA has been expressed in tobacco and potato under the control of the 35S promoter. Two forms of the HSA protein were expressed, with different signal sequences to ensure secretion of the HSA. One form had the human prepro sequence, while the other had the signal sequence from the tobacco extracellular PR-S protein. Although both forms of HSA were secreted, analysis showed that the form with the human prepro sequence was not properly processed, leading to proHSA being secreted. However, the form with the PR-S signal sequence was correctly processed, and mature HSA indistinguishable from the human protein was produced.

Arginine vasopressin

The examples we have looked at demonstrate that both plant and non-plant pharmaceutical proteins, which are of high value both clinically and economically, can be successfully produced in plants. However, in what is a typical situation in developing technologies, not all attempts at protein production in plants have been successful.

Attempts to express arginine vasopressin in tobacco cultures, for example, have proved unsuccessful due to protein instability and retarded cell growth.

Economic and regulatory considerations for molecular farming

Although there has been much discussion in this chapter of suitable targets for molecular farming, few of the examples described have yet been produced commercially. This may result from problems with the technology itself (see the section on cyclodextrins), or it could reflect the long time between initial laboratory-based demonstration and subsequent optimization and regulatory approvals for field release and approval for food or drug use. However, it may be the case that levels of production that represent major scientific breakthroughs are nonetheless insufficient to be economically viable. This may seem surprising, because it is tempting to make the naïve assumption that the production of compounds in a field, with free energy from the sun, must be cheaper than production in a chemical factory or industrial fermenter (Box 11.3 gives an interesting example of the economics involved in petroselenic acid

BOX 11.3 *The cost of producing petroselenic acid in transgenic soybeans*

A number of factors need to be taken into consideration when determining the economics of molecular farming. These can be exemplified by a calculation of the cost of producing a petroselenic acid-rich oil in transgenic soybean. A major factor that makes soybean an attractive crop for molecular farming is the current high volume and low cost of production. Thus, in the USA, about 6.5×10^{10} kg of soybeans are produced annually, at a cost of about \$0.24 kg^{-1}. The low cost of soybean production permits soybean oil to be priced in the range of \$0.41–0.53 kg^{-1}, which compares favourably with many petrochemical products. There are therefore considerable benefits in using a high-volume/low-cost crop like soybean as a platform for molecular farming. However, using a highly developed crop to produce a speciality product also incurs costs.

One factor to be taken into account is that the productivity of major crops like soybean is maintained by a continuous breeding programme that constantly improves traits like yield and pest resistance, and develops different cultivars for different environments. It is not then sufficient to introduce a transgenic trait, even in an elite background, without considering the effects of this constant improvement. The transgenic crop must either accept an ever-decreasing relative yield, or try to keep pace with the ongoing breeding programmes. Traits like yield are generally polygenic, so they need to be selected empirically from large breeding populations. Since each unlinked

(Continued overleaf)

BOX 11.3 *Continued*

locus that is required to be present in the crop increases the size of the breeding population by a factor of two, the introduction of three genes (to produce bioplastics, for example) would need an eightfold increase in the size of the breeding programme. Even if the transgenic crop keeps pace with the rate of improvement, the initial introduction of the transgene into a particular elite line is estimated to result in a yield penalty of 3% by the time that outdated cultivar reaches the field. This yield penalty will need to be paid to the crop producer to compensate for the loss of yield from growing normal elite soybean. At a 3% reduction in yield, the farmer would need to be paid an extra 3% of \$0.24 kg^{-1}. The speciality crop therefore costs an extra \$0.007 kg^{-1} to produce.

Another cost associated with the production of a transgenic compound in a major commodity crop is the cost of what is called identity preservation; that is, keeping the speciality crop separate from the major bulk crop. This is estimated to be about 6% of the price of the bulk crop (for simple identity preservation), so for soybean this would be about \$0.015 kg^{-1}. The total cost is therefore \$0.24 + 0.007 + 0.015 = \$0.262 kg^{-1}.

Since the seed oil content of soybean is 20%, and since only the oil in this speciality crop is of increased value, these increased direct costs will be added only to the oil. Thus, with an added production cost of \$0.022 kg^{-1} for the unprocessed bean, the oil will cost an added 5 × \$0.022 = \$0.11 kg^{-1}. For example, if the normal cost of soybean oil was \$0.58 kg^{-1}, the speciality oil would cost \$0.69 kg^{-1} to produce.

The next consideration is the cost of extraction and processing. Petroselenic acid (18:1 Δ^6) can be cleaved by ozonolysis to form adipic acid (which is a potential precursor to the nylon 6,6 monomers) and lauric acid. Assuming that petroselenic acid could be engineered to the same level as high-oleic-acid soybeans (85%), it is estimated that 1 kg of oil would produce 0.4 kg of adipic acid, 0.55 kg of lauric acid, 0.1 kg of glycerol, and 0.26 kg of other fatty acids. Given a market value of \$1.48 kg^{-1} each for adipic acid and lauric acid, \$1.01 kg^{-1} for glycerol, and a minimal value for the other fatty acids (\$0.07 kg^{-1}), the total return is \$1.52 kg^{-1} oil (\$0.59 for adipic acid + \$0.81 for lauric acid + \$0.10 for glycerol + \$0.02 for the other fatty acids).

However, the cost of similar ozonolysis processes is calculated to be \$1.29 kg^{-1}. This means that with a starting material cost of \$0.69 kg^{-1}, the total production costs would be \$1.98 kg^{-1} compared with a total return of \$1.52 kg^{-1}. Thus, production of petroselenic acid in soybeans, even at levels of 85% of total fatty acids, would not be profitable based on these figures. (Calculations from Hitz (1999).)

production in transgenic soybean). This is often not the case, because farming is not a cost-free activity, the extraction of the product from the plants may be a difficult and expensive process, and the price of the product must be significantly higher than that of the crops it has replaced. This may mean that the current levels of synthesis in transgenic plants coupled with the costs of production make a number of the examples described above uneconomic when compared with current production methods. On the other hand, continuing improvements in plant biotechnology, coupled with the inevitable rise in costs of non-renewable resources, such as petrochemicals, make it likely that many of these products will eventually be made in plants.

The economics of producing medically related proteins in plants has been much discussed, particularly because of the high unit price of many of these proteins and the expansion of the market for medically related proteins. It is in this area that many companies and people have invested money, in the hope of achieving economically viable molecular farming in plants. Fortunately for these companies and investors, the figures seem to suggest that there is a future for molecular farming. It has been estimated that the cost of producing immunoglobulins in alfalfa, in a 250-m^2 greenhouse, is US$500–600 g^{-1} compared with $5000 g^{-1} for a hybridoma-produced antibody. The company Planet Biotechnology has compared the costs of producing an IgA antibody in four different systems. The comparison was between production in a cell culture system, transgenic goats, grain (assuming a yield of 7.5 t ha^{-1}), and green biomass, such as leaves (assuming a yield of 120 t ha^{-1}). The cost of an IgA produced in leaves was estimated to be below $50 g^{-1}. This compared very favourably with the costs for production in a cell culture system ($1000 g^{-1}) and in a transgenic animal ($100 g^{-1}). It must be remembered that that these figures are fairly crude, being based on projected yields etc. However, they do illustrate that costs of protein production in plants are at least comparable with other systems of production, and further developments are now coming on stream. These include the use of the chloroplast as a production system. The advantages and disadvantages of using the organelle approach are discussed in Box 11.4.

The first three plant-produced proteins that made it to market were avidin, trypsin, and GUS (ProdiGene maize system). These were mainly for laboratory use so did not suffer too much regulation. Other proteins are moving towards commercialization and the market place. These include proteins such as the interferons α-2a and α-2b (Large Scale Biology and Sigma-Aldrich Co.), which are in a scale-up phase, and a number that are in clinical trials. Plants, as production systems for pharmaceuticals, require several different forms of regulation and research studies, particularly if they are to be released into the environment. Like food crops, they are controlled in Europe by various European Union (EU) regulations and in the USA by the US Department of Agriculture (USDA)/Animal and Plant Health Inspection Service (APHIS) process. The products are also governed by agencies that oversee the production of pharmaceuticals, the European Agency for the Evaluation of Medicine (EMEA) in Europe and the FDA in the USA. As a consequence, the pharmaceuticals are required to go through various clinical trials. A number are at the phase I level, which involves a small-scale trial (20–80 people) looking at a number of points, such as safety, tolerability, and dose optimization. Products at this stage include HBsAg (hepatitis B) vaccine from lettuce and potato, Norwalk virus capsid protein vaccine from potato, and a rabies glycoprotein produced using viral vectors in spinach. Also in phase I trials are some dozen or so personalized antibodies for the treatment of non-Hodgkin's lymphoma (see Case study 11.5). Some products have made it to phase II trials: these trials involve larger groups of people and are designed to assess the clinical efficacy of the therapy. They also continue to consider the parameters of the phase I trials.

At this level are proteins such as CaroRx™ manufactured by Planet Biotechnology. This is the sIgA for tooth decay discussed in detail above. It binds to *S. mutans*, the

● sIgA near to commercialization.

BOX 11.4 *Chloroplasts: a clean, high-level expression system*

Chloroplast systems have many advantages over other approaches.

As we have seen in Chapter 3, transformation systems are now available for the introduction of foreign genes into chloroplasts, giving an estimated 10 000 copies per cell. Using such systems, it is claimed that recombinant protein content has reached as much as 47% of the total soluble protein. An example given in Table 11.7, somatotrophin (growth hormone), produced in chloroplasts, reached about 7% of the total soluble protein. This compares favourable with that of a nuclear transgene (0.01% total soluble protein). Using chloroplast-based plant systems has other potential advantages over existing systems. Despite the very high number of gene copies, there have been no reports of gene silencing that could affect expression levels. Also, protein processing is one of the most expensive parts of producing proteins in existing systems such as *E. coli*. It seems that chloroplasts can assemble and fold complex foreign proteins, but whether they can cope with the other post-translational modification problems remains to be seen.

Chloroplasts are also able to cope with genes that have unusual structure. The tetanus toxin fragment C (TetC) has been difficult to express in other recombinant systems. This gene has a very high AT content (73%), which *E. coli* was unable to cope with. Expression was improved when a gene with an engineered codon bias was used. With the potential for toxic cell-wall pyrogens from *E. coli*, the use of yeast systems has been investigated. This was not a success either, as the gene contained several fortuitous polyadenylation sites that led to truncated mRNAs being made. To get expression in yeast, a high GC sequence lacking the polyadenylation sites was produced. Then the protein synthesized did not act as an immunogen because it was glycosylated. Both high AT and the high GC versions of the *TetC* gene have been introduced into tobacco chloroplasts where levels of 25 and 10% total soluble cellular protein were obtained for the different genes. Nasal immunization of mice with concentrated protein solution induced protective levels of TetC antibodies.

Another advantage of chloroplasts is that they are maternally inherited. This means that they will not be spread via pollen to non-transgenic systems. This type of clean-gene technology is most important for getting approval for growth in non-contained facilities.

Health-related proteins produced in chloroplasts include human proinsulin for the treatment of autoimmune disease, interferon-α 2b, human serum albumin, and synthetic human haemoglobin genes and antibodies such as the Guy's 13 monoclonal antibody.

One disadvantage or challenge for any molecular farming system is low-cost purification/processing. Being encased within the organelle may help the process of removing the protein away from the other cellular material. Again, with some proteins or enzymes, purification may not be necessary. This is the case for vaccines. Several potential vaccines have been produced, including the cholera toxin B subunit, *Bacillus anthracis* protective antigen, a fusion between the F1 (capsular protein) and the V antigen (a component of the type III protein-secretion system) from *Y. pestis*, and the canine parvovirus VP2 antigen. These have all been expressed well and in some cases have given some evidence of correct folding and immunogenicity. Instead of making them in tobacco leaf chloroplasts, some work has been done towards expressing the proteins in edible material such as the carrot root. Expression levels are not as high as in leaf material, but sufficient organelles are present to make it effective.

bacterium that causes tooth decay, preventing them binding to the teeth. (The company has also produced another product in plants called RhinoRx™ in which they have fused a human receptor for the rhinovirus to a Protected-SIgA™, which gives the complex greater stability when sprayed into the throat on infection. The idea is that the virus will bind this molecule and not the host throat.) Also in phase II trials are several proteins, such as gastric lipase for use in cystic fibrosis and pancreatitis, and lactoferrin for gastrointestinal infections (Meristem Therapeutics), and a vaccine for coccidiosis in poultry (Guardian Biotechnologies).

Several proteins have recently made it to market. Another protein for diagnostic use, human intrinsic factor, has been commercialized by Cobento, a Danish firm. Dow Chemicals has USDA approval for a vaccine against Newcastle disease virus of poultry. Interestingly, this vaccine was produced in a non-nicotine tobacco plant cell culture. This approach gets over some of the problems associated with release of engineered plants. It is vitally important that there is no cross-contamination of feed or food crops with these plants. Measures must be in place to prevent this happening pre- and post-harvest. ProdiGene accepted a civil penalty of US$250 000 (plus clean-up costs) after volunteer transgenic maize was found growing in a soybean crop 1 year after the pharmaceutical maize was harvested.

● Will the investors get a return?

SUMMARY

This chapter has illustrated the wide range of compounds that have already been produced in transgenic plants and those that might be valuable targets for plant biotechnology in the future. The chapter has concentrated on three major classes of biological macromolecule: carbohydrates, lipids, and proteins. It should be borne in mind that there is an extremely broad range of secondary metabolites that could also be produced in transgenic plants, but these are somewhat beyond the scope of this book.

In the case of carbohydrates, two types of product were identified. On the one hand, existing bulk carbohydrate products such as starch may be modified to enhance their utility. On the other hand, there is the potential to produce novel, high-value carbohydrate products in high-yielding crops. The most successful of these to date has been the development of the fructan beet, but many of the other attempts to produce novel carbohydrates have met with less success. Nevertheless, many of these studies have been useful in terms of the growing body of knowledge concerning the regulation of carbohydrate metabolism and the partitioning of carbon between different compartments and tissues in the plant.

The precise manipulation of lipid metabolism requires a detailed understanding of the role of different enzymes in producing the profile of different fatty acids in different plants. There is a considerable amount of fundamental research still required in this area before designer oils with a uniform composition become commonplace. On the other hand, the progress towards the production of bioplastics in plants has been impressive, partly because this can be achieved by importing well characterized pathways from bacteria that do not exist in the plant, rather than having to fine-tune complex plant pathways. Nevertheless, the import of this pathway into plants does have a significant impact on the host-plant metabolism, and it is still proving difficult to produce commercially significant amounts of this product in crop plants.

A wide variety of proteins has been expressed in plants, although the majority of these have not been produced on a commercial scale, and therefore many questions about the viability of using plants for producing proteins remain unanswered. Improvements in the technology of protein production in plants will undoubtedly come and, like lipid metabolism, fundamental research into protein processing, post-translational modification, and stability will make a contribution. These developments should see the costs of protein production in plants being reduced, although costs of development and regulatory approval will still be significant. It has been estimated that a plant-derived pharmaceutical will cost US$50 million merely to develop.

Although the future of protein production in plants may be uncertain, the production of recombinant proteins in plants provides some of the best examples of where plant biotechnology can make a positive impact on human (and animal) health and quality of life.

FURTHER READING

Visit www.oxfordtextbooks.co.uk/orc/slaterplants2e/ for hyperlinked references that take you directly to online abstracts for the journal articles listed below:

Plant biochemistry

Buchanan, B. B., Gruissem, W., and Jones, R. L. (2000) *Biochemistry and Molecular Biology of Plants*. American Society of Plant Physiologists, Rockville, MD.

Heldt, H.-W. (1997) *Plant Biochemistry and Molecular Biology*. Oxford University Press, Oxford.

Molecular farming

Cappell, T. and Christou, P. (2004) Progress in plant metabolic engineering. *Current Opinion in Biotechnology* **15**, 148–154.

Goddijn, O. J. M. and Pen, J. (1995) Plants as bioreactors. *Trends in Biotechnology* **13**, 379–387.

Hitz, B. (1999) Economic aspects of transgenic crops which produce novel products. *Current Opinion in Plant Biology* **2**, 135–138.

Carbohydrates

Andersonn, M., Melander, M., Pojmark, P., Larsson, H., Bulow, L., and Hofvander P. (2006) Targeted gene suppression by RNA interference: an efficient method for production of high-amylose potato lines. *Journal of Biotechnology* **123**, 137–148.

Caimi, P. G., McCole, L. M., Klein, T. M., and Kerr, P. S. (1996) Fructan accumulation and sucrose metabolism in transgenic maize endosperm expressing a *Bacillus amyloliquefaciens SacB* gene. *Plant Physiology* **110**, 355–363.

Cairns, A. J. (2003) Fructan biosynthesis in transgenic plants. *Journal of Experimental Botany* **54**, 549–567.

Ebskamp, M. J. M., Van der Meer, I. M., Spronk, B. A., Weisbeek, P. J., and Smeekens S. C. M. (1994) Accumulation of fructose polymers in transgenic tobacco. *Biotechnology* **12**, 272–275.

Hellwege, E. M., Czapla, S., Jahnke, A., Willmitzer, L., and Heyer, A. G. (2000) Transgenic potato (*Solanum tuberosum*) tubers synthesize the full spectrum of inulin molecules naturally occurring in globe artichoke (*Cynara scolymus*) roots. *Proceedings of the National Academy of Sciences USA* **97**, 8699–8704.

Jobling, S. (2004) Improving starch for food and industrial applications. *Current Opinion in Plant Biology* **7**, 210–218.

Oakes, J. V., Shewmaker, C. K., and Stalker, D. M. (1991) Production of cyclodextrins, a novel carbohydrate, in the tubers of transgenic potato plants. *Biotechnology* **9**, 982–986.

Schwall, G. P., Safford, R., Westcott, R. J., Tayal, A., Shi, Y. C., Gidley, M. J., and Jobling A. S. (2000) Production of very-high-amylose potato starch by inhibition of SBE A and B. *Nature Biotechnology* **18**, 551–554.

Sevenier, R., Hall, R. D., Van der Meer, I. M., Hakkert, H. J. C., van Tunen, A. J., and Koops, A. J. (1998) High level fructan accumulation in a transgenic sugar beet. *Nature Biotechnology* **16**, 843–846.

Slattery, C. J., Kavakli, H., and Okita, T. W. (2000) Engineering starch for increased quantity and quality. *Trends in Plant Science* **5**, 291–298.

Tanaka, N., Fujita, N., Nishi, A., Satoh, H., Hosaka, Y., Ugaki, M., Kawasaki, S., and Nakamura, Y. (2004) The structure of starch can be manipulated by changing the expression levels of starch branching enzyme IIb in rice endosperm. *Plant Biotechnology Journal* **2**, 507–516.

Van der Meer, I. M., Ebskamp, S. C. M., Visser, R. G. F., Weisbeek, P. J., and Smeekens, S. C. M. (1994) Fructan as a new carbohydrate sink in transgenic potato plants. *Plant Cell* **6**, 561–570.

Lipids and bioplastics

Bohmert, K., Balbo, I., Kopka, J., Mittendorf, V., Nawrath, C., Poirier, Y., Teschendorf, G., and Trethewey, R. N. (2000) Transgenic *Arabidopsis* plants can accumulate polyhydroxybutyrate to up to 4% of their fresh weight. *Planta* **211**, 841–845.

Bohmert, K., Balbo, I., Steinbuchel, A., Tischendorf, G., and Willmitzer, L. (2002) Constitutive expression of the β-ketothiolase gene in transgenic plants. A major obstacle for obtaining polyhydroxybutyrate-producing plants. *Plant Physiology* **128**, 1282–1290.

Conrad, U. (2005) Polymers from plants to develop biodegradable plastics. *Trends in Plant Science* **10**, 511–512.

Houmiel, K. L., Slater, S., Broyles, D., Casagrande, L., Colburn, S., Gonzalez, K., Mitsky, T. A., Reiser, S. E., Shah, D., Taylor, N. B., Tran, M., Valentin, H. E., and Gruys K. J. (1999) Poly(β-hydroxybutyrate) production in oilseed leucoplasts of *Brassica napus*. *Planta* **209**, 547–550.

John, M. and Keller, G. (1996) Metabolic pathway engineering in cotton: biosynthesis of polyhydroxybutyrate in fiber cells. *Proceedings of the National Academy of Science USA* **93**, 12768–12773.

Mittendorf, V., Robertson, E. J., Leech, R. M., Kruger, N., Steinbuchel, A., and Poirier, Y. (1998) Synthesis of medium-chain-length polyhydroxyalkanoates in *Arabidopsis thaliana* using intermediates of peroxisomal fatty acid β-oxidation. *Proceedings of the National Academy of Sciences USA* **95**, 13397–13402.

Napier, J. A., Michaelson, L. V., and Stobart, A. K. (1999) Plant desaturases: harvesting the fat of the land. *Current Opinion in Plant Biology* **2**, 123–127.

Neumann, K., Stephan, D. P., Ziegler, K., Huhns, M., Broer, I., Lockau, W., and Pistorius, E. K. (2005) Production of cyanophycin, a suitable source for the biodegradable polymer polyaspartate, in transgenic plants. *Plant Biotechnology Journal* **3**, 249–258.

Poirier, Y., Nawrath, C., and Somerville, C. (1995) Production of polyhydroxyalkanoates, a family of biodegradable plastics and elastomers, in bacteria and plants. *Biotechnology* **13**, 142–150.

Scheller, J. and Conrad, U. (2005) Plant-based material, protein and biodegradable plastic. *Current Opinion in Plant Biology* **8**, 188–196.

Thelen, J. J. and Ohlrogge, J. B. (2002) Metabolic engineering of fatty acid biosynthesis in plants. *Metabolic Engineering* **4**, 12–21.

Topfer, R., Martini, N., and Schell, J. (1995) Modification of plant lipid synthesis. *Science* **268**, 681–685.

Wrobel, M., Zebrowski, J., and Szopa J. (2004) Polyhydroxybutyrate synthesis in transgenic flax. *Journal of Biotechnology* **107**, 41–54.

Proteins

Boothe, J. G., Saponja, J. A., and Parmenter, D. L. (1997) Molecular farming in plants: oilseeds as vehicles for the production of pharmaceutical proteins. *Drug Development Research* **42**, 172–181.

Daniell, H., Streatfield, S. J., and Wycoff, K. (2001) Medical molecular farming: production of antibodies, biopharmaceuticals and edible vaccines in plants. *Trends in Plant Science* **6**, 219–226.

Daniell, H., Chebolu, S., Kumar, S., Singleton, M., and Falconer, R. (2005) Chloroplast-derived vaccine antigens and other therapeutic proteins. *Vaccine* **23**, 1779–1783.

Fischer, R., Schumann, D., Zimmermann, S., Drossard, J., Sack, M., and Schillberg, S. (1999) Expression and characterization of bispecific single-chain Fv fragments produced in transgenic plants. *European Journal of Biochemistry* **262**, 810–816.

Giddings, G., Giddings. G., Allison, G., Brooks, D., and Carter, A. (2000) Transgenic plants as factories for biopharmaceuticals. *Nature Biotechnology* **18**, 1151–1155.

Gil, F., Brun, A., Widgorovitz, A., Catalá, R., Martinez-Torrecuadrada, J. L., Casal, I., Salinas, J., Borca, M. V., and José, M. E. (2001) High-yield expression of a viral peptide vaccine in transgenic plants. *FEBS Letters* **488**, 13–17.

Gleba, Y., Marillonnet, S., and Klimyuk, V. (2004) Engineering viral expression vectors for plants: the 'full virus' and the 'deconstructed virus' strategies. *Current Opinion in Plant Biology* **7**, 182–188.

Goddijn, O. J. M. and Pen, J. (1995) Plants as bioreactors. *Trends in Biotechnology* **13**, 379–387.

Hemming, D. (1995) Molecular farming: using transgenic plants to produce novel proteins and other chemicals. *AgBiotech News and Information* **7**, 19N–29N.

Hiatt, A. (1990) Antibodies produced in plants. *Nature* **344**, 469–470.

Kusnadi, A. R., Nikolov, Z. L., and Howard, J. A. (1997) Production of recombinant proteins in transgenic plants: practical considerations. *Biotechnology and Bioengineering* **56**, 473–484.

Ma, J. K.-C. and Hein, M. B. (1995) Immunotherapeutic potential of antibodies produced in plants. *Trends in Biotechnology* **13**, 522–527.

Ma, J. K.-C., Hiatt, A., Hein, M. B., Vine, N. D., Wang, F., Stabila, P., van Dolleweerd, C., Mostov, K., and Lehner, T. (1995) Generation and assembly of secretory antibodies in plants. *Science* **268**, 716–719.

Ma, J. K.-C., Chikwamba, R., Sparrow, P., Fischer, R., Mahoney, R., and Twyman, R. M. (2005) Plant-derived pharmaceuticals—the road forward. *Trends in Plant Science* **10**, 580–585.

Miele, L. (1997) Plants as bioreactors for biopharmaceuticals: regulatory considerations. *Trends in Biotechnology* **15**, 45–49.

Peeters, K., De Wilde, C., and Depicker, A. (2001) Highly efficient targeting and accumulation of a Fab fragment within the secretory pathway and apoplast of *Arabidopsis thaliana*. *European Journal of Biochemistry* **268**, 4251–4260.

Santi, L., Giritch, A., Roy, C. J., Marillionnet, S., Klimyuk, V., Gleba, Y., Webb, R., Arntzen, C. J., and Mason, H. S. (2006) Protection conferred by recombinant *Yersinia pestis* antigens produced by a rapid and highly scalable plant expression system. *Proceedings of the National Academy of Sciences USA* **103**, 861–866.

Staub, J. M., Carcia, B., Graves, J., Peter, T. J., Hunter, P., Nehra, N., Paradkar, V., Schilittler, M., Carroll, J. A., Spatola, L., Ward, D., Ye, G., and Russell, D. A. (2000) High-yield production of a human therapeutic protein in tobacco chloroplasts. *Nature Biotechnology* **18**, 333–337.

Thanavala, Y. Huang, Z., and Mason, H. S. (2006) Plant-derived vaccines: a look back at the highlights and a view to the challenges on the road ahead. *Expert Review of Vaccines* **5**, 249–260.

Zhang, X., Buehner, N. A., Hutson, A. M., Estes M. K., and Mason, H. S. (2006) Tomato is a highly effective vehicle for expression and oral immunization with Norwalk virus capsid protein. *Plant Biotechnology Journal* **4**, 419–432.

12

Science and society: public acceptance of genetically modified crops

INTRODUCTION

In this chapter, we will briefly look at some of the concerns surrounding the use of genetically modified (GM) crops, or **biotech crops** as they are now often known. What are the concerns of the public? This is not a difficult question to answer as surveys have been undertaken in a number of countries over the last decade. What is difficult is to know how to interpret the feedback. For most people, their only source of information has been, and still is, sensational headlines in tabloid newspapers. Therefore, views are being formed from a very limited knowledge base. Some authors go further and suggest that these views are formed from misinformation spread by interest groups that are prejudiced or hold some extreme moral or ethical views that are anti-biotechnology. To this should also be added a lack of trust in information given out by governments and the big corporations that control agriculture, both of whom the public see as having their own agendas.

The aim of the chapter is not to prescribe any particular view but to highlight some of the difficult issues so that the reader can consider their own position in the debate and be stimulated to get involved. The first part of this chapter will consider some of the data on the public concerns and then the current status of GM crops in world agriculture. We will look at some of the measures that can be taken to minimize the risk associated with the crops and attempts to get this over to the public and then see how GM crops and food are regulated to ensure safety. Chapter 13 will take this further by looking at future developments in genetic modification that may bring further public support.

Public concerns

Surveys have been done in the USA by a number of different organizations (including Texas A&M University, Harris, Gallup, and The Pew initiative on Food and Biotechnology; in 2004) and in the UK as part of the GM Nation initiative (2003). References are given later in the chapter where more complete analysis can be viewed, but here a few headline points will be discussed. In the UK, the aim of the survey was to gauge public opinion in 2003 about GM food and crops and to identify any shifts in opinion due to the GM Nation public debate. This was supposed to be a national event that allowed everyone in the UK to express their opinion on GM foods and to become more educated on the subject. It was part of 'The national dialogue' set up by the UK Government that had three main strands: a public debate, a review of the science around genetic modification, and an economics study. Some of the information from these studies is available on the Internet at the UK GM science review website (see web-link 12.16).

The first issue was to put genetic modification is context: how important was it to people compared with other personal or social issues? In the survey, personal issues scored highly, with social issues—and in particular GM crops—among the least important. This is not surprising as in the UK GM foods are not available and are not therefore a major worry to most people. Attitudes to GM food were also evaluated. Only 14% of responders thought GM food a good thing, while 40% said it was neither good nor bad, and 40% thought it was a bad thing. A similar pattern is found when asking whether GM food should be opposed: more than half of the public were not sure whether GM food should be promoted or not. For benefit versus risk, 42% of the sample thought the risks of GM food outweighed the benefits, 23% that they are about the same, and 20% that benefits outweigh the risks. The survey indicated that the public who responded had complex views on the subject and not simple yes/no beliefs. It showed that there had been an increase in the level of concern in the UK and that direct support seemed to have fallen over the decade that GM food had been in the public domain but that the public were more ambivalent toward the complex issues. This was reinforced by investigation of specific issues. The majority, 85% of the public sampled, thought that we do not know enough about the potential long-term effects of GM food on our health. They were also concerned about the potential negative impact on the environment of GM crops. However, a substantial number thought that there could be a range of future benefits, for the environment, consumers, and developing countries.

In the USA, the surveys highlight the lack of knowledge about GM foods and engineered animals. Where there are issues, these seem greatest about cloning animals and the safety of GM foods. Most people don't want the technology stopped; rather, they want robust regulation in place to ensure safety and that moral and ethical points are covered in the regulations. The apparent risks of such foods are not as great as with the chemical or nuclear industries. The majority of North Americans covered in these surveys say they avoid eating GM food but a majority also said they would buy GM food if it was cheaper, tasted better, or was produced using fewer chemicals. This is interesting as over half the foods in supermarkets have some GM component. The majority wished research to continue for the benefits that would occur in the future. Most surveys indicated a 50/50 split for and against GM food.

In both sets of surveys, the biggest issue along with food safety is trust. Many people see the introduction of GM plants as being the prime example of the big corporation dominating the rights of the individual. It is difficult to concede that private companies should be able to claim ownership of or exploitation rights of genetic

- The main issues over biotech food are trust and safety.

material, even whole organisms. Yet in the USA they were given this right in the early 1980s by the Supreme Court and laws were introduced that gave seed companies the exclusive rights on material they developed. It was also bad publicity for companies to then impose contracts to prevent farmers from planting harvested seeds in subsequent years or to introduce terminator technology that leads to infertile seed. This is particularly important to the 1.4 billion people that depend on saved seed to grow the next season's plants. Another issue that the public has had difficulties with is that the first round of plants produced for the market seemed to be just for the benefit of the big corporations. Herbicide-resistant plants were produced that required the herbicides produced by the same companies. Some of these and further issues will be discussed below. One final point is that during the British survey, the point that the Government was driving the campaign for GM plants was highlighted. The Government was not seen as an honest broker: it was felt that information on genetic modification from the Government was tainted by their agenda. This has also been linked with a loss of trust for the UK Government that related to the Iraq war. Despite these reservations, the amount of GM plantings continues to increase; this will be discussed below.

The current state of transgenic crops

It seems appropriate to spend a little time reflecting on the current status of GM plants in broad terms. Despite public opposition and political difficulties in some regions of the world, particularly in Europe, the area of land cultivated with transgenic crops continues to increase. Figures from the International Service for the Acquisition of Agri-Biotech Applications (ISAAA) and summarized on the website GMO compass (web-link 12.15; a site funded by the European Union (EU) but written by independent science journalists) clearly show the importance of transgenic crops in both developed and developing countries (Box 12.1). It was estimated that in 2001 over 52 million ha, or 130 million acres, of land were planted with transgenic crops compared with only 1.7 million ha in 1996 when the crops were first being exploited. Since then, the area planted has continued to increase. In 2006, 102 million ha were cultivated. The USA has the greatest amount of land dedicated to transgenic crops, where for instance over 90% of the soybean crop is transgenic. Other countries including Brazil, Argentina, Canada, India, and China also grow significant amounts of GM crops and a further 16 countries are listed as growing crops at commercial levels or for large-scale field trials. Another interesting point is that to quote ISAAA, 'Biotech crops were grown by approximately 10.3 million farmers in 22 countries in 2006. . . . Notably, 90% or 9.3 million of the beneficiary farmers were small resource-poor farmers from developing countries.'

Who has benefited from these first-generation GM crops?

An analysis of the figures in Box 12.1 shows that a large area is planted with GM crops. However, just two traits, herbicide resistance and insect resistance (singly or in

BOX 12.1 *Synopsis of commercial transgenic crop utilization*

Transgenic crop cultivation by country

Six countries are listed as the main adopters of biotech crops. These are listed below.

Country	Area (millions of hectares)		
	1996	2001	2006
USA	1.53	5.7	54
Argentina	0.1	11.8	18
Brazil	–	5.7	11.5
Canada	0.1	3.2	6.1
India	–	–	3.1
China	–	1.5	3.5

Transgenic crop cultivation by crop

Four crops account for nearly 100% of the commercially grown transgenic crops. In the table below, the percentage figure is the proportion of the total area dedicated to transgenic crops made up of that particular crop in 2006.

Crop	Area (millions of hectares)			Proportion of total area (%)
	1996	2001	2006	
Soybean	0.5	33.3	58.6	68
Maize	0.3	9.8	25.2	18
Cotton	0.8	6.8	13.4	39
Canola	0.1	2.7	4.8	21

Other GM crops grown include squash, papaya, rice, and alfalfa.

Transgenic crop cultivation by trait

Herbicide resistance is the trait found in by far the largest area of transgenic crops. With insect resistance, these two traits account for effectively all the area cultivated with transgenic crops. There is also a move towards plants containing more than one trait.

Trait	Area (millions of hectares)		
	1996	2001	2006
Herbicide resistance	0.6	40.6	69.9
Insect resistance (*Bt*)	1.1	7.8	19
Insect resistance (*Bt*) + herbicide resistance	–	4.2	13.1

(Continued overleaf)

● Despite the public worries, the amount of GM crops being grown continues to increase.

BOX 12.1 *Continued*

GM crops growing in EU Member States

Country	Area (millions of hectares)	
	2006	2007
Maize		
Spain	54 000	70 000
France	500	20 000
Czech Republic	1 290	5 000
Portugal	1 250	3 000
Germany	950	2 650
Slovakia	Data not given	
Soybean		
Romania	100 000	

Transgenic crop cultivation by proportion of total area cultivated for each crop in the US in 2007

Crop	Area (millions of hectares)	%
Soybean	23.6	91
Maize	27.4	73
Cotton	3.9	87

The figures in this box can give a misleading picture of the current status of GM crops. Although they are grown on a large scale in countries like the USA, and for some crops make up the majority of the area grown, other countries refuse to allow GM crops to be grown or the products of GM to be imported. This difference in the status of GM crops is particularly pertinent if the situation in the EU is considered. GM maize and soybean are grown in the EU but the amounts are very small compared with the worldwide production of GM crops. This is in part due to the moratorium on licensing new GM products in the EU that was in place from 1998 and due to the activity of various pressure groups. GM crops and products made from GM ingredients still face considerable public antagonism in many EU countries. It is now legal to grow GM plants in all EU countries but few countries do. Within EU regulations, it is legal to prohibit GM crops on sound scientific grounds, for instance in the UK the growth of some crops has been shown to be detrimental to the environment (see below). However, some countries seem to have bent these rules and are being pursued legally. The main GM crop to be used in the EU at present is soybean. The EU imports some 40 million t of meal or soybean annually for processing for oils and other food additives and animal feed. Not all is GM but with nearly 70% of the world crop being transgenic, it is difficult to avoid. Products should be labelled as containing GM material unless it is less than 0.9% admixture.

combination), and four crops, grown predominantly in North America, account for the vast majority of this area. The agribiotech multinationals, like Monsanto and Syngenta, drove the development of these crops. The main beneficiary, other than the agribiotech multinationals, appears to have been the farmer, who has generally enjoyed increased yields and revenues from growing GM crops, despite a sometimes considerable premium on GM seed. Monsanto puts this increase in revenue at US$40.67 per ha ($16.46 per acre) from *Bacillus thuringiensis* (*Bt*) GM maize in areas where the European corn borer is a problem (see Chapter 6). This increased revenue is, in many cases, associated with a decrease in the use of herbicides and/or insecticides (although the figures are disputed). This reduction in the use of potentially harmful chemicals was trumpeted as one of the real benefits of these first-generation GM crops. However, it is interesting to note that surveys in the USA have shown that environmental benefits are not a driver for the adoption of GM technology by farmers. In many cases, but not all, there does seem to be a real reduction in the use of herbicides or pesticides associated with growing GM crops. For example, the introduction of herbicide-tolerant cotton resulted in increased yields but no decrease in herbicide usage, whereas the adoption of herbicide-resistant soybeans led to significant reductions in herbicide use and only small increases in yield (see Chapter 5). Improved weed control can also lead to reduced harvest costs and cleaner crops that can generate higher prices. The herbicides used on GM crops are also considered to be more environmentally friendly than are some of the alternatives. There does therefore appear to be a real environmental benefit associated with these GM crops.

For the period 1996 to 2005, the ISAAA summarizes the net economic benefits to biotech crop farmers as $27 billion ($13 billion for developing countries and $14 billion for industrial countries). For chemicals, the accumulated reduction in pesticides over the same time period is estimated at 224 300 t of active ingredient, equivalent to a 15% reduction in the associated environmental impact of pesticide use.

● There have been financial and environmental benefits from GM crops.

Other environmental benefits from growing GM crops might also be apparent with herbicide-tolerant crops. For instance, there should be less incorporation of herbicides in the soil, more environmentally friendly cultivation methods (non-tilling) can be used, which reduce soil erosion, there should be less crop damage due to alternative spraying patterns, and biodiversity may also be increased as weeds can be allowed to grow post-emergence if herbicide-resistant crops are being grown. However, agricultural management has to respond to the use of these crops if negative effects on the environment are not to materialize. Some farmers in Canada have encountered problems with multiply herbicide-resistant volunteer canola for instance, and the altered herbicide-application regime used with some GM crops can also bring about its own problems (increased spread from later sprayings, when the crop is taller, for instance). Also, major field trials carried out in the UK (reported in a series of major scientific reports in 2003 and later) indicated that if GM herbicide-tolerant beet and spring-sown oilseed rape were to be grown and managed as in these farm-scale evaluations, there would be adverse effects on the arable weed populations. This would likely result in adverse effects on organisms at higher trophic levels (e.g. birds) compared with conventionally managed crops. In other words, these GM crops could not be grown commercially in the UK until further work had been done to reduce these effects. For instance, other management systems should be investigated as they may not have detrimental effects. Interestingly, GM herbicide-tolerant maize grown in the field trials was not found to have any associated adverse effects.

Insect-resistant crops should also mean less exposure for the farmer and the consumer to dangerous pesticides. Both types of crop should lead to a reduction in the use of fuel when spraying and the reduction of energy in the chemical synthesis process. Both of these lead to a reduction in CO_2 emissions.

What will drive the development of the future generations of GM crops?

The noticeable feature of the first-generation crops is the absence of engineered traits that are not primarily for agronomic benefit. The agribiotech multinationals invested a large amount in the development of these crops with the expectation that the investment would eventually be repaid and produce a profit. The technology was relatively simple and there was obvious commercial gain to be had. However, other traits that could considerably improve the general well being of farmers and consumers in both the developing and developed worlds are conspicuous by their absence. This commercially driven development has implications for the future of GM crops in general. Distrust of the large multinationals, who control both the GM seed and the herbicide, for instance, contributes significantly to the distrust of GM technology, particularly in Europe. The perception, at least, is that GM crops have failed to live up to the considerable claims that were made for them. Many people therefore view future developments in GM crops, ones that could have significant benefits for the people of the world and the environment, with suspicion. They are seen by many as being unnecessary (at the very least) and commercially driven.

● Global warming may have unexpected effects on the use of GM crops.

One of the claims made is that GM crops have lead to a reduction in CO_2 emissions and contributed towards the fight against global warming. This may have really been a side issue but it has now become central to the image of GM crops. As part of his belated attempts to be seen to be doing something about global warming, US President George W. Bush has challenged the USA to remove the dependency on imported oil. The means to do this is to turn the bread basket of the world into the source of biomass that can be converted into biofuels. Instead of growing crops such as maize for food, they are now being converted into biofuels. This has done wonders for the shares of some biotech companies but it has lead to major hikes in the global prices of some basic foodstuffs such as maize and wheat. It may seem more acceptable to some that the GM crops are not being produced for food but it leaves the problem that there will be less actual food available globally. In the longer term, genetic modification may be used to increase the biomass produced by fuel plants such as maize (clearly moves to use trees as a fuel source are too long-term under this scheme), but in the short term it will be another issue that shows GM in a bad light.

Commercial concerns will undoubtedly continue to drive the development of some GM crops, but what is being seen now is the development of GM crops with traits other than agronomic ones. We have already looked at the example of Golden Rice (Chapter 10), which is a pioneering example of a GM crop developed for the developing world and without the backing of a commercial company. Like many of the new developments in GM crops, it was developed primarily in academia with backing from public funding and charitable institutions. On the same theme, medical products produced from GM plants are also seen as a way of convincing the consumer of the technology's validity. However, until the benefits of these so-called new generation GM crops become apparent, the technology faces some considerable hurdles to widespread acceptance.

In the next section, we will look at how GM crops are regulated and some of the concerns surrounding them. To some extent, this regulation reflects the concerns of the general public; as we will see, European legislation in particular is being modified to try to allay some of those concerns.

Concerns about GM crops

There is a considerable degree of public concern in many parts of the world about the real and perceived issues associated with GM crops. Some of these are based on fears about the potential impact of the technology on food safety, human health, and the environment. At the other end of the spectrum, there is moral or ethical opposition to GM crops on the basis that they are wrong in principle. These public concerns can have a massive impact on plant biotechnology and its adoption. In this section, we will look at some of the issues of public concern about GM crops, and examine what plant biotechnologists are doing to allay those public fears.

Antibiotic-resistance genes

Previous chapters have looked at the use of antibiotic-resistance genes as selectable markers for transformed plants. Plant transformation is a very low-frequency event and some means of selecting the transformed cells from the untransformed cells is required. The use of antibiotics as selective agents was already well established as a fundamental tool of molecular biology and cloning in other organisms, particularly in *Escherichia coli*. Their use in plant biotechnology was therefore logical. However, the use of antibiotic marker genes has proved to be one of the hurdles to the widespread acceptance of GM crops.

● Many antibiotic-resistance genes will not be allowed in transgenic crops.

In 1996, Novartis sought approval for a maize variety that carried an ampicillin-resistance gene. The UK Advisory Committee on Novel Foods and Processes (ACNFP) blocked this approval for a considerable length of time, but eventually the maize line was approved for cultivation in France. However, concerns about GM crop safety and lack of market acceptance meant that this GM maize was never cultivated widely. This episode was in some ways unusual, as ampicillin-resistance genes are not found widely in the current generation of GM crops. It did, however, help to set the tone for much of the discussion about the safety of antibiotic-resistance genes in GM crops. Ampicillin is an antibiotic of the penicillin family that is widely used to treat a variety of human infections. The presence of a resistance gene in a genetically modified organism (GMO) released into the environment was perhaps bound to raise fears about creating antibiotic-resistant bacteria, and particularly human pathogens. However, there are several scientific arguments that indicate that this scenario is unlikely and here we will look at two.

First, the antibiotic-resistance genes used in creating GM crops were originally isolated from bacteria. Resistance to antibiotics, which can confer a selective advantage, occurs naturally in bacteria, and is often carried on plasmids that can be readily transferred from one strain to another. The development of antibiotic-resistant strains of bacteria is therefore a real problem in hospitals, where antibiotics are used routinely and create a selective pressure for the development of resistance. Against this background, the transfer of antibiotic-resistance genes from plants to bacteria (for which there is no known mechanism) will not significantly alter the pool of antibiotic-resistance genes in the environment. The transfer of intact, functional, antibiotic-resistance genes to gut flora from ingested plant material is also highly unlikely, though not impossible. There is certainly little evidence for the transfer of DNA from food to gut flora, and it is difficult to see why a transgene would behave differently from the DNA in which it is integrated. Even if resistance genes were transferred, no real selective advantage would accrue and the resistant bacterium would be unlikely to survive. The one report of gene transfer in humans that has entered the literature is perhaps wanting in a number of key pieces of data but it is worth reporting here. In this work, 12 healthy volunteers were compared with seven ileostomists in feeding experiments involving transgenic food. DNA transfer of a target gene sequence was not detected in the former group but it was detected in the latter. This statement is not strictly true as the presence of the target genes were detected in gut flora but it was not clear whether the transfer had occurred during the tests or before. As indicated, this experiment left a lot of questions unanswered.

Second, many of the antibiotic-resistance genes commonly found in GM crops (such as *nptII*) confer resistance to antibiotics that are not used to treat disease in humans, their use having been superseded by less-toxic and/or more-effective alternatives.

Despite these explanations of the relative safety of antibiotic-resistance genes in GM crops, plant biotechnologists are seeking ways to obviate the need for their use or presence in GM crops. Various alternative selectable marker genes have been developed (see Chapter 4) and technologies that remove selectable markers, so-called clean-gene technology, are also being developed (see Chapter 4). The EU has changed its regulatory system to prevent the use of these resistance genes; this is discussed below.

Herbicide resistance and super-weeds

It is possible to use selectable marker genes that confer resistance to herbicides as an alternative to antibiotic-resistance genes (see Chapter 4). However, one problem associated with the use of herbicide resistance markers is the potential for the creation of super-weeds (see Chapter 5). Gene transfer of herbicide-resistance genes, predominantly via cross-pollination, to weedy relatives of GM crops could create such super-weeds. This problem is unlikely to occur if the herbicide-resistance gene was only used as a selectable marker during regeneration from tissue culture. However, the creation of GM crops engineered specifically to express herbicide resistance, as a desirable agronomic trait (see Chapter 5) to simplify crop production, is more problematic.

Gene transfer and the creation of herbicide-resistant weeds is not a theoretical consideration. The usefulness of some herbicides has already been reduced by the presence of naturally occurring herbicide-resistant weedy relatives of several major crops. In 2005, it was reported that at least 15 weedy species, including ryegrass, bindweed, and goosegrass, were either naturally resistant to glyphosate or had developed resistance to it. Whereas this herbicide has been used for some time, recently its use has escalated: 5000 t per year in 1995 to more than 30 000 t in 2002, and use has increased since then. The monoculture effect could lead to major problems with these weeds. The transfer of herbicide-resistance genes from GM crops to further species could exacerbate this problem. Practically every major crop species has weedy relatives that could be cross-pollinated by a GM crop. An example of this was noted in the UK after farm-scale trials. The transfer of herbicide-resistance genes from GM oilseed rape to a distantly related plant, charlock, was found to have occurred. The new form of charlock was growing in a field that had been used to grow GM rape. When scientists treated it with herbicide, it showed no ill-effects. In this case, the herbicide was glufosinate ammonium. The issue is exacerbated as charlock seeds can survive for many years in the soil.

The transfer of herbicide resistance to weeds is already an issue, although at present it is more a question for agricultural management than widespread public concern. The transfer of herbicide-resistance genes to weedy relatives may result in the weed becoming resistant to one or more herbicides, but it will still be susceptible to other chemicals. This may, of course, be of major concern to the farmer (or others), as the herbicide of choice may be rendered ineffective against some weeds. However, herbicide-resistance genes confer no selective advantage on weeds that are not subject to treatment with the herbicide, and therefore the trait is unlikely to spread throughout the population.

Plant biotechnologists can also decrease the use of herbicide-resistance genes as selectable marker genes by again either employing alternatives or by using clean-gene technology. A particularly attractive alternative is to engineer the chloroplast genome, which is discussed later in this chapter. Chloroplasts are, in most but not all cases, inherited maternally, so negating the chance of gene transfer by cross-pollination.

Gene containment

Preventing the transfer of foreign genes from GM crops to other plants is a wider environmental issue, a concern that does not only apply to herbicide-resistance genes. A great variety of foreign genes are being introduced into GM crops, but the environmental impact of these genes is currently difficult to predict. In view of these concerns and to prevent what has been described as genetic pollution, several strategies for preventing transfer of foreign genes are being developed.

Gene transfer usually occurs through pollen, although GM crops, if a wild relative has pollinated them, could also serve as a female parent for hybrid seeds. The dispersal of seeds from GM crops among weedy relatives could also produce mixed populations, with introgression of, for example, a herbicide-resistance gene resulting in herbicide-resistant weeds. The potential for foreign genes from GM crops to be transferred to weedy relatives depends on a great many variables. If gene transfer is through pollen, the amount produced, longevity, dispersal mechanism, and distance, as well as the proximity of weedy relatives that are sexually compatible with the GM crop, all influence the frequency of transfer. Dispersal by seeds obviously depends on whether the GM crop is allowed to set seed, whether the seeds are dispersed among or close to weedy relatives, and the viability and longevity of those seeds.

Table 12.1 Techniques for gene containment

Technique	Advantages	Disadvantages
Chloroplast transformation[a]	Maternally inherited; high level of transgene expression	Expressed proteins may not be processed properly
Male sterility[b] (disrupt anther development)	No pollen formed	Crops must be propagated by pollination from a non-GM crop or by other means; male sterile plant could serve as female parent
Terminator technology[c]	Seeds are sterile; inducible	Silencing of transgene results in viable seeds and introgression; terminator transgenes and desirable trait must remain linked
Apomixis[d] (seeds produced without fertilization)	Can be used to fix heterosis	Complex trait, not usual in crop plants; a plant may not be 100% apomictic
Cleistogamy[e] (fertilization occurs within unopened flower)	Self-pollination prevents outcrossing	Complex trait, not usual in crop plants; introgression still occurs
Transgenic mitigation[f]	Introduced trait is advantageous or neutral for the GM crop but deleterious for weeds	Does not prevent gene transfer; may cause extinction of weedy relatives, thus reducing biodiversity

[a]See Chapter 3 and Boxes 11.4 and 13.1.
[b]Male sterility can be engineered by interfering with the development of the tapetum, which is part of the anther and is important in pollen development. Tapetum-specific promoters have been used to drive the expression of ribonucleases in the tapetum, resulting in targeted destruction of the tapetum and no pollen development.
[c]See Box 12.2.
[d]See the apomixis section in this chapter.
[e]In cleistogamy, self-pollination and fertilization occur within the unopened flower. The genes responsible for the trait have yet to be identified and characterized.
[f]The deleterious traits need not be drastic for transgenic mitigation to work, as competition between weeds is intense. An example of transgenic mitigation is preventing seed-pod shatter (preventing the release of seeds), which is deleterious to weeds, but can actually be an advantage in crops.
Data from Daniell (2002), with permission.

Techniques for gene containment

A wide variety of techniques (see Table 12.1) are capable of preventing or reducing gene transfer. Some of the techniques listed in Table 12.1 are, at present, theoretical or have only been demonstrated in laboratory situations. It is also interesting to note that at least one of the techniques, so-called terminator technology (Box 12.2), has itself been the cause of much opposition to GM crops as seed cannot be collected and saved by the farmer for planting in subsequent years.

BOX 12.2 *Terminator technology terminated?*

Although by no means the only patented technology that has been labelled as terminator technology by its opponents (for instance, Astra Zeneca were issued US Patent 5,808,034 in September 1998 to cover their own terminator-type technology, and currently over 30 patents covering similar technology have been issued), it is the technology developed by the US Department of Agriculture (USDA) and the Delta and Pine Land Co. that has attracted most publicity. Much of this publicity, the vast majority of which was unfavourable to say the least, stemmed from the involvement of Monsanto in the saga following their attempts to acquire Delta and Pine Land Co. and rights to the technology.

Terminator technology is described in US Patent 5,723,765 (Control of Plant Gene Expression), issued on 3 March 1998, to the USDA's Agricultural Research Service (ARS) and Delta and Pine Land Co., a cotton and soybean breeder. The patent covers technology that was referred to as the Technology Protection System (TPS). Three genes (two from bacteria and one from plants) are introduced into self-pollinating plants in TPS. The key to TPS technology is treatment of the seeds, before sale, with a compound (the inducer) that activates what has been called a molecular switch, which initiates a chain of events, culminating with the introduced plant gene being activated shortly before the seed matures fully. The plant gene prevents the seeds from completing their development (after all the marketable elements of the seed, such as oil, have developed) and becoming fertile.

Various modes of action have been proposed for TPS, but one possible method can be illustrated as follows. A repressor gene normally prevents activation of the plant terminator gene (therefore allowing the seed company to use the seed as normal), but after treatment with the inducer (such as tetracycline) the repressor is itself repressed. This results in the second gene (a recombinase, like the Cre enzyme that was also looked at in Chapter 4, for instance) being activated, which leads to a piece of DNA (a blocker flanked by *lox* sites) that prevents expression of the plant gene being removed. The terminator gene (a ribosome-inhibitor protein, for example) is then under the control of a promoter that is only activated very late in seed development. The advantage of TPS is that, in the absence of treatment with the inducer, the seeds can be used as normal. It is only after treatment with the inducer that the seeds produced are destined to be sterile.

TPS was developed to protect the interests of seed companies and, it is claimed, to protect the environment by preventing gene spread (Table 12.1). The seed company's investment in developing and breeding new crops would be protected by preventing farmers from saving seed

(Continued overleaf)

BOX 12.2 *Continued*

from one year to the next, thereby avoiding paying for seed (or selling seed illegally). The developers claim that this protection of the interests of seed companies will, in the long term, benefit agriculture because it encourages the development of new strains and varieties.

The future of terminator technology is uncertain. Following widespread public concern and adverse publicity, the then chairman of Monsanto, Robert B. Shapiro, wrote to the president of the Rockefeller Foundation, Gordon Conway, in 1999 informing him that Monsanto had no intention of marketing seeds containing terminator technology. Astra Zeneca followed suit shortly after, and no terminator technology seeds have been marketed, although this may, in part, be due to difficulties with the technology. Companies like Monsanto are at pains to point out that terminator technology was not one of their main development interests, but it is interesting to note that many companies have continued to work on the technology after 1999. Recently, various developments have occurred in this area. In 2000, the United Nations Convention on Biological Diversity recommended that governments not approve Terminator for field tests or commercial use. This created what is now recognized as an international moratorium. It was upheld and strengthened in March 2006 thanks to pressure from people across the world. However, on 1 June 2007, Monsanto concluded its deal to acquire Delta and Pine Land Co., the US seed company conducting greenhouse trials of Terminator and multinational corporation Syngenta has applied for a Canadian patent on its Terminator potatoes. It seems this technology has not gone away and the answer to the question above is *no*.

Big business

The threat of terminator technology being introduced led to protests in many parts of the world, and was seen as yet another example of big business imposing its wishes on farmers and consumers alike. It is informative to look at the example of terminator technology as it illustrates some interesting points about the plant biotechnology industry and public perception (Box 12.2).

Terminator technology has not been introduced commercially, although most of the big plant biotechnology firms are thought to have the technology. Terminator technology has the potential to make GM crops safer by reducing gene transfer to weedy relatives. The publicity has, however, focused on its potential use to prevent farmers from saving a portion of their grain for use as seed in subsequent years, thus tying them to an expensive source of fresh seed each year. Concerns that terminator technology had already been introduced led to GM cotton crops being burnt in India amid wide-scale unrest.

For terminator technology to become a problem for farmers in developing countries (and even in the USA, where 20–30% of soybean farmers reuse seed), they would already have to be growing seed, with terminator technology engineered in, from a commercial company like Monsanto. Otherwise seed can continue to be kept as normal. There is concern, though, that future development will mean that the highest-yielding varieties (which governments or banks will promote) will come with technology-protection schemes attached. Farmers will, therefore, either be forced to pay for seed every year or continue to grow the low-yield varieties.

Apomixis

In May 1998, in what may well come to be heralded as the beginning of a new era for GM crops, leading researchers in apomixis declared that the technology needs to remain available to developing countries. The Bellagio Apomixis Declaration recognized that apomixis offers too great a benefit for it to be controlled by a small number of concerns (which inevitably would be Western-based biotechnology companies protecting their investment with patents). The needs of commercial concerns (that will undoubtedly play a major role in technology development) are not ignored, but the emphasis is put on novel approaches to patenting, licensing, and technology development to ensure 'open access' to the technology.

So what is apomixis, and why is it such an important technology? In plants, seeds are usually produced as a result of fertilization in sexual reproduction. In some plants, however, seeds can be produced without fertilization, a process known as **apomixis**. Natural apomixis occurs in a limited number of plant species (about 400 in 40 families), but is not found in any major crop plant. Plants from apomictic seeds are genetically identical to the mother plant and are therefore clones. Any desirable features of the mother plant are retained in subsequent generations.

- Are apomixis and terminator technology two technologies that should not be in the hands of seed companies?

Apomixis, if it could be introduced into cereal crops like rice, wheat, and maize, would provide major advantages to both plant-breeding and biotechnology companies and farmers, particularly those in the developing world. Much of the seed sown by farmers is high-performance hybrid seed, produced from parents that have different, desirable characteristics. Hybrid seed is expensive, though, because of the continuous effort needed to produce seed from the parents, as hybrid plants do not breed true and the desirable characteristics are lost. If apomixis were introduced into the hybrid, it could be cultivated indefinitely, as the offspring would be genetically identical and therefore retain the desirable traits. This is known as **fixing heterosis** (where the desirable characteristics of both parents are combined in one plant).

Plant biotechnology companies are interested in genetically engineering apomixis into crop plants because it can be used to considerably speed up the development of new varieties. The advantage to farmers is that seed, even from high-performance hybrids, can be saved and reused, without the loss of desirable characteristics. This frees the farmer from having to buy expensive seed every year, and would make expensive, high-performance hybrids available to farmers in the developing world. The advantage of apomixis, removing farmers from dependence on expensive seed, is, of course, one of the potential hurdles to its development, as seed companies cannot protect their investment.

Several ways round this problem have been suggested, such as coupling apomixis and terminator technology, so that farmers who could afford to buy new seed each year are supplied with seed with the terminator technology activated. Seed for farmers in the developing world would not have the terminator technology activated and seed could therefore be reused. The problem, according to some from the plant biotechnology industry, may not be all that great, as farmers in the developed world (the natural market for hybrid seed) prefer to grow the latest hybrids anyway, and would not save and reuse on a large scale.

Food safety

Concerns about GM food safety

The surveys of public feeling highlight the central issue as being the influence of biotech crops on health, rather than effects on the environment or food security. Will these so-called 'frankenfoods' make me ill? Will I be allergic to them? This is a natural position to take and one that is really no different to that which we all have for all types of food. The organic food industry is based on some people's concerns about the use of chemicals in agriculture.

The Pusztai affair, and its effect on public confidence in GM foods, was looked at in Box 6.4. Whatever the scientific limitations of the work carried out by Pusztai and his colleagues, the revelations certainly contributed to the general anti-GM-food feeling in the UK. Several scientists have supported Pusztai's findings, and, as was pointed out in Chapter 6, the area of GM food safety clearly needs further investigation. One of the contributory factors to the public reaction to the Pusztai report was undoubtedly the loss of public confidence in food safety following the bovine spongiform encephalopathy (BSE) outbreak.

It is important to remember there is no evidence that GM foods are any less safe than non-GM foods. However, the issue of GM food safety has now become much broader. How do plant biotechnologists assess the safety of GM foods, and how do these data get fed into a meaningful debate involving all interested parties? Does the public trust either the information or, indeed, the scientists? If GM foods are to accepted in Europe, it is incumbent upon plant biotechnologists and governments to engage in a real dialogue with the public, and not simply criticize or ignore public attitudes in a post-BSE Europe as certainly was the case when GM tomatoes were first commercialized in the UK. The public feeling, particularly in the UK, against GM foods was badly underestimated then by some government spokespeople, and the subsequent use of the full weight of the UK academic hierarchy to investigate Pusztai's findings only served to create the impression that Pusztai was being silenced by government, academia, and big business. Thus, a bad situation for plant biotechnology was made even worse.

Further issues have hit the media that have not necessarily been presented in a balanced way. In 2004, the population of a village in the Philippines was reported to have been struck down by a serious allergic reaction to *Bt* corn that caused a range of symptoms. A debate developed to the validity of this claim, some suggesting the villagers were being manipulated by the anti-GM lobby and that the cause of the sickness was a virus. Other examples of concern in the media have been the contamination of human food in 2000 with StarLink maize (*Bt* insect-resistant variety) that had been licensed only for animal feed and not for human consumption because of the possibility of allergic reactions. Recently there has been a scare over non-transgenic rice being contaminated with GM rice from trial crops. This has happened with both Chinese and US crops that were being imported into the EU, and in China there seems to be an illegal trade in GM rice seed, presumably due to its popularity with farmers, which is being grown and is finding its way into the food chain. Such problems have led to major front-page stories such as 'GM: The cover up' as it has been deemed safe by government officials to allow the rice with minor amounts of contamination to be eaten.

The dialogue around GM foods must deal with the way safety is measured and what features are tested. For instance, it must deal with the very real scientific issues surrounding concepts like substantial equivalence (see Box 12.4, below). What sort of compositional analysis, if any, is sufficient, or more than sufficient, to prove substantial equivalence? What alternative methods (like transcriptomics and metabolomics) could be used in analysing GM foods? Is substantial equivalence a scientifically valid and, more importantly, publicly acceptable concept? Part of the dialogue also needs to emphasize the real benefits (e.g. improved nutritional or health values) that GM crops and food could bring to the consumer, rather than to the agribiotech multinationals. Governments have had significant problems dealing with biotech crops and food, the public has not seen any benefit, companies are seen as self-interested capitalists, and governments are not trusted. One approach to allay public concern in the EU has been to revisit the regulatory process, and this is discussed in the next section.

● Are GM foods safe to eat? Increased testing and tighter regulations will help to sort this out.

The regulation of GM crops and products

The future of GM crops is, to a large extent, determined by the regulatory framework that is applied to their growth and processing. As a minimum expectation, regulations should ensure that the GM crops and products derived from them are safe (both to the environment and to human and animal health). However, regulations pertaining to GM crops, particularly in the EU, cover additional aspects and reflect a range of public concerns. These additional regulations cover aspects such as the labelling of GM foods and GM-derived products so that consumers can exercise choice over whether to eat GM ingredients. Regulations are also designed to ensure that the public is reassured and confident that any GM crops or products are safe and that they are not being consumed without their knowledge.

The EU

In EU countries, national laws implement various Council Directives and Regulations governing the regulation of GM crops and products. Many of the early biotechnology developments were covered by the key legislative elements, such as Directive 90/220/EEC on the deliberate release into the environment of GMOs (which was replaced by Directive 2001/18/EC in late 2002; see below) and Regulation (EC) 258/97 concerning novel foods and novel food ingredients. Directive 90/220/EEC (and subsequent amendments) required an extensive environmental risk assessment

(ERA) to be carried out before any authorization to place the GM crop on the market under the Directive could be given.

The company responsible for the GM crop (the notifier) submitted a technical dossier and a summary (the notification) to the competent authority of a Member State. The technical dossier was examined, with particular reference to the ERA, to ensure compliance with the Directive. If the outcome was favourable, then the process progressed, with the dossier being forwarded to the European Commission. The competent authorities of all Member States evaluated the notification and raised any objections. If no objections were received, the product could then be placed on the market under the consent of the Member State where the original notification was made. If objections were raised, the proposal was subjected to further scrutiny. A Commission proposal was submitted to the Regulatory Committee of Member States and was subjected to a vote. This proposal took advice from the Scientific Committee on Plants. If a qualified majority of the Member States supported the proposal, then it was accepted. If no qualified majority was achieved, then the proposal was referred to the Council of Ministers of the Environment, where it could be accepted, rejected, or sent back to the Commission. As will be described below, elements of this process have remained the same despite the changes in regulation. What has changed are the requirements of the process: the regulations bring more rigour into ensuring the safety of the food chain. Before we consider the new system in any detail, we will consider how GM crops have been dealt with in the UK as this illustrates the rigour of the process (see Box 12.3).

BOX 12.3 *How European legislation worked at a national level in the UK*

In the UK, the Genetically Modified Organisms (Deliberate Release) Regulations 1992 (amended 1995 and 1997), drawn up under several Acts of Parliament (the Environmental Protection Act 1990, the European Communities Act 1972, and the Health and Safety at Work Act 1974), implemented the relevant pieces of EU legislation, particularly EU Directive 90/220/EEC. These regulations applied to all GMOs, but by far the most interest has been shown in GM plants.

A step-by-step approach was adopted in the regulation and release of GMOs. Initial developments took place in a contained environment (such as a laboratory or greenhouse). The safety of the GMO was assessed in this contained environment before small scale release into the environment was permitted. The scale of these releases was gradually increased if safety assessments indicated it was safe to do so.

The Health and Safety Executive regulated the initial, contained use of GMOs in the UK. Releases into the environment were subject to regulation under the terms of the EU Directive 90/220/EEC. Release into the environment was divided into two categories: small-scale (research) trials and release for commercial purposes. A detailed environmental risk assessment (ERA), prepared by the applicant, underpinned all releases of GMOs into the environment, irrespective of the scale of release. This ERA was subjected to scientific scrutiny, and release was only allowed if any risks were considered to be of a very low level. Low-level risks generally meant that the GMO posed no greater risk than a non GM equivalent.

BOX 12.3 *Continued*

Approval for a research trial

Applicants had to submit a detailed dossier, which included an ERA, to the Joint Regulatory Authority (JRA), who advise the Secretary of State at the Department for Environment, Food and Rural Affairs (DEFRA). The JRA reviewed the dossier, which was also forwarded to experts in other government departments and other organizations, such as English Nature. The Advisory Committee on Releases to the Environment (ACRE) also reviewed all applications for release into the environment. ACRE, a committee of independent scientists, advised on both the human health and environmental aspects of the proposal. If ACRE and the other assessors (the Advisory Committee on Novel Food and Processes, for instance, if the GMO is to be used as food) were satisfied that the proposal represented a very low risk, then the release of the GMO into the environment, on a small scale, could proceed. The release approval may have been subject to conditions, and scientists from the Central Science Laboratory closely monitored these initial releases. If the JRA, ACRE, and relevant ministers were satisfied that no problems were connected with the release of the GMO, the scale of the release could be increased.

Approval for marketing releases

Approvals for large-scale releases, aimed at commercialization, had to be sought at a European level. Some of the process was similar to that required for approval of a small-scale release. The applicant had to submit another dossier to the Member State, which contained—in addition to all the previous information—details on the diversity of sites where it may have been used, ecosystems that might have been affected, and proposals for labelling. The Member State reviewed the proposal, if satisfied that the proposal represented a very low risk, and the dossier was forwarded to the European Commission for consideration by the Member States as outlined above in the introduction to EU regulations.

Farm-scale evaluations

In a unique programme to attempt to put the evaluation of GMOs on a sound scientific basis, the UK decided that no GM crops would be grown commercially in the UK until the results of a series of farm-scale evaluations were known. Only if these farm-scale evaluations showed that there were no significant risks to the environment would the commercial planting of GM crops be allowed. As we have seen, this was not found to be the case for a number of crops.

To address public concerns, the EU has revisited the regulations on biotech crops and brought in a new directive to cover deliberate release into the environment of GMOs (Directive 2001/18/EC) and the regulation of GMO food and feed (1829/2003). Also, in 2003, an amendment to 2001/18 was introduced; this regulation (1830/2003) concerns the traceability and labelling of GMOs and food and feed products produced from GMOs. The aims of this change in legislation are to tighten up the safety of biotech plants and foods and to give the consumer freedom of choice; this will be achieved by more scientific testing, labelling, and traceability. The provisions of the ERA have been strengthened to cover direct or indirect and immediate or delayed effects of GMOs. Post-marketing monitoring of unanticipated effects of GMOs will also be introduced. In moves designed, in part, to allay public fears about some of the dangers of GMOs, antibiotic-resistance marker genes that pose a risk to human and animal life will be phased out, GMO products will be labelled clearly and will be traceable, and there will be public consultation over proposed releases.

The scientific changes are brought in to make the GMOs and the products safe and not damage human and animal health or the environment. The important issue is that the GMOs and products must be considered as safe as their conventionally derived counterparts and that this must be proved by doing tests using the most advanced knowledge and technology available. This is a real attempt to overcome the problems associated with the use of substantial equivalents and the precautionary principle (see Box 12.4). Without this information, a product cannot be authorized. Freedom of choice deals more with people's rights to direct their own lives and choose whether to use or reject GMOs or their products. This includes farmers, consumers, and businesses. Non-transgenic and transgenic crops must co-exist, but GM crops should be handled so there is no uncontrolled mixing with conventional crops. The management of these proposals rests with the Member States. To help with this, labelling and the concept of traceability have been introduced. In forensic terms, there must be a clear chain of evidence. If GMOs are intentionally used, the consumer

BOX 12.4 *Substantial equivalence and the precautionary principle in risk assessment*

Few issues have generated such a wealth of conflicting data and opinion as the use of substantial equivalence in GM-crop risk assessment. It is not our intention to review the mass of information that surrounds these issues, but simply to make the reader aware of these concepts and some of the issues associated with their use in risk assessments.

Substantial equivalence

The substantial equivalence dogma, at least in the context of GM crops, stems from an Organization for Economic Co-operation and Development (OECD) document *Safety Evaluation of Foods Derived by Modern Biotechnology: Concepts and Principles*, which reported the findings of a group of experts in the safety of food biotechnology. It maintained that the most effective way of assessing the safety of novel foods and food products was to consider whether the novel food product was substantially equivalent to an existing analogous food product (if indeed one existed). Substantial equivalence is based on observation of phenotypic and agronomic characteristics and a compositional analysis of key components. This, of course, makes perfect sense on one level. After all, if it looks like a potato, grows like a potato, and tastes like a potato, the chances are it's a potato (even if it is a GM potato) and has, therefore, been consumed safely for several hundred years. Substantial equivalence and its application to GM foods has, though, been criticized by a number of scientists. First, substantial equivalence has not been defined strictly, and therefore remains open to interpretation. The use of compositional analysis in determining substantial equivalence has also been criticized as limited and imprecise by some scientists. In the USA in particular (where regulations focus on the product and not the technology

BOX 12.4 *Continued*

used for production), if a GM food is considered to be substantially equivalent then it is exempt, in many cases (self-cloning or if the gene product has been approved previously, for instance), from the necessity of safety evaluation. Plant transformation can lead to unexpected results or unforeseen changes in the product, just like any technology (and conventional breeding is certainly not immune from unexpected results). These potentially deleterious effects may not be detected before marketing due to the lack of comprehensive safety testing. It has also been argued that GM crops must be substantially different from their predecessors, by virtue of the way they are produced and/or by virtue of the transgene that is introduced, if they are to be patented.

Substantial equivalence, properly applied, may provide a sound framework for safety assessment (it was never intended that the concept of substantial equivalence should be used as a substitute for a safety assessment). However, its use to justify marketing GM foods and products without comprehensive testing has generated a great deal of public resentment, particularly in Europe.

The precautionary principle

Another approach to guide risk assessment is the precautionary principle. Like the concept of substantial equivalence, it is not a new idea (it was first recognized in the 1970s) and has become a general principle of international environmental law. It engenders what is considered by some to be a fundamentally different approach to risk assessment from the concept of substantial equivalence.

The precautionary principle is invoked in the event of a potential risk, even if this risk cannot be fully demonstrated or quantified, or its effects determined. Appropriate measures, designed to anticipate and prevent harm, should then be taken. The precautionary principle is considered by many to be appropriate when considering GMOs, because of the complexity and novelty associated with GM technology. Basically, in using the principle it is accepted that it is impossible to precisely predict every possible outcome from GM technology, even if at a superficial level the end product appears no different from its natural counterpart. The precautionary principle has, like substantial equivalence, also been widely criticized in some quarters, being described as unscientific and antitechnology.

Elements of both of these approaches are likely to feed into the regulatory process.

should be able to read this fact on the product label. Traceability is required to allow the consumer to know that even small amounts of GM product may be found. Clear documentation and good record keeping are needed to follow GMs from field to product. These regulations are applied on a case-by-case basis and are not seen in generic terms.

Authorizations made under the previous regulations will stand, and other ongoing applications will be dealt with under the new regulations. If the intention is to apply to grow a biotech crop in the EU and then produce a food from it, then both the release and food and feed regulations must be satisfied. However, if the crop is grown outside the EU, then the product itself may only need authorization. This will depend on the nature of the product; even if the intention is not to grow a crop in the EU but the product is capable of cultivation or propagation (such as maize kernels or rapeseed), then the release regulations need to be satisfied. If the product was sugar or oil, then only the food regulations need to be dealt with.

- Is the new process for getting approval in the EU for GM plants and derived products a solid, science-based set of regulations? Or is it a smoke screen to benefit big business?

Directive 2001/18 was supposed to have been adopted by Member States by October 2002. The directive covered the commercial use of a GM plant that is able to reproduce, release into the environment for cultivation, or importing plant material. The application should include a scientific safety assessment that shows there will be no harmful effects on humans or the environment through an environmental impact assessment. There also needs to be supplied a standardized method for detecting the GMO and a system for its monitoring. The process is similar to that outlined above; in the first instance, the proposal would normally be submitted to national authorities to give an initial scientific assessment. The documents are then forwarded to the European Commission and national authorities of all Member States to give an opinion. However, the applicant may submit directly to the European Food Safety Authority (EFSA) for scientific evaluation. Where there are open questions or objections, the EFSA carries out a scientific review. This is used by the European Commission to generate a draft decision, which is subjected to a vote in the Standing Committee for the Food Chain and Food Safety. The draft may be accepted or rejected with a qualified majority; that is, a certain number of votes that also represents a certain percentage of the EU population. As qualified voting is politically sensitive and a target for reform in the EU, this may change. If no majority is obtained, then the draft is submitted to the Council of Ministers who have 90 days to approve or reject the draft with a qualified majority. If the draft is rejected, it goes back for redrafting. If it is accepted or no decision can be reached, then the draft comes into effect and is valid for 10 years after which the authorization expires.

GM food and GM-derived food ingredients were regulated under Regulation (EC) 258/97. Notification included information on the proposed labelling of the GM food so that consumers could make informed choices about whether to eat GM food. One point worth noting is that foods derived from GMOs (but no longer containing them) that are substantially equivalent to existing foods could be regulated by a simplified procedure (the concept of substantial equivalence is discussed in Box 12.4). In such cases, the European Commission was simply notified (with scientific justification as to why the product is substantially equivalent, or with an opinion to the same effect from the competent authority of a Member State) when such a product was placed on the market. The notification was passed to Member States and the product could then be marketed throughout the EU.

The new regulations (1829/2003) needed to be more scientifically vigorous and the substantially equivalent approach has been strengthened such that much more evidence is needed to prove the point. An application for authorizing food or feed consisting of or made from a GMO must be submitted to the authorities of the Member State and must include a scientific safety assessment showing studies confirming that the GM food is not dangerous to health or the environment, analysis showing that the food is substantially equivalent to conventional counterparts, methods and sample material for detecting GM content, a proposal for post-market monitoring, and suggestions for product labelling. The proposal dossier is forwarded to the EFSA who have 6 months to provide an opinion on safety. They also make the application available to all Member States. The EFSA provides an opinion based on a scientific evaluation carried out by a panel of experts who confirm the substantially equivalent status. The EFSA also has the EU reference laboratory evaluate the detection methods in the dossier. If the product requires approval under 2001/18, that should also be confirmed. The EFSA provides an opinion to the European Commission that should include the scientific safety assessment, product-labelling suggestions, suggestions for post-market monitoring, and validated detection methods. An environmental monitoring plan for GM plants may be included if necessary. The final part of the process is as for 2001/18 so it is possible to submit a proposal for deliberate release and for food and feed for an integrated safety assessment.

The first key issue to be highlighted here is that the use of antibiotic-resistance genes for selection is discouraged. They have not been completely banned, but rather they have been divided into three groups by the EFSA. The first group includes resistance genes to antibiotics such as kanamycin that are seldom or never used in medicine. These marker genes can still be used in GM plants. The second group are not to be used in commercial plants but can still be used in plants that get as far as field trials. These are resistance genes to antibiotics that are used to treat specific conditions. The resistance genes are widespread in microorganisms and it is unlikely that the use of the plants in field trials will increase the occurrence of them in nature. The third group of genes are those for antibiotics that are valuable in medicine, such as amikacin (*nptIII*). The use of these genes is banned so that the value of the antibiotics is not compromised. Alternative selectable markers have been presented in Chapter 4.

The second key issue is what scientific tests need to be carried out to ensure safety. This is quite a difficult question to answer here but the EFSA would expect proof that the novel GM trait is safe and that there are no unforeseen changes in plant metabolism as a result of gene transfer. Transgenes may make proteins that are novel to humans so some tests will need special consideration. To prove that a protein product is safe will involve looking at toxicity, usually by animal feeding tests; however, the effects may be different in humans. The protein/food should be tested for allergenic properties. Novel proteins should be compared with known allergens for similarities in structure and for stability during digestion. This test is based on observations that for well-known allergens, such as peanuts, the allergenic proteins are not digested completely in the gut, which results in peptide fragments that carry immunodominant IgE-binding epitopes. Test are carried out with blood from individuals that are sensitive to allergies (looking for cross-reactions) and animals can be tested to see whether they develop a hypersensitive response to the proteins. In a recent paper (Goodman *et al.*, 2008), the validity of the tests used has been reviewed and in some cases challenged. The reference is listed in the Further reading section and makes interesting reading. Another important analysis is to see whether there have been unforeseen changes in plant metabolism as a result of the gene transfer. Again animal tests can be used—for example long-term feeding studies to look for changes in the health of the animal—but this is an area where -omics can play a role. Proteomic analysis of protein content, metabolomic analysis of chemical content, and measurements of nutritional value and vitamin content should be made. Many plants produce toxins at low levels, and analysis is required to ensure there are no increases in the levels of these proteins due to the transgenes.

As we have seen, there have been claims that feeding animals with transgenic material can lead to health problems. Often these have been flawed experiments and other studies have found no evidence for health effects. Inclusion of these stringent tests in the health and safety review is necessary to build the public's confidence. Any negative findings in these studies would prevent the food from being marketed.

A lot more information can be found on the GMO compass website and sources such as the Department for Environment, Food and Rural Affairs (DEFRA) in the UK (see the web-links subsection of the Further reading list).

The USA

In the USA, regulations pertaining to GM crops are implemented by three federal agencies. The US Department of Agriculture (USDA) and, in particular, the Animal and Plant Health Inspection Service (APHIS) under the authority of the Federal Plant Pest Act, determines whether a transgenic crop is likely to become an agricultural or environmental pest. The USDA regulates the transport, import, and testing of transgenic plants. In many cases, field tests of GM crops can be conducted if APHIS are simply notified and six criteria regarding safety of the GM crop are met (see Box 12.5) and if the applicant meets certain standards to ensure biological confinement.

If these criteria and standards are not met, a more involved process is required to obtain a permit for release from APHIS. The field tests and the subsequent reviews by federal authorities can take many years to ensure the safety of the GM crop to human health and the environment.

GM crops that are to be commercialized can be granted non-regulated status by APHIS upon petition. This process requires that extensive data dealing with the nature of the modification and its effects on the crop and the wider ecosystem are presented to APHIS. If non-regulated status is granted, the product and any offspring can be released in the USA without APHIS review. APHIS can prevent the continued use of a GM crop if it is thought to present a risk of becoming a pest. However, the USDA regulatory process has been criticized recently (in a series of National Academy of Sciences reports) for not being rigorous, inclusive, and transparent.

The US Environmental Protection Agency (EPA) ensures the safety of pesticides, including those produced biologically. Under the authority of the Federal Insecticide, Fungicide, and Rodenticide Act, the EPA regulates the testing, distribution, marketing, and use of plants that have been genetically engineered to be pest-resistant. The EPA also sets tolerances for herbicide residues on herbicide-resistant crops. The EPA issues permits for field testing and registrations for the commercialization of GM crops that are pest-resistant.

BOX 12.5 *The six criteria for notification in the USA*

The six criteria that must be met if the Animal and Plant Health Inspection Service (APHIS) is to be notified of environmental release can be summarized as:

1 the plant species is maize, cotton, potato, soybean, tobacco, or tomato (or any other plant species that has been so determined);

2 the transgene(s) must be stably integrated;

3 the function of the transgene(s) must be known and expression must not be detrimental to plant health;

4 the transgene does not: (a) result in the production of an 'infectious entity' like a virus; (b) encode substances known, or likely, to be toxic to non-target organisms likely to feed or live on the plant;

5 introduced sequences that are derived from plant viruses must not pose the risk of the creation of new plant viruses;

6 the plant must not contain certain genetic material derived from an animal or human pathogen.

The US Food and Drug Administration (FDA), under the authority of the Federal Food, Drug and Cosmetic Act, regulates foods and animal feed derived from transgenic plants. Genetically engineered foods have to meet the same safety standards as all other foods. However, many GM crops do not need pre-market approval from the FDA because they do not contain substances that are significantly different from those already present in other foods (they are deemed to be substantially equivalent, see Box 12.4). Some US pressure groups want to bring in similar regulations to the EU that require rigorous scientific testing and labelling.

One or more of the federal agencies may regulate a particular GM plant, depending on the nature of the modification and its intended use. Thus, herbicide-tolerant crops would be regulated by the USDA (to ensure the GM crop would not become a pest), the EPA (in the event of the new use of a herbicide), and the FDA (to ensure the crop was safe to eat). If no food or feed use is intended, the FDA need not be involved in the regulatory process.

As well as federal regulation, some states require additional regulation of GM crops. The acceptance and regulation of GM crops has been a source of considerable tension between the EU and the USA. The EU moratorium has, for instance, cost the USA over $200 million annually in corn sales to EU Member States. The new regulations and the lifting of the moratorium should open up the EU market to (properly assessed) GM imports from the USA. However, the USA claims that some aspects of EU regulation put imports from the USA at an unfair disadvantage. With these two different approaches to the regulation of GM crops and food safety, continuing discussions and a sensible common approach to the resolution of these problems is necessary for the benefit of all.

SUMMARY

In this chapter, we have looked at GM crops in a wider context, considering their acceptance, regulation, and some elements of their future prospects (more of this in Chapter 13). It is not meant to be an exhaustive survey, but examples have been given to stimulate debate and illustrate how GM crops are no different from many other new technologies, in that they will be challenged. What we have seen is that GM crops do have great potential for improving the quality of life of many people, but fully realizing this potential depends on acceptance. The lack of public acceptance of GM crops, particularly in Europe, has proved to be a major stumbling block in the development of GM crops. For much of the developed world, this lack of acceptance effectively means lack of consumer choice and dwindling export markets for GM producers. Some countries from the developing world, like China, have begun to develop an independent strategy to the development of GM crops, relying on local solutions to local problems. The consequences for the developing world, where countries may not have the resources available for an independent approach like that being adopted in China, could be much worse.

Attempts are being made to persuade the public that GM crops are safe and offer previously unknown opportunities for a better life and a better environment. However, studies show that the public in Europe is still sceptical about the involvement of companies like Monsanto in GM crop production and is not convinced by some of the claims made for GM crops.

The challenge for plant biotechnology is to convince public opinion that GM crops offer real advantages, which cannot be realistically achieved any other way, and that the technology is safe. Only then will the world market be fully open to GM crops and products and the full potential realized.

FURTHER READING

Visit www.oxfordtextbooks.co.uk/orc/slaterplants2e/ for hyperlinked references that take you directly to online abstracts for the journal articles listed below:

Benefits, risks, acceptance, and safety of GM crops and GM foods

Dale, P. J., Clarke, B., and Fontes, E. M. G. (2002) Potential for the environmental impact of transgenic crops. *Nature Biotechnology* **20**, 567–574.

Daniell, H. (2002) Molecular strategies for gene containment in GM crops. *Nature Biotechnology* **20**, 581–586.

Fletcher, L. (2001) GM crops are no panacea for poverty. *Nature Biotechnology* **19**, 797–798.

Fu, X., Sudhakar, D., Peng, J., Richards, D. E., Christou, P., and Harberd, N. P. (2001) Expression of *Arabidopsis GAI* in transgenic rice represses multiple gibberellin responses. *Plant Cell* **13**, 1791–1802.

Gaskell, G., Bauer, M. W., Durant, J., and Allum, N. C. (1999) Worlds apart? The reception of genetically modified foods in Europe and the US. *Science* **285**, 384–387.

Goodman, R. E., Vieths, S., Sampson, H. A., Hill, D., Ebisawa, M., Taylor, S. L., and Uan Ree, R. (2008) Allergenicity assessment of genetically modified crops—what makes sense, *Nature Biotechnology* **26**, 73–81.

Johnson, B. and Hope, A. (2000) GM crops and equivocal environmental benefits. *Nature Biotechnology* **18**, 242.

Junk, B. and Gaugitsch, H. (2001) Assessing the environmental impacts of transgenic plants. *Trends in Biotechnology* **19**, 371–372.

Kleter, G. A., van der Krieken, W. M., Kok, E. J., Bosch, D., Jordi, W., and Gilissen L. J. W. J. (2001) Regulation and exploitation of genetically modified crops. *Nature Biotechnology* **19**, 1105–1110.

Kuiper, H. A., Kleter, G. A., Noteborn, P. J. M., and Kok, E. J. (2001) Assessment of the food safety issues related to genetically modified foods. *Plant Journal* **27**, 503–528.

Marris, C. (2001) Public views on GMOs: deconstructing the myths. *EMBO Reports* **2**, 545–548.

Moffat, A. S. (1998) Toting up the early harvest of transgenic plants. *Science* **282**, 2176–2178.

Smyth, S., Khachatourians, G. G., and Phillips, P. W. B. (2002) Liabilities and economics of transgenic crops. *Nature Biotechnology* **20**, 537–541.

Wolfenbarger, L. L. and Phifer, P. R. (2000) The ecological risks and benefits of genetically engineered plants. *Science* **290**, 2088–2093.

Web-links

Advisory Committee on Releases to the Environment (ACRE), web-link 12.1: www.defra.gov.uk/environment/acre/index.htm

AgBioForum, web-link 12.2: www.abioforum.org

AgBiotechNet, web-link 12.3: www.agbiotechnet.com

AgBioWorld, web-link 12.4: www.agbioworld.org

Agricultural Biotechnology, web-link 12.5: www.usda.gov/agencies/biotech

Agriculture and Environment Biotechnology Commission, web-link 12.6: www.aebc.gov.uk

Biotechnology Industry Organization, web-link 12.7: www.bio.org

Council for Agricultural Science and Technology, web-link 12.8: www.cast-science.org

Department for Environment, Food and Rural Affairs (DEFRA), web-link 12.9: www.defra.gov.uk

Essential Biosafety, web-link 12.10: www.essentialbiosafety.info/main.php

European Federation of Biotechnology, web-link 12.11: www.efbweb.org

European Union On-Line, web-link 12.12: http://europa.eu.int

Genetically modified organisms: the regulatory process (DEFRA report), web-link 12.13: www.defra.gov.uk/environment/fse/leaflet/pdf/gmcrop2.pdf

GM crops: the first 10 years—global socio-economic and environmental impacts, web-link 12.14: www.pgeconomics.co.uk/pdf/global_impactstudy_2006_v1_finalPGEconomics.pdf

GMO compass, web-link 12.15: www.gmo-compass.org/eng

GM science review, web-link 12.16: www.gmsciencedebate.org.uk

Information Systems for Biotechnology, web-link 12.17: www.isb.vt.edu

Institute of Science in Society, web-link 12.18: www.i-sis.org.uk

International Service for the Acquisition of Agri-Biotech Applications (ISAAA), web-link 12.19: www.isaaa.org/

National Agricultural Biotechnology Council, web-link 12.20: www.cals.cornell.edu/extension/nabc

NCBE GM Food Information, web-link 12.21: www.ncbe.reading.ac.uk/NCBE/GMFOOD/menu.html

New developments in crop biotechnology and their possible implications for food product safety, web-link 12.22: www.rikilt.wageningen-ur.nl/Publications/Tekstrapport2000%20004.htm

New Scientist GM Food Special Report, web-link 12.23: www.newscientist.com/hottopics/gm

Pew Initiative on Food and Biotechnology, web-link 12. 24: http://pewagbiotech.org/

The Plant Journal. Plant GM technology, web-link 12.25: www.blacksci.co.uk/~cgilib/jnlpage.asp?=tpj&File=tpj&Page=plantGM

USDA Animal and Plant Health Inspection Service (APHIS), web-link 12.26: www.aphis.usda.gov/ppq/biotech

USDA Economic Research Service (ERS), web-link 12.27: www.ers.usda.gov

US Department of Agriculture (USDA), web-link 12.28: www.usda.gov

US Environmental Protection Agency (EPA), web-link 12.29: www.epa.gov

US Food and Drug Administration (FDA), web-link 12.30: www.fda.gov

US Mission to the European Union, web-link 12.31: www.useu.be

13

Beyond genetically modified crops

INTRODUCTION

One of the themes running through this book is a consideration of the social context within which genetically modified (GM) crops have been developed. Apart from describing the technological development, several chapters have considered the economic forces that have determined which GM traits were commercialized. The social and political pressures that have influenced the development of GM technology have also been discussed along with the potential health and environmental impacts of GM crops. This chapter attempts to predict the future trajectory of GM crops by looking not only at recent technological developments but also at how these may respond to social and political pressures.

'Greener' genetic engineering

● The removal of antibiotic-resistance marker genes addresses some of the health and environmental concerns about GM crops.

Although many of the products of plant biotechnology on the market today already offer real environmental and health benefits (from reduced pesticide use or with products such as Golden Rice, for instance), considerable opposition to GM crops remains. However, new approaches to GM crop production are becoming apparent and might help to make the technology more acceptable. These new approaches, which make use of data from genome-sequencing projects (Chapter 1) together with our increased understanding of the fundamental biology underlying agronomically important traits, rely on plant genes to produce the desired effects. We have already seen that some of the concerns about GM crops stem from a perception of the potential dangers of inserting foreign genes into plants. One particular concern relates to the use of antibiotic marker genes, and the possibility that the widespread dispersal of such genes may contribute to the development of antibiotic-resistant pathogens. Consumer rejection of food containing foreign genes has driven the technology to concentrate on the manipulation of endogenous genes and to explore methods to reduce the amount of undesirable genetic material in transgenic plants. Thus, alternatives to antibiotic-resistant marker genes are discussed in Chapter 3, and techniques for the removal of such genes after transformation are described in Chapter 4. Targeted modification of endogenous genes by homologous recombination provides

● Gene targeting would permit the precise manipulation of endogenous genes, but the low rate of homologous recombination in higher plants precludes this strategy.

a more precise mechanism for the genetic manipulation of plants that could overcome some of the objections to current GM strategies. Homologous recombination and gene targeting are discussed in Chapter 5, as are the obstacles that prevent this technology being widely adopted by the plant community.

Another concern about the environmental impact of GM crops is the potential for spread of transgenes in pollen (Chapter 6). Chloroplast engineering offers an alternative approach to the delivery of transgenes that offers a number of advantages while reducing this environmental impact, and is described in Box 13.1.

BOX 13.1 *Chloroplast engineering*

Engineering of the chloroplast genome provides one of the most exciting developments in modern plant biotechnology, and holds out the promise of being a clean, efficient, and environmentally friendly way of genetically engineering plants. Chloroplasts are normally inherited maternally, in the same way that organelles such as mitochondria are in animals. This means that in the majority of cases, chloroplasts are not found in pollen (the pollen of some plants does contain active chloroplasts, although the chloroplast DNA is usually lost during maturation of the pollen. Some plants also exhibit biparental inheritance of chloroplasts). This predominantly maternal inheritance has two beneficial effects. First, gene transfer to weedy relatives, which occurs principally through pollen, is reduced. Second, pollen from plants in which the chloroplast genome has been transformed does not represent a toxic risk to non-target species (Chapter 6 highlighted the Monarch butterfly controversy), as the transgene products are not found in pollen. This has proved to be the case, even when the transgene product normally accumulates to a very high level.

Chloroplasts are also ideal targets for multigene engineering (see below). Chloroplast genes are, like prokaryotic genes, often found in operons, where one promoter controls the expression of a group of genes, with polycistronic mRNA being produced. There is therefore the possibility of introducing a group of foreign genes, as an operon, under the control of a single promoter. This has the added advantage that all the genes in the operon are expressed at similar levels. Expression of the *Bacillus thuringiensis* cry2Aa2 operon (insecticidal Cry proteins were discussed in Chapter 6) in chloroplasts results in extremely high levels of the transgene product accumulating as stable crystals. This is thought to be due partly to the presence of a chaperonin gene on the cry2Aa2 operon, which mediates the folding of the Cry protein.

High levels of transgene expression and foreign protein accumulation are relatively common with chloroplast transformation, due to each chloroplast having multiple copies (>50) of the genome and each cell having a number of chloroplasts (50 or more, although this figure varies). Thus, tremendous amplification is achieved if each copy of the genome in each chloroplast carries the transgene.

Gene silencing does not appear to be a problem encountered with chloroplast transformation, and homologous recombination allows transgenes to be inserted precisely into the chloroplast genome. Antibiotic-resistance genes need not be used as selectable markers or they can be easily excised. Toxicity of the foreign protein appears to be less of a problem, in many cases, than with nuclear transformation. Chloroplasts apparently seem to offer one of the best hopes for finding an acceptable system for foreign-gene expression in plants.

Table 13.1 Examples of environmentally beneficial GM crops

Substance/trait	Gene	Benefit	Plant
Heavy metals	Mutated *GUS* gene	Mutation assay (which restores activity) used to detect toxins in environment	*Arabidopsis*[a]
TNT (trinitrotoluene)	Nitroreductase	Phytoremediation; reduces the contamination of land by residues from explosives	Tobacco[b]
Lignin	Cinnamyl alcohol dehydrogenase/caffeate 5-hydroxy-ferulate O-methyltransferase	Altered lignification; more environmentally benign paper-making processes	Poplar[c]

Sources: [a]Kovalchuk *et al.* (2001); [b]French *et al.* (1999) and Hannink *et al.* (2001); [c]Pilate *et al.* (2002).
GUS, β-glucoruronidase.

It is also worth re-stating many of the examples of genetic modification discussed in earlier chapters where there is a clear environmental benefit when compared with conventional agricultural practices, notably pest-resistant (Chapter 6) and disease-resistant (Chapters 7 and 8) traits that reduce reliance on chemical sprays. Positive environmental effects could also be derived from stress-resistant (Chapter 9) and higher yielding (Chapter 10) crops and crops grown for biofuels (Chapter 11). More obvious environmental benefits could be derived from other examples that have not been covered previously, and are shown in Table 13.1. These include plants engineered for various phytoremediation purposes and as environmental indicators.

- Certain GM traits are targeted directly at environmental improvement.

Genetic manipulation of complex agronomic traits

The removal of undesirable marker genes has been coupled with an increased emphasis on using plant genes in preference to bacterial or animal genes. (Compare the reliance of herbicide- and pest-resistance strategies on bacterial genes outlined in Chapters 5 and 6 with the strategies for the manipulation of the tomato ripening described in Chapter 10). Gene transfer from widely differing organisms and the creation of publicly perceived 'frankenfoods' is avoided, and the emphasis shifts towards modification of plant growth, development, and physiology to bring about the desired changes. This new strategy could bring benefits to the developed and developing world alike. The approach involves identifying plant genes that control or influence agronomically important traits and manipulating these to produce agronomic benefits. These improvements could be applied to a range of crops, benefiting producers and consumers from a range of economic backgrounds.

- Future GM crops will focus on the use of plant genes that control agronomically important traits.

For instance, rice takes only a little more than 6 months to produce a crop in some areas. Introduction of the *Arabidopsis LEAFY* gene (a regulatory gene that promotes floral meristem identity) into rice accelerates flowering just enough to allow two

- Regulatory genes can be used to accelerate flowering.

crops per year, with only a small reduction in yield per crop. Similar approaches have been taken to speed up the reproductive development of citrus trees. The constitutive expression of either *LEAFY*, or another *Arabidopsis* floral meristem identity gene known as *APETALA*, shortens the juvenile phase of citrus trees, so that they crop as early as their first year, as opposed to having a delay of more than 6 years.

Dwarf rice plants can be obtained by transformation with the *gai* transgene.

Another example was described in Chapter 10. The development of dwarf strains of cereals like wheat and rice led to the Green Revolution of the 1960s and 1970s, which transformed countries like India into net wheat exporters. So great an achievement was the development of these dwarf strains that Dr N. E. Borlaug was awarded the Nobel Prize for Peace for his pioneering work in this area. Dwarf strains are generally higher-yielding as less of the photoassimilate goes into growth and more goes into the grain. They also have the advantage of being less prone to damage from wind and rain. Dwarf strains can, however, take a considerable time to produce by conventional breeding, and in many cases breeding in dwarfing characteristics has led to the loss of other, desirable characteristics. Scientists are now beginning to manipulate the mechanism (which involves the plant hormone gibberellin) responsible for stem growth directly, using GM technology. Both a dominant-mutant form of the *Arabidopsis gai* (gibberellin insensitive; a gibberellin derepressible repressor of plant growth) gene and the wild-type gene, when introduced into rice, confer reduced responses to gibberellins and result in dwarf rice plants. New, dwarf cereal strains could, therefore, be produced relatively quickly using this approach. The ability to directly manipulate elite varieties also means that the other desirable characteristics might not be lost.

These examples are interesting, not only because of the hope they hold out for increasing crop yield and productivity, but because they also highlight the background information that is required for this strategy to be universally successful. Note in particular how information and resources from fundamental investigations conducted in model plant species, like *Arabidopsis*, have been combined with the accumulated knowledge from crop breeding to benefit plant biotechnology.

The pressures to make GM technology more acceptable combined with the scientific opportunities presented by the genomics revolution point the way forward for a new paradigm of crop improvement: molecular breeding. This describes a set of approaches and technologies that integrate genetic/genomic technologies with conventional breeding. To fully appreciate molecular breeding, it is necessary to consider the nature of the information and technologies that might contribute to this new approach, as follows.

1 The identification of genes that play a role in determining agronomic traits. To date, a large number of genes have been characterized in terms of the class of protein they encode and the biological function the protein may have. However, there is still a gap between this type of molecular knowledge and understanding the genetic factors that influence the performance of crops in the field.

2 The elucidation of pathways and systems that underpin complex traits. This is important not only for identifying targets for genetic manipulation (see examples of metabolic engineering in Chapters 9–11), but also for understanding the consequences

of modifying complex systems (examples of unforeseen consequences of modifying the shikimate pathway are discussed in Chapter 5 and of the carotenoid pathway in Chapter 10).

3 The ability to transfer multiple genes into crop plants. Many of the complex traits of interest will require modification of several genes. This is discussed in Box 13.2.

BOX 13.2 *Multigene transformation*

In the earlier chapters of this book, we looked at several examples of what might be termed first-generation GM crops. These are the products of fairly simple, single-gene transformations, with insect- and herbicide-resistant GM crops being prime examples. However, many traits that might be desirable in a GM crop are the products of complex biochemical pathways and introducing the biosynthetic capability into GM crops could require multiple transgenes to be introduced. Various examples of the introduction of multiple transgenes into crops have been met in different chapters, and it is useful to review some of the strategies involved. The development of stacked herbicide- and insect-resistant traits in commercial crops has been mentioned in Chapters 5 and 6. The advantages of pyramiding more than one pest-resistant gene was also discussed in Chapter 6. The strategy for stacking or pyramiding used in most of these examples involved crossing homozygous breeding lines containing each trait and selection for inheritance of both genes. While appropriate for large breeding companies with a small number of well-characterized GM lines, this strategy is slow and labour-intensive, the major problem being that the genes are not linked and may therefore segregate in future generations. The more transgenes that are stacked, the more generations that are required to produce homozygous lines, and the larger the population required to maintain and screen for segregation.

In Chapters 10 and 11, examples of multigene engineering included Golden Rice and bioplastic (polyhydroxybutyrate, PHB) production. Golden Rice was a remarkable achievement, not only because of the potential health benefits to consumers in the developing world, but because it took 7 years to introduce just three genes into rice. The approach used is instructive, in that two different strategies are combined: the use of linked genes in one construct, and co-transformation with more than one construct. These approaches are both widely used and have their advantages and disadvantages. Linked genes are positioned together on the same construct and so generally remain linked after integration into the genome. Normally only two or three genes are linked in this way, but larger constructs with four to six genes have been tried, the main limitation being the ability to engineer sufficient unique restriction sites into the cassette. New vector systems have been designed to overcome this problem (see Chapter 4). Co-transformation does not require the engineering of new constructs, but uses multiple vectors during the same transformation event (T-DNA or biolistics). One of the interesting and useful observations from such procedures is that the transgenes are often co-integrated, meaning that they are unlikely to segregate in subsequent generations (although this is less helpful if the aim is to segregate the transgene from a marker gene).

One of the problems encountered with multiple gene transformations is gene silencing (see Chapter 1), which may result from complex integration patterns of multiple transgene copies.

(Continued overleaf)

BOX 13.2 *Continued*

Alternative strategies to overcome this problem have been to express multiple genes as a polycistronic mRNA. The strategy is well adapted to chloroplast engineering (see Box 13.1), where the translational apparatus normally translates polycistronic mRNAs. For nuclear-integrated genes translated on cytoplasmic ribosomes, internal ribosome-entry sites can be inserted to recruit ribosomes to internal positions and initiate translation at internal start codons. The main drawback of this method is that internal ribosome-entry sites are relatively inefficient, and the downstream open reading frames are not translated as efficiently as the one at the 5′ end.

An alternative strategy is to produce a polyprotein, which is cleaved into its component peptides after translation. Cleavage at the required positions in the polyprotein can be obtained by inserting the recognition sequence for specific proteases present in the host plant, or by incorporating an enzyme such as the plant potyvirus NIa protease into the same polyprotein. An alternative strategy has been to incorporate the 20-amino acid 2A protein from foot-and-mouth disease virus, which causes the ribosome to skip during translation, leaving a gap between the 19th and 20th amino acid of the viral sequence.

Identification of genes associated with desirable traits

Genetic mapping

One of the most useful concepts when dealing with the large genomes of crop plants is that of the genetic map: effectively a series of milestones along the chromosomes that provide positional information. Genetic maps are of value for conventional plant breeding as well as providing a scaffold for genome-sequencing projects. The milestones or markers on a map are generally of two types.

First are genetic markers identified by specific phenotypes. The map location is determined by conventional methods of genetic crossing exactly as used by Morgan in the 1940s to map the position of eye colour and wing shape in *Drosophila*. (The genetic distance apart of two markers is still measured in centiMorgans, cM.) Genetic markers are typically single-gene mutations defining two alleles with readily distinguishable phenotypes. These are often not traits that are particularly valuable for crop-breeding purposes, and mapping complex agronomic traits directly is much more difficult (see the section on Quantitative trait loci, below).

● Polymorphic sites in the genome can be used as molecular markers.

Second are physical markers identified by molecular biological techniques. The benefits of using physical markers is that the marker can be observed directly, rather than via the expression of a phenotype, removing the time required for expression and complications that may affect the interpretation of phenotype, such as incomplete dominance, conditional expression, and epistasis. Like genetic markers, the only requirement is that the molecular marker is polymorphic (i.e. there is more than one form of the marker) and that the different alleles are distinguishable, albeit by molecular rather than phenotypic methods. One advantage is that the tests can be carried

Allele 1	**Allele 2 substitution**	**Allele 3 deletion**
CCGCTAAGACTAG	CCGCTGAGACTAG	CCGCT | AGACTAG
GGCGATTCTGATC	GGCGACTCTGATC	GGCGA | TCTGATC

Figure 13.1 Examples of two single-nucleotide polymorphisms (SNPs) caused by a base substitution (allele 2) or by a base deletion (allele 3).

Sequence	**SSR**
---CCCCCCCCCCCCCCCC---	$(C)_{16}$
---TATATATATATATATATA---	$(TA)_9$
---GGAGGAGGAGGAGGAGGA---	$(GGA)_6$
---CTATCTATCTATCTATCTAT---	$(CTAT)_5$

Figure 13.2 The repeated motifs that are characteristic of a number of common simple sequence repeats (SSRs). The number of tandem repeats of each sequence that is typically seen in a particular SSR is shown in the right-hand column.

out as soon as DNA can be isolated from the progeny. Another advantage for many molecular markers is that they are not dominant or recessive and both alleles can be detected equally well. To understand molecular markers, it is necessary to consider how polymorphic sites arise in a genome.

One of the important areas of information to flow from the large genome projects is the extent and type of polymorphisms between individual genomes. Polymorphisms can be classed according to type, as follows.

First are single-nucleotide polymorphisms (SNPs). As the name suggests, these are single-base mutations. They are often base substitutions (e.g. A/T→G/C in alleles 1 and 2 shown in Figure 13.1) but can also include insertion/deletions (indels; compare alleles 1 and 3 in the figure). Extensive mapping of SNPs in the human genome indicates that they occur throughout the genome at a frequency of about one every few hundred bases and are estimated to arise from mutations at a rate of approximately 10^{-9} per nucleotide. Most SNPs in higher eukaryotes occur in non-coding parts of the genome and have no effect on genetic function.

Second, simple sequence repeats (SSRs) are tandem repeats of short sequence motifs (one to six bases) that occur throughout the genome. Polymorphisms in the number of repeats accumulate much more frequently than SNPs ($\approx 10^{-3}$) due to slippage during replication or crossing over. The polymorphism in SSRs (microsatellites) forms the basis of genetic profiling of human DNA for forensic purposes. Some exemplar SSRs are shown in Figure 13.2.

Finally, there are a number of other types of sequence—larger-scale insertions, deletions, and transposition—that contribute to genetic variation in the genome, including longer tandem repeats (minisatellites or variable-number tandem repeats), different categories of transposable element (retrotransposons and DNA transposable elements), and larger breaks and transpositions of chromosome fragments. However, these are not generally used for genetic mapping.

The molecular biological methods that are used to detect these different types of polymorphisms for mapping are described in Box 13.3.

BOX 13.3 *Methods for the detection and mapping of polymorphic markers*

Randomly amplified polymorphic DNA

Randomly amplified polymorphic DNA (RAPD) is based on the use of an arbitrary short (≈10-base) primers for the amplification of genomic DNA. Due to the short size, this primer will bind to several sites in the genome, some of them close enough together and in the correct orientation to produce PCR products. The pattern of PCR products is determined by the random distribution of priming sites in the genome, and polymorphisms that affect priming sites will be detected by the alteration in PCR product size (see diagram). RAPD analysis is relatively cheap and straightforward and a series of different arbitrary primers can be tested to produce characteristic patterns for a particular genome. However, one problem with this method is its low reproducibility among laboratories, since the banding pattern obtained is highly dependent upon the PCR conditions (particularly annealing temperature) used.

Restriction-fragment length polymorphism

Restriction-fragment length polymorphisms (RFLPs) are mapped by restriction digestion of the genome and detection of hybridizing bands by Southern blotting using random DNA sequence probes. Differences in the pattern of bands on the blot reflect differences in the distribution of certain restriction sites between lines or cultivars. These are most likely to the result of single-nucleotide polymorphisms (SNPs) that create or destroy a restriction-enzyme cutting site, causing a marked shift in the restriction fragment length.

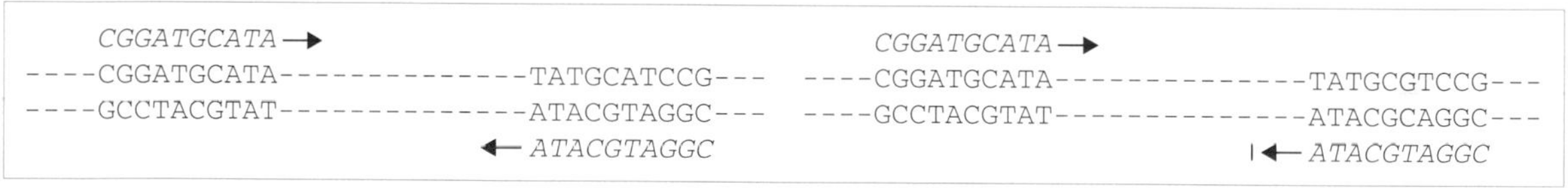

RAPD amplification of genomic DNA. The random distribution of an arbitrary 10-base sequence throughout the genome will permit amplification by the corresponding primer when two such sequences are close enough together and in the correct orientation (left-hand panel). A single base mutation in a primer site (shown in green) will prevent amplification from that position (right-hand panel), altering the pattern of amplified bands.

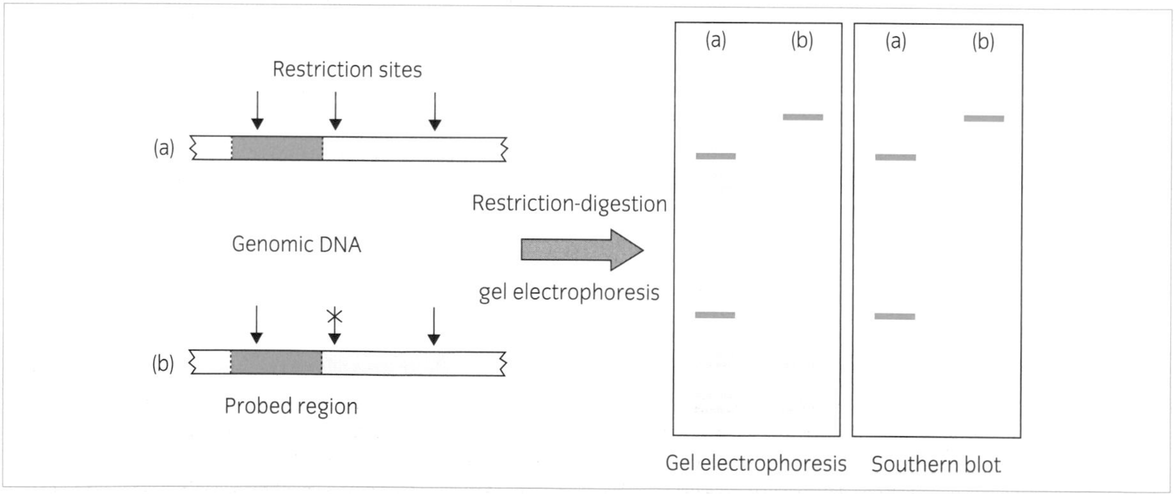

RFLP mapping. Restriction-enzyme digestion of genomic DNA yields a large number of fragments. Specific fragments can be detected by Southern blotting using either a probe designed to hybridize to a region of interest, or an arbitrary-sequence probe. In this example, an SNP at the middle restriction site prevents digestion, altering the pattern of fragments detected on the Southern blot. The pattern is therefore a direct indicator of which polymorphic allele is present in individual genomes.

BOX 13.3 *Continued*

Amplified-fragment length polymorphism

Amplified-fragment length polymorphism (AFLP) combines PCR amplification with the basic RFLP concept. Rather than relying on the detection of restriction fragments by Southern blotting, the digestion products are ligated to standard primer sequences that are used to amplify the fragments. A second, more specific, PCR is then performed to amplify a subpopulation of the total PCR

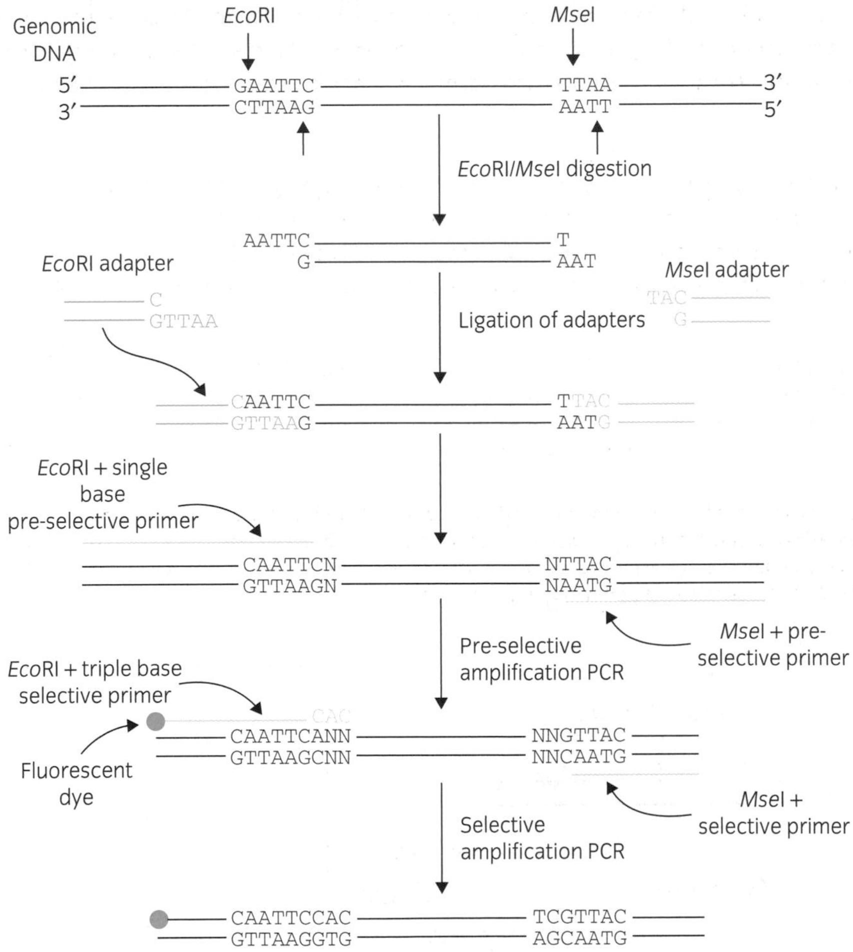

AFLP mapping. AFLP detects RFLPs by PCR rather than Southern blotting. The basic strategy involves attaching adapters onto the restriction fragments of genomic DNA. A pair of enzymes is used, one a four-base cutter such as *Mse*I and the other a six-base cutter such as *Eco*RI. Only fragments containing an *Mse*I site at one end and *Eco*RI at the other have the correct pair of primer sequences for amplification. (Note that the majority of fragments will have an *Mse*I site at both ends, so only a small subpopulation will be amplified.) In order to reduce the complexity of the amplified population even further, two PCR amplifications are carried out. The first pre-selection stage uses primers with one additional base at the 3′ end, whilst the selective primers have three specific bases. The pattern of fluorescently tagged fragments is separated by sequencing gel or capillary electrophoresis.

(Continued overleaf)

BOX 13.3 *Continued*

products. The fragment subset is then separated by sequencing gel or capillary electrophoresis. The method is reproducible among laboratories (although there is not yet a standardized database of AFLP patterns) but compared with RAPDs, it is time-consuming, technically demanding, and expensive. Polymorphic AFLP bands can also be isolated and sequenced, producing single-locus sequence-characterized amplified region (SCAR) markers.

Inter-simple sequence repeats

In contrast to the previous methods, which generally detect SNPs, it is possible to target simple sequence repeat (SSR) polymorphisms. The DNA fingerprinting techniques perfected for human forensic applications rely on precise sequence information about the unique sequences up- and downstream of a specific SSR. Multiplex PCR is then used to obtain a fingerprinting profile. However, it is also possible to create patterns based on SSR polymorphisms without sequence information by inter-simple sequence repeats (ISSRs). ISSR uses PCR primers based on SSRs, for example $(AG)_{10}$, $(GGC)_7$, $(AGGC)_5$. As with RAPDs, the method detects two instances of the same sequence that are within an amplifiable distance from each other in the genome. However, it is less sensitive to annealing temperature than RAPD, although optimization of annealing temperature is required.

SNPs

There are many methods now available to detect individual SNPs, exploiting differences in hybridization, priming, ligation, digestion, or sequencing reaction between polymorphisms. They all require sequence information about the SNPs to be detected, so are only applicable to characterized SNPs. On a larger scale, there are systems in place based on high-throughput technologies like microarrays (see Box 13.4) capable of genome-wide SNP mapping.

Quantitative trait loci

One of the aims of molecular mapping is to produce a sufficiently fine-scale map to pinpoint the location of genes that play a role in determining important agronomic traits. Many of these traits are described as quantitative; that is, they are influenced by multiple genetic and environmental factors. However, it is possible to map genes that have a major effect on quantitative traits by evaluating the correlation between the value of the quantitative trait and the allelic states at linked genetic markers. A quantitative trait locus (QTL) is therefore a chromosomal location where there is considered to be a reasonable probability that functionally different alleles segregate and cause significant effects on a quantitative trait. QTL mapping requires a statistical analysis of molecular marker and phenotypic data from a large segregating population to determine those markers where allelic polymorphism correlates with the quantitative trait phenotype. This primary QTL mapping is coarse and locates the gene within a chromosome region (known as a QTL supporting interval) of approximately 10–30 cM, which could contain several hundred genes (Figure 13.3). To identify the gene involved in the quantitative trait, two methods are available: positional cloning and association mapping.

● Positional cloning of genes responsible for QTLs requires several stages.

Positional cloning requires further steps to map the QTL to a much finer resolution and relate this map position to the DNA sequence. One strategy is to cross nearly isogenic lines in which the only allelic variation occurs in the short region of the coarse-mapped QTL. In the resulting populations, the QTL is described as **Mendelized**

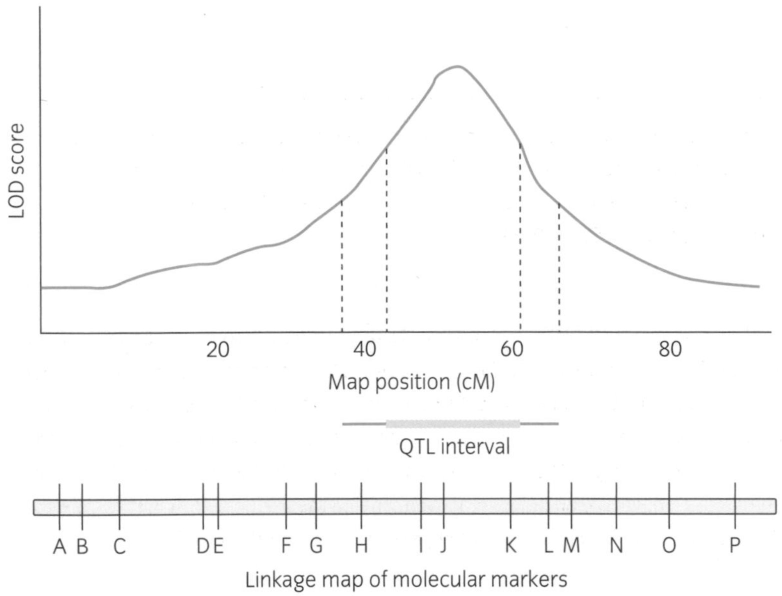

Figure 13.3 Quantitative trait locus (QTL) graph of LOD score against genetic distance (cM) along a particular chromosome. The LOD score is the logarithm (base 10) of the odds that the loci are linked (recombination fraction = θ) or not linked (recombination fraction = 0.5). The LOD score is therefore a measure of the likelihood of genetic linkage between the physical markers mapped below the graph and the quantitative trait phenotype being mapped. The QTL interval therefore defines a region of statistical probability of linkage, with the inner and outer limits of the interval corresponding to different levels of probability. Thus, markers I, J, and K show highly significant linkage with the quantitative trait, whereas linkage with H and L is less significant.

and a fine map with more precise cM distances between the QTL and adjacent molecular markers can be produced. This is relatively straightforward when there are many polymorphic markers mapped in the region, but this is currently possible only for those plants whose genome has already been, or is in the process of being, sequenced. At this stage, the markers closest to the QTL are used to anchor the genetic map to the physical map and it may then be possible to determine the gene responsible from candidate genes in the location by identifying the mutation responsible for the QTL effect (the QTN, or quantitative trait nucleotide polymorphism). Alternatively, it may be necessary to test each predicted coding sequence in the region functionally, by overexpressing or down-regulating the gene.

Association mapping avoids the requirement for producing nearly isogenic lines (which may require five to 10 generations of backcrosses) by calculating the statistical association between allelic variants at candidate loci within an existing set of genotypes (e.g. breeding lines, cultivated varieties and germplasm accessions). Ecotilling (see below) will provide a powerful tool for identifying SNPs within the target locus, allowing the association between haplotype variation at a candidate gene and the quantitative phenotype to be calculated.

While these strategies are being developed, only a relatively small number of genes responsible for QTLs have been identified to date (Table 13.2). It is interesting to note

Table 13.2 Positional cloning of genes from QTLs

Species	Trait	QTL	Gene	Function
Arabidopsis	Flowering time	ED1	*CRY2*	Cryptochrome
	Flowering time	FLW	*FLM*	Transcription factor
	Glucosinilate structure	GS-elong	*MAM*	Methylthioalkylmalate (MAM) synthase
	Root morphology	BRX	*BRX*	Transcription factor
Maize	Plant architecture	Tb1	*Tb1*	Transcription factor
Rice	Heading time	Hd1	*Se1*	Transcription factor
	Heading time	Hd3a	*Hd3a*	Unknown
	Heading time	Hd6	*αCK2*	Protein kinase
	Heading time	Ehd1	*Ehd1*	B-type response
Tomato	Fruit sugar content	Brix9-2-5	*Lin5*	Invertase
	Fruit shape	Ovate	*Ovate*	Unknown
	Fruit weight	Fw2.2	*ORFX*	Unknown

(Data from Salvi and Tubersosa (2005).)

that several of the identified genes encode regulatory proteins involved in flowering time. This can be contrasted with the experiments described above in which *Arabidopsis LEAFY* and *APETELA* genes were used to manipulate flowering time on the basis of molecular-biological information from the model organism. It can be seen that these completely different approaches to identifying genes of interest complement each other and accelerate the move towards a knowledge-based GM technology. One of the other examples in Table 13.2 shows the use of QTL mapping to elucidate the genes involved in tomato fruit quality, again in contrast to the case study on tomato fruit quality outlined in Chapter 10.

Investigating gene function by reverse genetics

Insertional mutagenesis

● T-DNA transformation can be used to create collections of insertional mutants.

One of the most direct ways of investigating gene function is to mutate the gene so that its function is altered or destroyed. This process has been described as reverse genetics: determining the effect on phenotype of changes in genotype (as opposed to conventional genetics where genotype is deduced from phenotype). The absence of a facile system for homologous recombination means that site-specific mutagenesis is not possible in plants. However, there are other ways of introducing mutations, the most simple involving the introduction of a piece of foreign DNA into the gene. Two different methods have been used for this: T-DNA insertion by *Agrobacterium*-mediated transformation (see Chapter 3) and transposon-induced mutagenesis. One advantage of this type of insertional mutagenesis is that it is relatively straightforward to locate the point of insertion, as the mutagen itself acts as a sequence tag. As this process is more or less random, large collections, or libraries of mutated plants, each of which has an insertion in a particular gene, have been established. The major limitation of these collections is that the majority of mutations are complete loss-of-function or knockout mutations. This means that the effect of more subtle changes in gene function cannot be determined easily, and a significant number of the mutations are lethal.

TILLING

Targeted induced local lesions in genomes (TILLING) is a mutation-detection tool that offers the potential to study gene function at a fundamental level, but also genetically modify crop plants in a specific, targeted fashion without using GM technology. The basis of the method is the mutagenesis of a population of plants by irradiation or treatment with a chemical mutagen such as ethylmethane sulphonate. The benefit of this strategy compared with gene-silencing or insertional-mutagenesis methods is that it produces a large number of recessive mutations by single-base substitution. The problem then is how to screen a large population to identify mutations in a candidate gene. The answer is to pool DNA from batches of plants and amplify the gene region of choice by polymerase chain reaction (PCR) with the 5′ and 3′ primers tagged with the fluorescent dyes IRD700 and IRD800, respectively (Figure 13.4). After the final cycle, the products are melted and allowed to re-anneal under conditions where single-base mismatches between strands will not prevent hybridization. The key to the method is then treatment with CEL-1, a nuclease enzyme from celery that cuts only at single-base mismatches. The PCR fragments are then analysed by separation on a LiCor sequencing gel. The bulk of the fragments will remain full length, but mismatch fragments will be detected as smaller bands. The size of the IRD700- and IRD800-labelled bands should add up to the full length, and indicate the position of the mutation within the amplified region.

● TILLING is a procedure for identifying specific mutants in a large mutagenized population.

To create and identify mutations in any gene of interest, a large population of mutagenized plants must be created and a high-throughput strategy for screening pools and subpools used to pick out desired mutants efficiently. For this reason, TILLING is not best used as an individual laboratory tool, but rather as a dedicated service for a particular crop community. There are now TILLING projects underway in a number of plant species, including *Arabidopsis*, *Brassica oleracea*, lotus, and maize. Currently, the emphasis is on the ability of TILLING to generate several single-base mutations within one gene and to analyse the effects on phenotype experimentally. In this regard, it provides much finer-detailed information about gene function than the general knockout effect produced by transposon or T-DNA insertional mutagenesis. However, it is clear that the system is also capable of identifying mutations that improve agronomic traits: not necessarily in the first round of screening, but in an iterative process in which specific mutations predicted to have the desired effect are sought. The first round of screening typically identifies about 10 mutations per gene, but further rounds can be screened subsequently.

● Several TILLING projects provide a service to their crop community.

The ability of the TILLING system to screen an artificially mutagenized population applies equally well to screening the variation within natural populations, a process

● Ecotilling is used to detect allelic variation between natural populations.

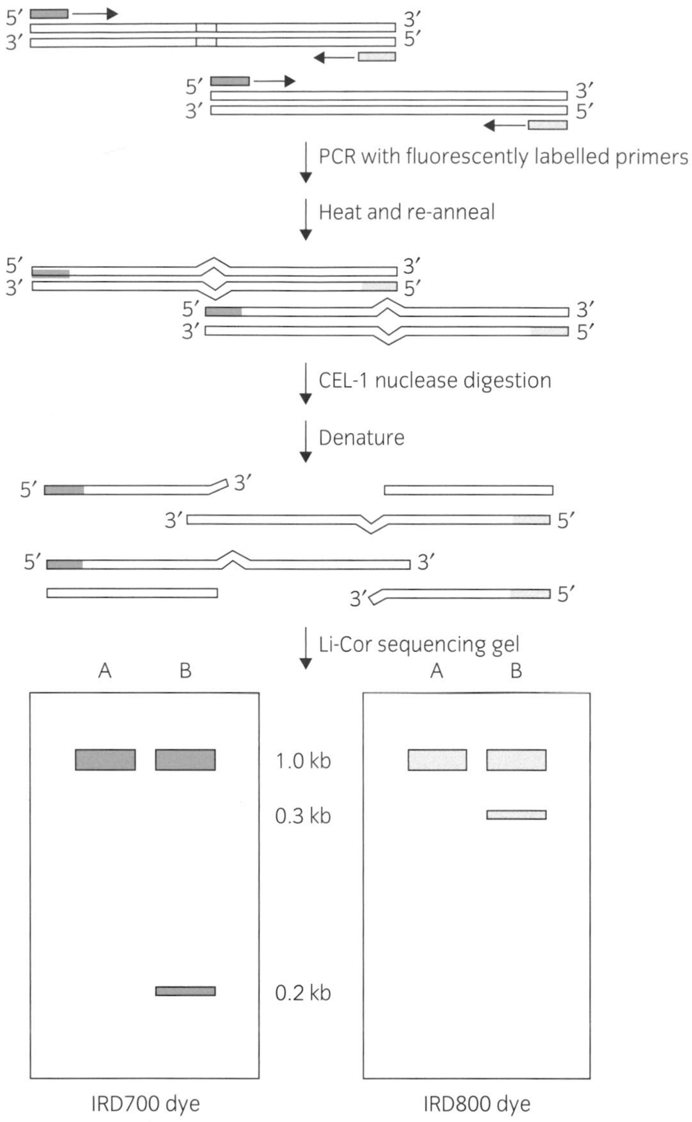

Figure 13.4 Targeted induced local lesions in genomes (TILLING) procedure for identification of specific mutants in a population. Pooled genomic DNA is amplified using primers specific to a gene of interest. PCR products are heated and allowed to anneal such that heteroduplexes can form between the wild-type and mutant DNA strands. Single-base mismatches are cut by the enzyme CEL-1, giving a minority of shorter fragments compared with the bulk of wild-type products. Each primer is labelled with a different fluorescent dye (IRD700 and IRD800), so the location of the mutation can be confirmed by the size of the band labelled with each dye. In this example, the mutation is 300 bp from the 5′ primer, confirmed by the fact that the sum of the IRD700 and IRD800 fragments add up to the full-length amplicon. A, no SNP; B, SNP.

described as Ecotilling. In a minor modification to the TILLING procedure, only a partial CEL-1 digest is performed, on both the mixed population and on each nat ural population alone, to allow for the more extensive heterogeneity between and within samples.

One of the most exciting prospects for the future is the possibility of incorporating mutants identified by TILLING directly into crop-breeding programmes without recourse to traditional GM technology. This presents the interesting prospect of genetic modification being consigned to the status of a specialized laboratory tool for confirming gene function, without the more controversial requirement of using GM technology directly for food production.

● TILLING may provide a GM-free method for genetic improvement of crops.

Understanding gene function within the genomic context: functional genomics

The sequencing of the *Arabidopsis* and rice genomes was described in Chapter 1 and the importance of using this sequencing information for determining the function of plant genes and genomes was discussed. The relevance of functional genomics to the future of crop improvement should now be apparent. Functional genomics provides a powerful tool to discover the multiple genes involved in complex traits and to understand their regulation and interactions within the plant. In addition to the reverse genetic tools discussed above, functional genomics is informed by several other important technologies.

Transcriptomics

Transcriptomics encompasses a number of technologies developed to enable the genome-wide analysis of gene-expression patterns at the level of the mRNA population: the transcriptome. These are high-throughput technologies designed to measure global changes in gene-expression profiles in different tissues or in response to different conditions or treatments (see Box 13.4). The detection of groups of genes

BOX 13.4 *Transcriptomic technologies*

Expressed-sequence-tag collections

One of the earliest resources that provided an overview of mRNA abundances was the expressed-sequence-tag (EST) sequencing data used to underpin the genome projects. EST programmes involve the large-scale sequencing of random cDNA clones from a library. The frequency of sequencing a particular gene sequence from an EST collection gives some indication of its abundance within the population. It is not a method that would be used directly for transcript profiling, but given the accumulation of EST data for other purposes, it can prove useful to interrogate this data for specific transcript profiles.

Serial analysis of gene expression

Serial analysis of gene expression (SAGE) is a method of EST sequencing designed specifically to provide transcriptomic information. Instead of producing normal ESTs of several 100 bases long, short tags of 9–14 bases are generated and joined into long concatemers that are then cloned

(Continued overleaf)

BOX 13.4 *Continued*

and sequenced. The sequence data is analysed to resolve the individual tags (they are separated by a short linker sequence) and the number of times each tag is represented indicates the abundance of the corresponding mRNA. (The size of the SAGE tag is just sufficient to provide an unambiguous identification of a specific mRNA or gene sequence.)

Microarrays and GeneChips®

Microarray hybridization is the most widely used method for global transcript profiling. The technology is similar in concept to earlier hybridization methods for studying gene expression (e.g. Northern blot hybridization) but is performed at a microscopic scale, enabling a large number of different sequences to be analysed in the same experiment. The simplest form of microarray consists of a series of micro-dots arranged on a solid support (typically a microscope slide), each one containing a different DNA sequence or probe. A fluorescent RNA (or cDNA) sample is hybridized to the slide and the intensity of fluorescence in each spot measured and analysed. In fact, in this type of experiment, it is common to hybridize two samples to the same microarray at the same time. These would typically be a test sample labelled with red Cy3 and a control labelled

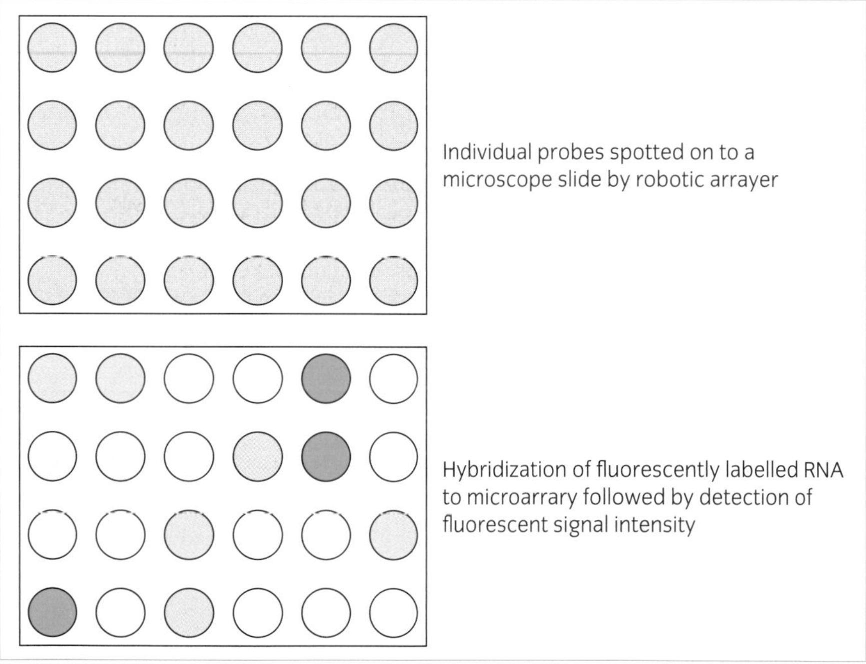

Microarray construction and use. Microarrays are constructed by spotting micro-spots of DNA probe on to a solid support such as a treated microscope slide. Different types of robotic devices for spotting are available. Typical probes are PCR-amplified cDNA clone inserts. Fluorescently labelled RNA samples are hybridized to the array and the intensity of fluorescence associated with each spot is measured and analysed. Cy3- and Cy5-labelled samples are generally hybridized together and the ratio of Cy3:Cy5 fluorescence calculated for each spot.

BOX 13.4 *Continued*

with green Cy5. The fluorescence at the red and green wavelengths is measured and the ratio of Cy3:Cy5 determined. This ratio gives an indication of whether the test condition has increased or decreased the level of gene expression for each mRNA compared with the control condition.

Although the term chip is often used interchangeably with microarray, the GeneChips® differ in a number of ways from the spotted array, in a process patented by Affymetrix. Oligonucleotide probes are designed *in silico* and synthesized directly on to a silicon wafer or chip using a process called photolithography. Separate probes are synthesized in adjacent squares on the chip, at densities such that many thousands of genes can be analysed on one chip. The normal format is to design 20 perfect-match 'features' per gene, with a corresponding set of single-base mismatch features as a control. The entire chip is enclosed in a black-box system, with all of the hybridization and washing fluidics, and the fluorescent reading carried out by proprietary instruments.

Massively parallel signature sequencing

Massively parallel signature sequencing (MPSS) is a recent variation on the SAGE concept but incorporating microarray-type technology to enable the parallel sequencing of millions of DNA-tagged microbeads. Although expensive and not widely used yet, this method has been used to study the entire *Arabidopsis* transcriptome.

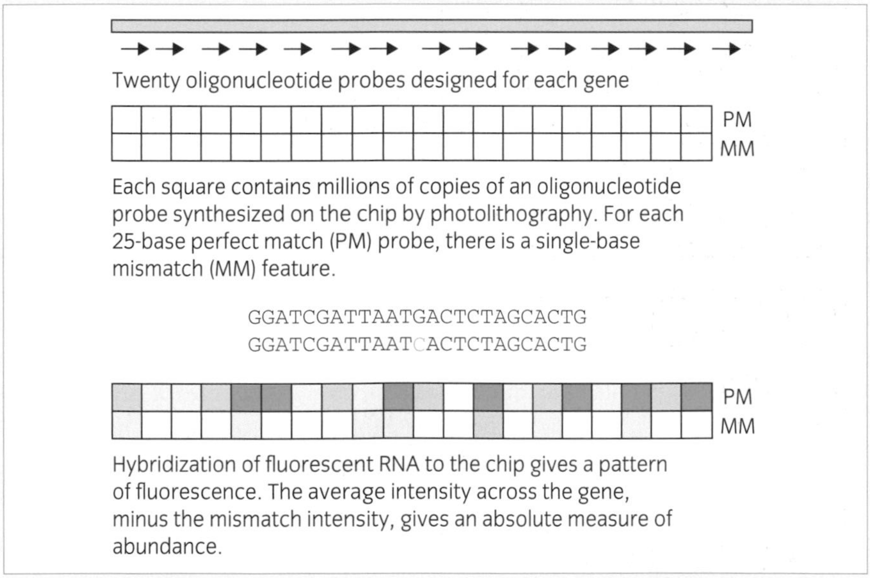

Construction and use of the Affymetrix GeneChip®. The GeneChip is one of a number of proprietary systems that differ significantly from spotted arrays. In this system, oligonucleotide probes are designed from the gene sequences to be analysed, and chemically synthesized on to the chip by photolithography. A typical design involves 20 different short (25-mer) probes spanning each gene, with each perfect match (PM) probe having a single-base mismatch (MM) probe as a control. A single sample is hybridized to the chip and corrected average fluorescent intensity calculated for each gene.

that show altered expression patterns under specific experimental conditions or at particular developmental stages provides a means of identifying likely multigene clusters involved in a particular response or developmental stage. Thus, the technology is one of the most powerful tools for gene discovery. However, it has wider applications in the development of GM crops, for example for analysing the effects of genetic modification upon the regulation of other genes in the genome.

One of the first global expression-profiling experiments in plants used a cDNA microarray to monitor the expression pattern of 1300 *Arabidopsis* genes in response to cold and drought stresses. Nineteen cold-induced and 44 drought-induced mRNAs were identified, 10 and 30 of which, respectively, were novel. Twelve of the stress-induced genes (six of them novel) were identified as being regulated by DREB1A/CBF3, 11 of which were shown to contain the dehydration-responsive element (DRE) in the promoter region. The relevance of this type of study to the strategies for development of stress-tolerant plants in Chapter 9 is apparent. Similar studies of transcriptional responses to a range of stimuli (e.g. wounding, elicitors, and pathogens) as well as growth and developmental processes relevant to crop performance will reveal similar regulatory networks amenable to genetic manipulation.

● There are extensive transcriptomic resources available for plant researchers.

There are now extensive transcriptomic resources available for researchers, from centralized microarray facilities that will analyse RNA samples to the wide range of expression data stored in various online databases. Although initially pioneered for the *Arabidopsis* community, these resources now extend to a number of important crop plant species.

Proteomics

Whereas the nature of transcriptomics lends itself to high-throughput technologies (since mRNAs differ only by nucleic acid sequence), the platforms available for proteomics are more limited in their capacity to analyse entire proteomes. However, proteomics has persevered against these technical limitations because of the importance for certain studies of being able to directly monitor the proteome. The important thing to remember is that whereas the transcriptome is often used as a shorthand for the gene-expression profile of a sample, it is really only a measure of the steady-state abundances of mRNAs. Although it holds true for many proteins that mRNA level is a good indicator of protein expression, this is not the case for all proteins. There is no substitute for knowing the amounts of each protein in a cell or tissue, and their activity (as judged by post-translational modification and interaction with other proteins). Box 13.5 outlines a number of the tools in the proteomic toolbox.

● Different aspects of the plant proteome are being investigated in several crop plants.

There have been a significant number of explorations of the plant proteome using these various tools. The information being derived can be classed into a number of broad areas, as listed below.

1 The plant cell proteome. In particular, there has been extensive characterization of the proteins found in different subcellular compartments, including the nucleus, chloroplasts, mitochondria, plasma membrane, and cell wall.

BOX 13.5 *Proteomic technologies*

Two-dimensional electrophoresis of proteins

For many years, it has been possible to separate relatively abundant proteins by two-dimensional polyacrylamide gel electrophoresis (2D-PAGE). Proteins are separated in the first dimension by isoelectric focusing according to their isoelectric point, and in the second dimension by SDS-PAGE according to size. More recent advances have improved the reproducibility of the technique by, for example, using stabilized pH gradients for the isoelectric focusing dimension. The other area of improvement has been in-gel-analysis software, such that the complex pattern of spots obtained for each sample can be documented and compared. The method still has some advantages over more recent innovations, one being that the position of the spot carries information about the molecular mass and amino acid composition of the protein, and it is also possible to detect post-translational modifications such as phosphorylation. Another reason that the method still has currency is that spots of interest can be cut out of the gel, digested with a proteolytic enzyme such as trypsin, and identified by mass spectrometry. The major limitation remains that it is only possible to resolve a few thousand proteins at most. The majority of low-abundance proteins are never seen on a two-dimensional gel.

Multidimensional liquid chromatography

Multidimensional liquid chromatography is in some ways analogous to two-dimensional gel electrophoresis in that proteins are first separated by one property, for example charge by ion-exchange chromatography and then size by reversed-phase high-performance liquid chromatography. The technique has advantages in that a relatively large amount of sample can be loaded, allowing detection of lower-abundance proteins. The proteins do not need to be stained to be detected and they can be passed directly to a mass spectrometer for peptide identification. Multidimensional protein identification (MudPIT) is a variation on this strategy, in that a biphasic column is used, with an ion-exchange resin packed on top of the reversed-phase resin. Peptide mixtures loaded on to the column are eluted with a stepped salt gradient and released into the reversed-phase resin. These second-dimension fractions are then eluted into a mass spectrometer with a gradient of organic solvent.

Mass-spectrometric methods

The importance of mass spectrometry in peptide identification after separation by 2D-PAGE or MudPIT has been mentioned above. One strategy is to perform peptide mass fingerprinting, using a mass spectrometer with an electron spray ionization (ESI) or matrix-assisted laser-desorption ionization (MALDI) source coupled with a time-of-flight (TOF) analyser. This allows the molecular mass of a population of trypsin digest fragments from a gel or liquid chromatography to be determined. These molecular-mass data are used to search protein databases for amino acid sequences that could generate the observed pattern of fragments.

Protein microarrays

Protein microarrays are being developed in a similar manner to nucleic acid microarrays (Box 13.4). However, the probes are either proteins themselves, or molecules that will bind specific proteins, such as antibodies or ligands. For example, antibody arrays have been developed, where each spot is a different antibody, and capture of labelled proteins from solution provides the signal. Variations on this method involve labelling a secondary antibody rather than the proteins. Conversely, antigen arrays have been used in which proteins are spotted on to an array and used to detect antibodies in solution. Protein arrays can also be used to screen for ligands, enzyme substrates, and interacting proteins.

2 The developing plant proteome. Several studies have focused on different cell and tissue types (seed, coleoptile, root, stem, leaf, petiole, xylem/phloem sap, trichomes, pollen tissues), developmental stages (germinated seeds, seedlings, and mature plant) and processes (seed germination, root formation, pollen maturation, somatic embryogenesis, programmed cell death, response to plant hormones).

3 The responsive proteome. The response of plants to a wide range of stimuli has been investigated, including growth conditions (nutrients, light), abiotic stresses (drought, salt, temperature, anoxia, heavy metals, herbicides, etc.), and biotic stresses (pathogens, herbivores, hypersensitive response).

4 Genetic variation in the proteome. Differences in protein profiles between genotypes, mutants, and transgenic plants have been studied.

What is also clear is that proteomics is no longer confined to model plants. Proteomic investigations already cover a wide range of plant species including many of the important crop species.

Interactomics

One of the most recent technologies aims to study the interactions between proteins: the interactome. Techniques involve the use of yeast two-hybrid assays and protein microarrays (Box 13.5) to determine protein–protein interactions. The interactome is displayed as a network of interactions, in which certain proteins form nodes with connections to many other proteins, and others are more peripheral with one or a few partners.

Metabolomics

Metabolomics describes attempts to obtain a global view of cellular chemical constituents using the tools of analytical chemistry. The aim is to study entire classes of metabolites, for example complete metabolic pathways and networks. As with the other -omics fields, differences in metabolic profiles can be correlated with desirable traits, providing information for metabolic engineering projects of the type described in Chapters 10 and 11. One of the important findings of metabolomic investigation is the demonstration that genetic engineering really is a quite precise science, with far fewer differences in metabolite concentrations between a transformed plant and a control plant than between different varieties of the same plant.

Systems biology

● Systems biology attempts to integrate the information from different strands of functional genomics.

If the array of -omic technologies is bewildering, so is the deluge of information flowing from the application of these techniques. One of the major problems facing genomic scientists is not only how to store and annotate this data in a reliable and accessible form, but also how to integrate the findings of disparate experiments. For example, is it possible to integrate the proteomic and metabolomic data accumulating from the study of drought stress with the type of transcriptomic data mentioned above? In other words, is it possible to make sense of the drought response of a plant by considering the regulation of gene expression, the effect this has on the cellular profile of proteins and their interactions, and the consequences in terms of metabolite concentrations? This problem is at least being considered under the heading of systems biology, an attempt to model how all of these systems are integrated in order to more fully explain biological phenomena. This area of genomic science is in its infancy, but it offers the hope that future developments in crop improvement will be much more precisely targeted and based on a much deeper understanding of the effects of manipulating plant systems.

SUMMARY

This final chapter has attempted to predict the future direction of plant biotechnology applied to crop improvement, and predicting the future is always a hostage to fortune. Nevertheless, some of the issues described here are incontrovertible. It is apparent that our knowledge of plant biology is accelerating as new genomic tools are applied and brought to bear not only on model species but also on crop plants. These tools are already enabling the discovery of genes that play a role in important agronomic traits, and it is inevitable that this knowledge will be used in some way for crop improvement. It is also apparent that for food crops in particular, the emphasis will be on using plant genes for the majority of manipulations. What is less clear is whether GM technology itself will remain the primary vehicle for delivery of these crop improvements (albeit a greener, more acceptable GM technology), or whether alternative methods will be found to manipulate the crop plant genome. Also, whereas this prediction may be valid for food crops, it may well be the case that the imperative to produce biofuels and cheap vaccines and pharmaceuticals will produce new generations of GM crops that are still based on the single, foreign-gene approach, yet are much more acceptable to the public.

FURTHER READING

Visit www.oxfordtextbooks.co.uk/orc/slaterplants2e/ for hyperlinked references that take you directly to online abstracts for the journal articles listed below:

Developments in genetic transformation

Daniell, H. and Dhingra, A. (2002) Multigene engineering: dawn of an exciting new era in biotechnology. *Current Opinion in Biotechnology* **13**, 136–141.

Daniell, H., Wiebe, P. O. and Millan, A. F. (2001) Antibiotic-free chloroplast engineering—an environmentally friendly approach. *Trends in Plant Science* **6**, 237–239.

Halpin, C. (2005) Gene stacking in transgenic plants—the challenge for 21st century plant biotechnology. *Plant Biotechnology Journal* **3**, 141–155.

Moffat, A. S. (2000) Can genetically modified crops go greener? *Science* **290**, 253–254.

Transcriptomics

Meyers, B. C., Tej, S. S., Vu, T. H., Haudenschild, C. D., Agrawal, V., Edberg, S. B., Ghazal, H., and Decola, S. (2004) The use of MPSS for whole-genome transcriptional analysis in *Arabidopsis*. *Genome Research* **14**, 1641–1653.

Rensink, W. A. and Buell, C. R. (2005) Microarray expression profiling resources for plant genomics. *Trends in Plant Science* **10**, 603–609.

Seki, M., Narusaka, M., Abe, H., Kasuga, M., Yamagucki-Shinzaki, K., Carninci, P., Hayashizaki, Y., and Shinozaki, K. (2001) Monitoring the expression pattern of 1300 *Arabidopsis* genes under drought and cold stresses by using a full-length cDNA microarray. *Plant Cell* **13**, 61–72.

Shoemaker, D. D. and Linsley, P. S. (2002) Recent developments in DNA microarrays. *Current Opinion in Microbiology* **5**, 334–337.

Venkatasubbarao, S. (2004) Microarrays—status and prospects. *Trends in Biotechnology* **22**, 630–637.

Proteomics

Aebersold, R. and Mann, M. (2003) Mass spectrometry-based proteomics. *Nature* **422**, 198–207.

Link, A. J. (2002) Multidimensional peptide separations in proteomics. *Trends in Biotechnology* **20**, S8–S13.

Phizicky, E. M., Bastiens, P. I. H., Zhu, H., Snyder, M., and Fields, S. (2003) Protein analysis on a proteomic scale. *Nature* **422**, 208–215.

Rabilloud, T. (2002) Two-dimensional gel electrophoresis in proteomics: old, old fashioned, but it still climbs up the mountains. *Proteomics* **2**, 3–10.

Zhu, H. and Snyder, M. (2003) Protein chip technology. *Current Opinion in Chemical Biology* **7**, 55–63.

Interactomics

Drewes, G. and Bouwmeester, T. (2003) Global approaches to protein–protein interactions. *Current Opinion in Cell Biology* **15**, 199–205.

Figeys, D. (2003) Novel approaches to map protein–protein interactions. *Current Opinion in Biotechnology* **14**, 119–125.

Metabolomics

Dunn, W. B., Bailey, N. J., and Johnson, H. F. (2005) Measuring the metabolome: current analytical technologies. *Analyst* **130**, 606–625.

Sumner, L. W., Mendes, P., and Dixon, R. A. (2003) Plant metabolomics: large-scale phytochemistry in the functional genomics era. *Phytochemistry* **62**, 817–836.

Systems biology

Blanchard, J. L. (2004) Bioinformatics and Systems Biology, rapidly evolving tools for interpreting plant response to global change. *Field Crops Research* **90**, 117–131.

Genomics and biotechnology

Gutterson, N. and Zhang, J. Z. (2004) Genomics applications to biotech traits: a revolution in progress? *Current Opinion in Plant Biology* **7**, 226–230.

Morgante, M. and Salamini, F. (2003) From plant genomics to breeding practice. *Current Opinion in Biotechnology* **14**, 214–219.

Paterson, A. H., Freeling, M., and Sasaki, T. (2007) Grains of knowledge: genomics of model cereals. *Genome Research* **15**, 1643–1650.

Tuberosa, R. and Salvi, S. (2006) Genomics-based approaches to improve drought tolerance of crops. *Trends in Plant Science* **11**, 405–412.

Varshney, R. K., Hoisington, D. A., and Tyagi, A. K. (2006) Advances in cereal genomics and applications in crop breeding. *Trends in Biotechnology* **24**, 490–499.

Molecular breeding

Dayteg, C., Tuvesson, S., Merker, A., Jahoor, A., and Kolodinska-Brantestam, A. (2007) Automation of DNA marker analysis for molecular breeding in crops: practical experience of a plant breeding company. *Plant Breeding* **126**, 410–415.

Meudt, H. M. and Clarke, A. C. (2006) Almost forgotten or latest practice? AFLP applications, analyses and advances. *Trends in Plant Science* **12**, 106–117.

Morandini, P. and Salamini, F. (2003) Plant biotechnology and breeding: allied for years to come. *Trends in Plant Science* **8**, 70–75.

Salvi, S. and Tuberosa, R. (2005) To clone or not to clone plant QTLs: present and future challenges. *Trends in Plant Science* **10**, 297–304.

Varshney, R. K., Graner, A., and Sorrells, M. E. (2005) Genic microsatellite markers in plants: features and applications. *Trends in Biotechnology* **23**, 48–55.

TILLING

Comai, S. and Henikoff, S. (2006) TILLING: practical single-nucleotide mutation discovery. *Plant Journal* **45**, 684–694.

Comai, L., Young, K., Till, B. J., Reynolds, S. H., Greene, E. A., Codomo, C. A., Enns, L. C., Johnson, J. E., Burtner, C., Odden, A. R., and Henikoff, S. (2004) Efficient discovery of DNA polymorphisms in natural populations by Ecotilling. *Plant Journal* **37**, 778–786.

Gilchrist, E. J. and Haughn, G. W. (2005) TILLING without a plough: a new method with applications for reverse genetics. *Current Opinion in Plant Biology* **8**, 211–215.

Henikoff, S., Till, B. J., and Comai, L. (2004) TILLING. Traditional mutagenesis meets functional genomics. *Plant Physiology* **135**, 630–636.

Innovative agronomic GM traits

He, Z. H., Zhu, Q., Dabi, T., Li, D. B., Weigel, D., and Lamb, C. (2000) Transformation of rice with the *Arabidopsis* floral regulator *LEAFY* causes early heading. *Transgenic Research* **9**, 223–227.

Niki, T., Nishijima, T., Nakyama, M., Hisamatsu, T., Oyama-Okubo, N., Yamazaki, H., Hedden, P., Lange, T., Mander, L. N., and Koshioka, M. (2001) Production of dwarf lettuce by over-expressing a pumpkin gibberellin 20-oxidase gene. *Plant Physiology* **126**, 965–972.

Pena, L., Martin-Trillo, M., Juarez, J., Pina, J. A., Navarro, L., and Martinez-Zapater, J. M. (2001) Constitutive expression of *Arabidopsis LEAFY* or *APETALA1* genes in citrus reduces their generation time. *Nature Biotechnology* **19**, 263–267.

'Green' GM traits

French, C. E., Rosser, S. J., Davies, G. J., Nicklin, S., and Bruce, N. C. (1999) Biodegradation of explosives by transgenic plants expressing pentaerythritol tetranitrate reductase. *Nature Biotechnology* **17**, 491–494.

Hannink, N., Rosser, S. J., French, C. E., Basran, A., Murray, J. A. H., Nicklin, S., and Bruce, N. C. (2001) Phytodetoxification of TNT by transgenic plants expressing a bacterial nitroreductase. *Nature Biotechnology* **19**, 1168–1172.

Kovalchuk, O., Titov, V., Hohn, B., and Kovalchuk, I. (2001) A sensitive transgenic plant system to detect toxic inorganic compounds in the environment. *Nature Biotechnology* **19**, 568–572.

Pilate, G., Guiney, E., Holt, K., Petit-Conil, M., Lapierre, C., Leple, J.-C., Pollet, B., Mila, I., Webster, E. A., Marstorp, H. G., Hopkins, D. W., Jouanin, L., Boerjan, W., Schuch, W., Cornu, D., and Halpin, C. (2002) Field and pulping performances of transgenic trees with altered lignification. *Nature Biotechnology* **20**, 607–612.

Index

A

D

H

I

J

K

L

M

Q

R

S

Plant Biotechnology